Indoor Climate and Energy Efficiency in Buildings

Indoor Climate and Energy Efficiency in Buildings

Guest Editors

Yanan Liu
Shen Wei
Lili Dong

MDPI

Basel • Beijing • Wuhan • Barcelona • Belgrade • Novi Sad • Cluj • Manchester

Guest Editors

Yanan Liu
College of Architecture and Urban Planning
Chongqing Jiaotong University
Chongqing
China

Shen Wei
The Bartlett School of Sustainable Construction
University College London
London
UK

Lili Dong
College of Architecture and Urban Planning
Chongqing Jiaotong University
Chongqing
China

Editorial Office
MDPI AG
Grosspeteranlage 5
4052 Basel, Switzerland

This is a reprint of the Special Issue, published open access by the journal *Buildings* (ISSN 2075-5309), freely accessible at: https://www.mdpi.com/journal/buildings/special_issues/WOC8K56919.

For citation purposes, cite each article independently as indicated on the article page online and as indicated below:

Lastname, A.A.; Lastname, B.B. Article Title. *Journal Name* **Year**, *Volume Number*, Page Range.

ISBN 978-3-7258-7829-1 (Hbk)
ISBN 978-3-7258-7830-7 (PDF)
https://doi.org/10.3390/books978-3-7258-7830-7

Contents

About the Editors

Yanan Liu

Yanan Liu is affiliated with the Department of Architecture at the School of Architecture and Urban Planning, Chongqing Jiaotong University, and the Institute of Future Cities and Carbon Neutrality Technologies. His academic and research interests focus on building performance, sustainability, HVAC systems, and carbon neutrality technologies. He has been involved in key projects such as the Chongqing Overseas Returnees Innovation Support Program and the Top-notch Young Talents Program of Chongqing Jiaotong University. His achievements include earning a Ph.D. through a joint program between Chongqing University and the Georgia Institute of Technology, becoming a National Registered Public Facility Engineer, and serving as a LEED Accredited Professional and Singapore Green Building Specialist. Additionally, he has delivered over 10 keynote speeches, chaired the International Conference on Energy and Environmental Science, and serves as an editor for several academic journals.

Shen Wei

Shen Wei is affiliated with the Bartlett School of Sustainable Construction, UCL. His main research areas include intelligent buildings, occupant behaviour in buildings and optimization control of building energy systems. Dr Wei has led and contributed to research projects funded by Innovate UK, British Council, EPSRC, Horizon Europe Funding, Ministry of Science and Technology, China, and Natural Science Foundation of China. Dr Wei has authored/co-authored over 140 journal articles, with an H-index of 41. He is currently the International Coordinator of CIB W98 - Intelligent and Responsive Buildings, Co-director of UCL Centre for Global Infrastructure Resilience, Committee Member of CIBSE Intelligent Buildings Group, CIBSE Resilient Cities Group, and ASHRAE UK, and the Associate Editor of *International Journal of Intelligent Buildings*.

Lili Dong

Lili Dong is affiliated with the School of Architecture and Urban Planning at Chongqing Jiaotong University. Her primary interests include green buildings, sponge cities, rural revitalization, and landscape garden planning and design. She has led over 10 teaching and research projects at the provincial and ministerial levels, built model courses in Civics, and guided 10 innovation and entrepreneurship training programs for college students. Her notable achievements include being a Class 1 registered architect in China, authoring 15 books, publishing 12 high-level journal papers, and delivering 12 lectures at international conferences. She has also won over 30 discipline competition awards and received honorary titles such as "Chongqing Women's Pioneer in Building Success" and "Chongqing Outstanding Young Architect."

buildings

MDPI

Article

CFD Analysis of Building Cross-Ventilation with Different Angled Gable Roofs and Opening Locations

Jingyuan Shi, Changkai Zhao and Yanan Liu *

School of Architecture and Urban Planning, Chongqing Jiaotong University, Chongqing 400074, China; 990201900031@cqjtu.edu.cn (J.S.); changkaizhao@mails.cqjtu.edu.cn (C.Z.)
* Correspondence: lyn3620@cqjtu.edu.cn

Abstract: The geometric shape of the roof and the opening position are important parameters influencing the internal cross-ventilation of buildings. Although there has been extensive research on natural ventilation, most of it has focused on flat or sloping roofs with the same opening positions. There is still limited research on the impact of different opening positions and sloping roofs on natural ventilation. In this study, computational fluid dynamics (CFD) was used to investigate the air exchange efficiency (AEE) in general isolated buildings. These buildings encompassed three distinct opening configurations (top–top, top–bottom, and bottom–top) and six varying slope angles for gable roofs (0°, 9°, 18°, 27°, 36°, and 45°). Computational simulations were carried out using the SST k-omega turbulence model, and validation was performed against experimental data supplied by the Japanese AIJ Wind Tunnel Laboratory. Grid independence validation was also conducted to ensure the reliability of the CFD simulation results. The study revealed that the highest AEE was 48.1%, achieved with the top–bottom opening configuration and a gable roof slope angle of 45°. Conversely, the lowest AEE was 31.4%, attained with the bottom–top opening configuration and a gable roof slope angle of 27°. Furthermore, it was observed that when the opening configuration was set to top–top and bottom–top, the slope angle of the gable roof had minimal influence on AEE, with an average AEE of only around 33%. When the opening configuration was top–bottom, it was found that there was a positive correlation between the gable roof slope angle and AEE.

Keywords: gable roof; natural ventilation; computational fluid dynamics (CFD); air exchange efficiency (AEE)

1. Introduction

The global demand for energy has been continually increasing, with the construction industry playing a significant role. Therefore, enhancing its energy efficiency becomes especially important. An overreliance on mechanical and electrical equipment to provide a comfortable environment has led to a disproportionate energy consumption in buildings, particularly from heating, ventilation, and air conditioning (HVAC) systems [1]. In China, between 2005 and 2020, the national total energy consumption from buildings increased from 930 million tce to 2.233 billion tce, with an average annual growth rate of 6.0%. By the end of 2020, the building stock in China had reached 69.6 billion square meters [2]. With such an immense inventory, reduction in building energy consumption has remained a key concern for both the government and society. With mild weather conditions, it is highly recommended to use natural ventilation to replace air conditioning for cooling [3]. Compared with air conditioning systems, the cooling ability of natural ventilation seems to be limited and its performance is also not quite stable. Therefore, many researchers have moved to studying how to maximize the performance of natural ventilation in cooling buildings.

Regarding opening positions, natural ventilation can be generally categorized into three types: (1) SSV—single-sided ventilation (one or more vents are placed on one side of the building); (2) CV—cross-ventilation (allowing natural airflow through the building by having windows positioned on opposite facades), and (3) SV—stack ventilation (utilizing temperature difference between indoor and outdoor environments, with vents located at different heights) [4]. Compared with SSV, CV generally gives higher ventilation efficiency [5], hence it was selected in this study.

In terms of opening positions in the vertical direction, according to Karava [6], apart from wall porosity, the relative locations of openings and exits on building facades are significant factors affecting cross-ventilation. Zhang [7] conducted a study on two different-sized external openings and found that the most effective configuration for enhancing cross-ventilation was to have a larger opening on the leeward side than on the windward side. The aforementioned studies primarily focused on the impact of opening size and position on cross-ventilation, with limited exploration of the building's form. However, as the roof is an integral component of the building's form, its significance is self-evident. Roof forms can be categorized into gable roofs [8–10], arched roofs [11], mono-pitched roofs [12], sawtooth roofs [13] and various other roof types [14,15]. Previous studies have focused on analyzing the surface wind-pressure coefficients of these different types of roofs and providing some information on the potential for natural ventilation, but they have lacked an analysis of airflow inside buildings.

In recent years, the effects of various types of roofs and openings on the natural ventilation of buildings have been studied. Vaishnani [16] conducted numerical simulations and used the PMV model to assess the impact of roof slope angles on cross-ventilation under winter, summer, and monsoon weather conditions in Delhi, India. The results showed that the PMV values decreased with increasing roof slope angles in winter, while in summer, the PMV values increased with greater roof slope angles. Esfeh [17] enhanced natural ventilation capacity by adjusting the geometric shape of a semi-circular arc-shaped roof. The research revealed that the ventilation performance of the arched roof is highly sensitive to the wind angle α, with the best performance at α equals 0 and the poorest performance at 75° and 90°. Kosutova [18] conducted a study on the impact of louvers with varying positions and fixed angles in buildings with flat roofs on natural ventilation. Building upon this foundation, Tai [19] conducted research on the impact of louvers with different positions and angles in buildings with flat roofs on air exchange efficiency. He also proposed optimization factors to measure the relationship between AEE and DFR. Starting from roof types and roof angles, Peren [20] conducted simulations using computational fluid dynamics (CFD) and found that, compared to buildings without eaves, the maximum increase in internal air volume flow occurred when the windward eave had a slope angle of 27°. Then Peren [21] investigated a single-sloped roof and several other concave and convex roofs, and the results showed that convex roofs could maximize the negative pressure of the wake near the building's exhaust vents, thereby enhancing cross-ventilation inside the building. Leite [22], on the other hand, combined the slope angle of a single-sloped roof with the window opening position and found that the roof slope angle had a significant impact on ventilation airflow, while the vertical position of the outlet opening had a minor effect. Atmaca [23] conducted wind-tunnel experiments and computational fluid dynamics (CFD) simulations on three differently sloped gable roofs, and found that the lowest negative pressure at all roof slopes occurred when the wind direction angle was 90 degrees. Tominaga [24] conducted numerical simulations (CFD) to study the airflow around isolated gable-roofed buildings with different roof slopes and found significant differences in the flow fields for slope ratios of 3:10 and 5:10. Additionally, there are also studies related to wind towers, wind catchers, Mansard roofs, and other structures on rooftops [25–34].

A review of the literature indicates that there is almost no research on the combination of the slope angle of gable roofs and the vertical position of outlet openings. Therefore, the innovation of this study was to analyze the effects of gable roof slope angle and vertical opening position on indoor air exchange efficiency (AEE), wind-pressure coefficient, and wind speed of independent houses by using CFD method. In the remaining part of this paper, Section 2 introduces the research methods adopted in this study. Section 3 analyzes the results from the CFD simulation with relevant discussions. A conclusion can then be found in Section 4.

2. Methodology

2.1. Wind-Tunnel Experiment

To calibrate the CFD model used in this study, data of wind loads on low-rise buildings, which were collected by wind-tunnel experiments at the Tokyo Institute of Technology in Japan [35], were used. Due to the prevalence of low-rise buildings in suburban areas, we selected suburban terrain, categorized as Terrain Category III in accordance with AIJ (2004), as the test wind field [36]. This particular category features a mean wind velocity profile exponent of 0.20 and a gradient height of 450 m. To replicate it, we utilized turbulence-generating spires, roughness elements, and a carpet on the upstream floor of the wind tunnel's test section. The wind velocity profile and turbulence intensity profile of the simulated wind field are presented in Figure 1. At a height of 10 cm, the turbulence density measured approximately 0.25. The wind velocity during testing at this height reached approximately 7.4 m per second, which corresponds to roughly 22 m per second at a height of 10 m at full scale. The baseline model selected for this study was the flat-roof model retrieved from the database. Wind pressure measurement taps were uniformly distributed across the surfaces of the tested models, as illustrated in Figure 2. The spacing between the taps was set at 20 mm, equivalent to 2 m at full scale. However, due to limitations in the capacity of wind pressure measurement scanivalve to measure a large number of taps simultaneously, certain inner points on models with larger surfaces were not measured. Each tap was connected to a pressure measurement scanivalve via synthetic resin tubes, measuring 80 cm in length and 1.2 mm in internal diameter. These scanivalves were capable of synchronously recording fluctuating wind pressures at nearly 384 points.

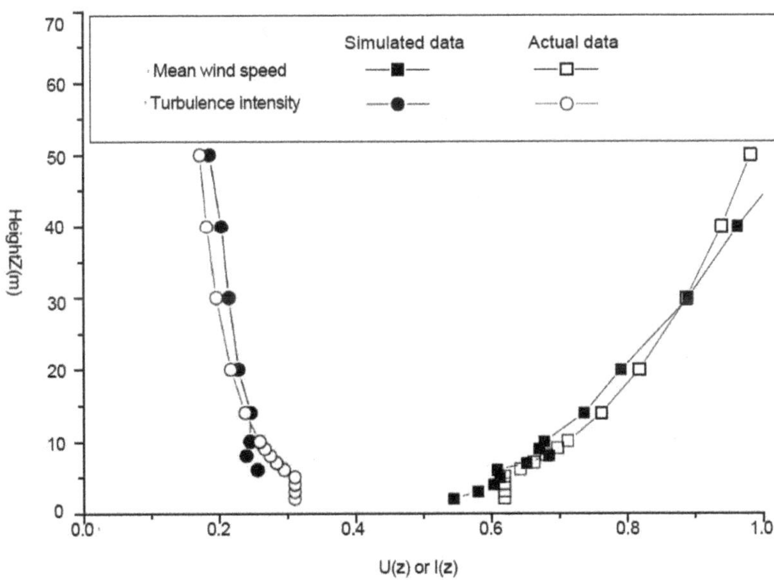

Figure 1. Simulated wind field of suburban terrain.

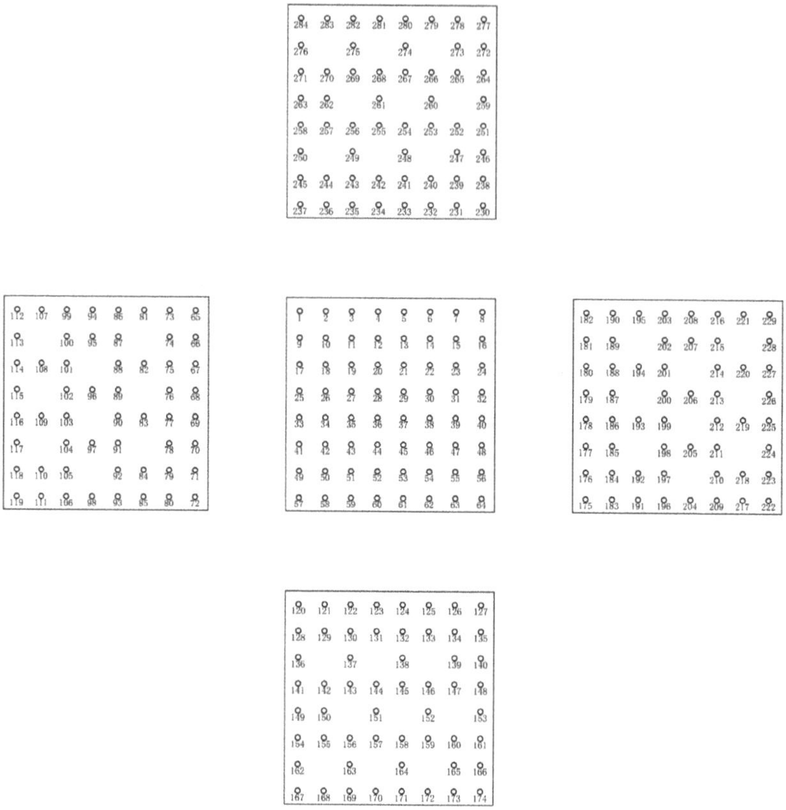

Position of measured points of a flat-roofed low-rise building

Building geometetrical parameters: H = 16.0 m, B = 16 m, D = 16 m, β = 0°

Figure 2. Arrangement of wind pressure measurement taps.

2.2. Model Geometry, Computational Domain, and Building Configuration

The dimensions of the building were 160 mm × 160 mm × 160 mm (L × W × H). The distance from the front of the building model to the entrance was 3H, from the side of the building model to the side and top wall of the fluid domain it was 5H, and from the back of the building model to the exit it was 15H, where H represents the height of the building model. The blockage ratio of the fluid domain was less than 3%. The above settings complied with the design specifications for the fluid domain [37].

The front view of the fluid domain is shown in Figure 3a, the longitudinal section in Figure 3b, and the transverse section in Figure 3c. ANSYS 2022 R1 was used for grid partitioning and CFD simulation, and local refinement of BOI was required, as shown in Figure 3b,c. The inlet and incident profile are shown in Figure 3e and the 3D mesh is shown in Figure 3f. Poly-hexcore cells were used in the meshing and 20 prism layers with a growth rate of 1.2 were applied on the ground and buildings, as shown in Figure 3d. The first cell height of the building was 35 μm, with a corresponding Y+ value of less than 0.60. Skewness of the grid was controlled to be no more than 0.65 in all numerical cases. Fine element sizes were applied at the model edges to achieve the desired grid quality and control the maximum Y+ value. The final grid had a maximum Y+ value of 1.26 for the walls and 1.13 for the ground. The respective average values of Y+ for the walls and ground were 0.48235 and 0.41380.

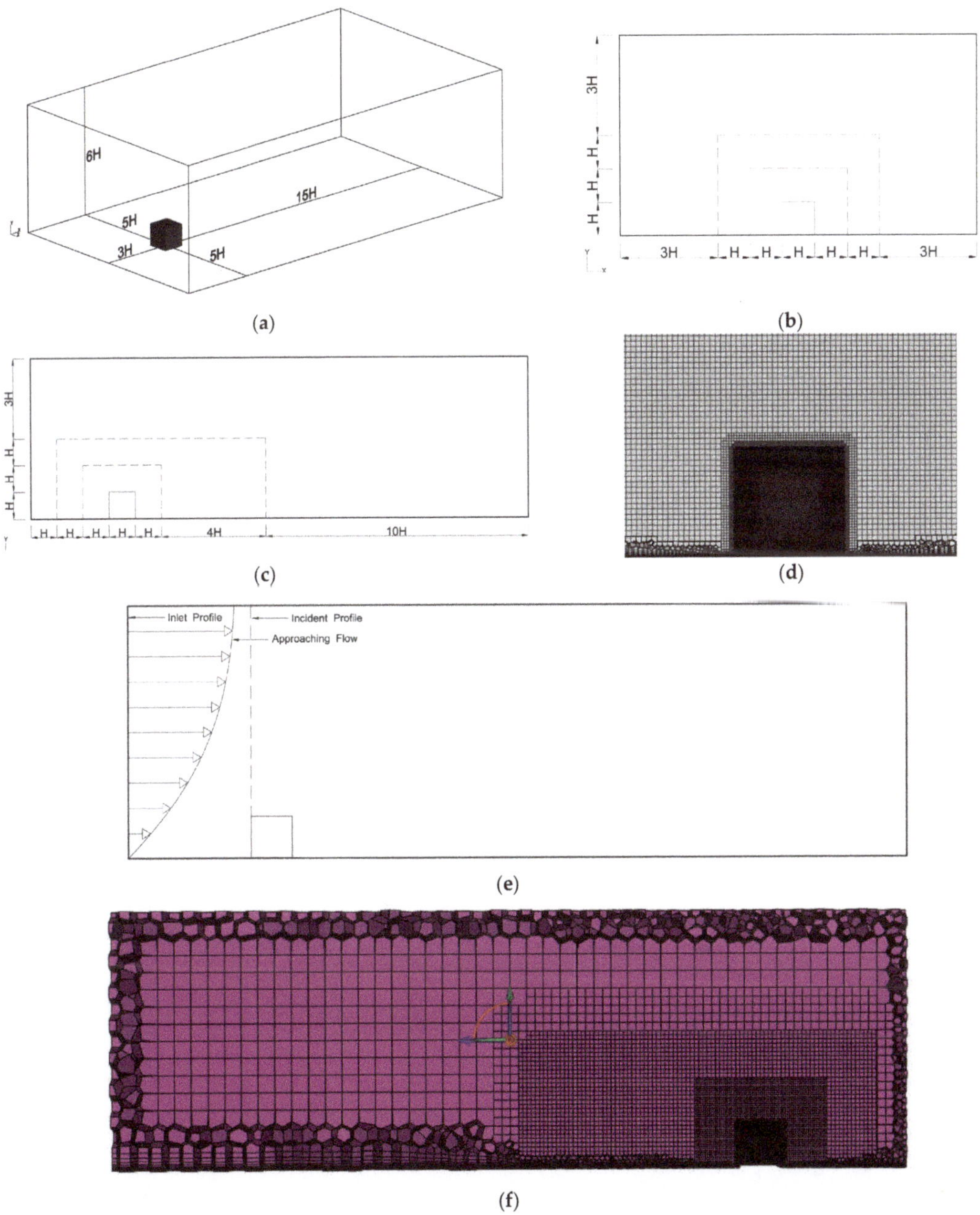

Figure 3. (**a**) The building in its computational domain, (**b**) longitudinal section, (**c**) transverse section, (**d**) prism layer around the house and ground, (**e**) inlet and incident profiles, and (**f**) 3D mesh.

2.3. Governing Equations

The shears-stress transport (SST) k-ω model transport equations are as follows:

$$\frac{\partial}{\partial t}(\rho k)+\frac{\partial}{\partial x_i}(\rho k u_i)=\frac{\partial}{\partial x_j}\left[\Gamma_k\frac{\partial k}{\partial x_j}\right]+G_k-Y_k+S_k \tag{1}$$

and

$$\frac{\partial}{\partial t}(\rho\omega)+\frac{\partial}{\partial x_j}(\rho\omega u_j)=\frac{\partial}{\partial x_j}\left[\Gamma_\omega\frac{\partial \omega}{\partial x_j}\right]+G_\omega-Y_\omega+D_\omega+S_\omega \tag{2}$$

where G_k represents the production of turbulence kinetic energy; G_ω represents the generation of ω; Γ_k and Γ_ω represent the effective diffusivity of k and ω, respectively; Y_k and Y_ω represent the dissipation of k and ω due to turbulence; D_ω represents the cross-diffusion term; and S_k and S_ω are user-defined source terms. The calculation process is as follows:

2.3.1. Production of k

According to the exact equation of the transport of k, the term can be defined as

$$G_k=-\rho\overline{u_i'u_j'}\frac{\partial u_j}{\partial x_i} \tag{3}$$

To evaluate G_k in a manner consistent with the Boussinesq hypothesis,

$$G_k=\mu_t S^2 \tag{4}$$

where S is the modulus of the mean rate-of-strain tensor, defined as

$$S\equiv\sqrt{2S_{ij}S_{ij}} \tag{5}$$

2.3.2. Production of ω

The production of ω is given by

$$G_\omega=\frac{\alpha\alpha^*}{V_t}G_k \tag{6}$$

where G_k is given by Equation (3) and α_∞ is given by

$$\alpha_\infty=F_1\alpha_{\infty,1}+(1-F_1)\,\alpha_{\infty,2} \tag{7}$$

$$\alpha_{\infty,1}=\frac{\beta_{i,1}}{\beta^*_\infty}-\frac{\kappa^2}{\sigma_{\omega,1}\sqrt{\beta^*_\infty}} \tag{8}$$

$$\alpha_{\infty,2}=\frac{\beta_{i2}}{\beta^*_\infty}-\frac{\kappa^2}{\sigma_{\omega 2}\sqrt{\beta^*_\infty}} \tag{9}$$

$$\kappa=0.41 \tag{10}$$

The coefficient is given by

$$\alpha=\frac{\alpha_\infty}{\alpha^*}\left[\frac{\alpha_0+Re_t/R_\omega}{1+Re_t/R_\omega}\right] \tag{11}$$

$$\alpha^*=\alpha^*_\infty\left[\frac{\alpha^*_0+Re_t/R_k}{1+Re_t/R_k}\right] \tag{12}$$

$$Re_t=\frac{\rho k}{\mu\omega} \tag{13}$$

$$R_k=6 \tag{14}$$

$$\alpha_0^* = \frac{\beta_i}{3} \tag{15}$$

$$\beta_i = 0.072 \tag{16}$$

2.3.3. Dissipation of k

The dissipation of k is given by

$$Y_k = \rho\beta^* f_{\beta^*} k\omega \tag{17}$$

where

$$f_{\beta^*} = 1 \tag{18}$$

and

$$\beta^* = \beta_i^* [1 + \zeta^* F(M_t)] \tag{19}$$

$$\beta_i^* = \beta_\infty^* \left[\frac{4/15 + \left(Re_t/R_\beta\right)^4}{1 + \left(Re_t/R_\beta\right)^4} \right] \tag{20}$$

$$\zeta^* = 1.5 \tag{21}$$

$$R_\beta = 8 \tag{22}$$

$$\beta_\infty^* = 0.09 \tag{23}$$

where Re_t is given by Equation (13).

2.3.4. Dissipation of ω

The dissipation of ω is given by

$$Y_\omega = \rho\beta f_\beta \omega^2 \tag{24}$$

where

$$f_\beta = 1 \tag{25}$$

$$\beta_i = F_, \beta_{i,1} + (1 - F_1)\, \beta_{i,2} \tag{26}$$

$$\beta = \beta_i \left[1 - \frac{\beta_i^*}{\beta_i} \zeta^* F\,(M_t) \right] \tag{27}$$

F_1 is obtained from Equation (38), β_i^* is obtained from Equation (20), and $F\,(M_t)$ is obtained from Equation (29).

2.3.5. Cross-Diffusion Modification

D_ω is defined as

$$D_\omega = 2\,(1 - F_1)\, \rho \frac{1}{\omega\sigma_{\omega,2}} \frac{\partial k}{\partial x_j} \frac{\partial \omega}{\partial x_j} \tag{28}$$

2.3.6. Compressibility Effects

The compressibility function, $F\,(M_t)$, is given by

$$F\,(M_t) = \begin{cases} 0 & M_t \le M_{t0} \\ M_t^2 - M_{t0}^2 & M_t > M_{t0} \end{cases} \tag{29}$$

where

$$M_t^2 \equiv \frac{2k}{a^2} \tag{30}$$

$$M_{t0} = 0.25 \tag{31}$$

$$a = \sqrt{\gamma RT} \tag{32}$$

In the high-Reynolds number form of the k–ω model, $\beta_i^* = \beta_\infty^*$. In the incompressible form, $\beta^* = \beta_i^*$.

2.3.7. Modeling the Effective Diffusivity

The effective diffusivities for the SST k–ω model are given by

$$\Gamma_k = \mu + \frac{\mu_t}{\sigma_k} \tag{33}$$

$$\Gamma_\omega = \mu + \frac{\mu_t}{\sigma_\omega} \tag{34}$$

where σ_ω and μ_t are the turbulent Prandtl numbers for k and ω, respectively. The turbulent viscosity is μ_t.

$$\sigma_k = \frac{1}{F_1/\sigma_{k,1} + (1 - F_1)\ /\sigma_{k,2}} \tag{35}$$

$$\sigma_\omega = \frac{1}{F_1/\sigma_{\omega_1 + (1-F_1)\ /\sigma_{\omega,2}}} \tag{36}$$

$$\mu_t = \frac{\rho k}{\omega} \frac{1}{max\left[\frac{1}{\alpha^*}, \frac{SF_2}{a_1\omega}\right]} \tag{37}$$

where S is the strain rate magnitude and α^* is defined in Equation (12).

The blending function F_1 is given by

$$F_1 = tanh\ (\phi_1^4) \tag{38}$$

$$\phi_1 = min\left[max\left[\frac{\sqrt{k}}{0.09\omega y}, \frac{500\mu}{\rho y^2\omega}\right], \frac{4\rho k}{\sigma_{\omega 2} D_\omega^+ y^2}\right] \tag{39}$$

$$D_\omega^+ = max\left[2\rho\frac{1}{\sigma_{\omega 2}}\frac{1}{\omega}\frac{\partial k}{\partial x_j}\frac{\partial \omega}{\partial x_j}, 10^{-10}\right] \tag{40}$$

The blending function F_2 is given by

$$F_2 = tanh\left(\phi_2^2\right) \tag{41}$$

$$\phi_2 = max\left[2\frac{\sqrt{k}}{0.09\omega y}, \frac{500\mu}{\rho y^2\omega}\right] \tag{42}$$

where Y is the distance to the next surface.

2.3.8. Continuity Equation

$$\partial p/\partial t + \nabla \cdot\ (pu) = 0 \tag{43}$$

where p represents the velocity of the fluid and u represents the velocity of the fluid.

2.3.9. Momentum Equation

$$\partial\ (pu)\ /\partial t + \nabla \cdot\ (puu) = -\nabla p + \nabla \cdot\ (\mu eff \nabla u) + \rho g + F \tag{44}$$

where ρ represents the pressure of the fluid, μeff represents the effective viscosity of the fluid, g represents the acceleration of gravity, and F represents the external force.

2.3.10. Turbulence Equation

$$\partial k/\partial t + \nabla \cdot (ku) = Pk - \varepsilon + \nabla \cdot (\mu efft \nabla k) \quad (45)$$

$$\partial \omega/\partial t + \nabla \cdot (\omega u) = P\omega - \beta\omega\omega + \nabla \cdot (\mu efft \nabla \omega) \quad (46)$$

where k represents turbulent flow energy, ω represents turbulent dissipation rate, Pk represents the generation term of turbulent flow energy, ε represents the dissipation term of turbulent flow energy, $\mu efft$ represents the effective viscosity of turbulence, $P\omega$ represents the generation term of turbulent dissipation rate, and $\beta\omega$ represents the dissipation coefficient of turbulent dissipation rate. The constant values used in Equations (1)–(46) are shown in Table 1.

Table 1. Values of the constants used in Equations (1)–(46).

$\sigma_{k,1}$	$\sigma_{\omega,1}$	$\sigma_{k,2}$	$\sigma_{\omega,2}$	α_1	$\beta_{i,1}$	$\beta_{i,2}$	α_∞^*	α_∞	$\alpha_\circ$	β_∞^*	R_β	R_k	R_ω	ζ^*	M_{t0}
1.176	2.0	1.0	1.168	0.31	0.075	0.0828	1	0.52	1/9	0.09	8	6	2.95	1.5	0.25

2.4. *Atmospheric Boundary Layer (ABL)*

The ABL is the velocity profile that needs to be created from the inlet. The vertical velocity profile U on flat terrain is typically given by Equation (47), where U_{ref} is the wind speed at the reference height of 7.4 m/s, Y_{ref} is the reference height of 0.1m, and α is the roughness coefficient of 0.2 [36].

$$U = U_{ref}\left(\frac{y}{y_{ref}}\right)^{\alpha} \quad (47)$$

Turbulent kinetic energy K is determined by Equation (48), in which, I_u represents turbulent intensity, equal to 0.25, and A is equal to 1.

$$k = A\left(I_U U\right)^2 \quad (48)$$

The value of turbulent dissipation rate ε is typically estimated using the assumption of local equilibrium [38].

$$\varepsilon = C_\mu^{1/2} k \frac{dU}{dy} \quad (49)$$

where C_μ is the model constant equal to 0.09. Equation (50) can be obtained by rearranging the above equation.

$$\varepsilon = C_\mu^{1/2} k \frac{U_{ref}}{y_{ref}} \alpha \left(\frac{y}{y_{ref}}\right)^{\alpha-1} \quad (50)$$

Therefore, the specific turbulent dissipation rate ω is defined by Equation (51) [39]. The specific values of U, k, ε, ω are shown in Table 2.

$$\omega = \frac{\varepsilon}{C_\mu k} = C_\mu^{-1/2} \frac{U_{ref}}{y_{ref}} \alpha \left(\frac{y}{y_{ref}}\right)^{\alpha-1} \quad (51)$$

Table 2. U, k, ε, ω values at the inlet boundary.

Y (m)	U (m/s)	k (-)	ε (-)	ω (-)
0.02	5.363	1.798	28.924	178.799
0.04	6.160	2.371	21.917	102.682
0.06	8.816	4.858	32.455	74.237
0.08	9.338	5.440	28.927	58.975

2.5. Fluent Solver Settings

All CFD simulations were performed using ANSYS FLUENT 2022 R1. The SST K-omega turbulence model was employed, and a production limiter was chosen to obtain more accurate wind-pressure coefficients. The COUPLE scheme was used for pressure–velocity coupling. Pressure was selected with a second-order interpolation scheme, while momentum, turbulent kinetic energy, and turbulent dissipation rate were discretized using a second-order upwind scheme; user-defined scalars (UDS) were chosen as first-order upwind to generate more accurate results. Standard initialization was used in this numerical simulation. Convergence criteria for x, y, and z velocities; turbulent kinetic energy (k); and turbulent dissipation rate (ε) were set to 1×10^{-4}. For user-defined scalars (UDS), the convergence criteria were set to 1×10^{-5}. The computation method employed was pseudo-transient, with a time step of 0.02 s and 900 timesteps. All solutions converged after 500 iterations.

2.6. Grid Sensitivity Analysis

In the same computational domain, the grid sensitivity analysis was conducted using the aforementioned boundary conditions and solution methods. By varying the grid size of the computational domain, grid sensitivity analysis was performed on three different grid schemes with 707,430 cells (Grid 1); 1,913,049 cells (Grid 2); and 2,665,529 cells (Grid 3). Twelve points were placed on the windward, top, and leeward sides of the building to measure the wind-pressure coefficients. As shown in Figure 4, there was a significant deviation between mesh 1 and mesh 2, while there was a slight deviation between mesh 2 and mesh 3. The average percentage deviation of wind-pressure coefficients between mesh 1 and mesh 2 was 7.47%, while between mesh 2 and mesh 3, it was 0.96%. From the grid sensitivity analysis, the simulation results were not sensitive to mesh 2 and mesh 3. Therefore, mesh 2 was used for all other CFD simulations.

2.7. Model Calibration

The results from the CFD simulation were compared with data from the Japanese wind-tunnel laboratory. Twelve measurement points with a radius of 1.2 mm were selected on the windward, roof, and leeward sides of the building for comparing pressure coefficients. The pressure coefficients at various locations were calculated using Equation (6). Figure 5 compares the pressure coefficients on the building surfaces, showing overall good consistency despite some differences in negative pressure areas.

$$C_P = \frac{(P - P_0)}{\left(0.5\rho U_{ref}^2\right)} \tag{52}$$

where P represents static pressure, P_0 is the reference static pressure, ρ is the air density, and U_{ref} is the reference velocity.

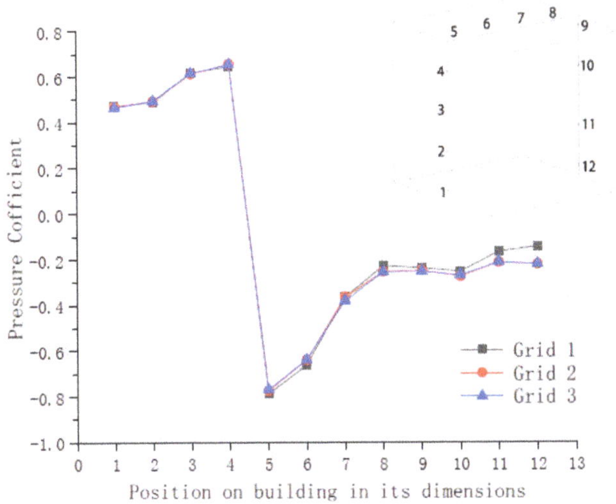

Figure 4. Results of grid sensitivity analysis.

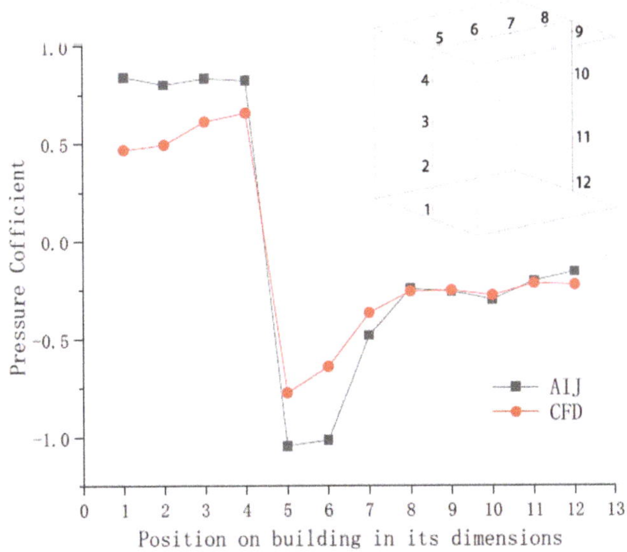

Figure 5. Comparison of CFD result with AIJ test.

3. Results and Discussion

To assess the impact of the roof inclination angle (RIA) and vertical opening position on ventilation airflow, a study was conducted on 18 different building models. These models encompassed three different opening positions (top–top, top–bottom, and bottom–top) and six different roof inclination angles (0°, 9°, 18°, 27°, 36°, and 45°). To facilitate a smooth comparison, the above models had the following characteristics: (1) identical internal volumes, (2) identical inlet and outlet opening sizes, and (3) different exit and entrance locations. Figures 6 and 7 illustrate three different opening positions and various angles of gable roofs.

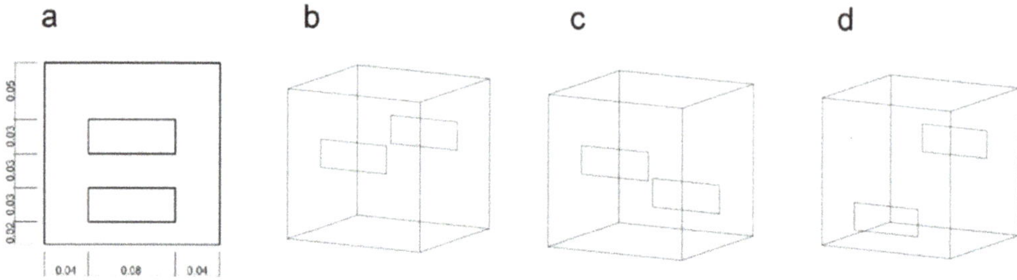

Figure 6. Opening positions and sizes. (**a**) The location and size of the top and bottom openings, (**b**) Top-Top opening configuration, (**c**) Top-Bottom opening configuration, (**d**) Bottom-Top opening configuration.

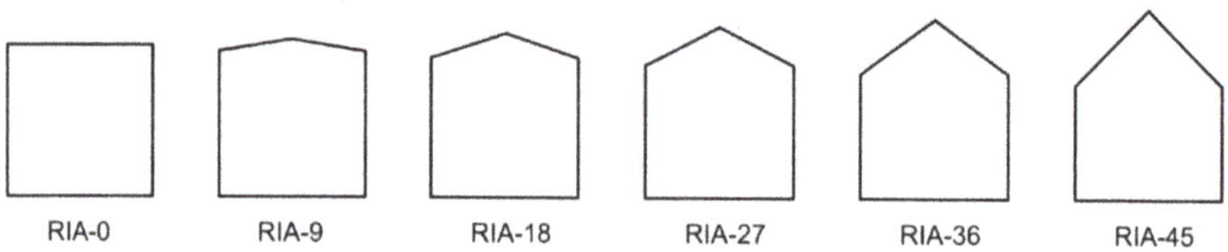

Figure 7. Gable roof inclination angles.

3.1. Pressure Coefficient, Cp

The pressure coefficient *Cp* is defined as the dimensionless ratio of static pressure at a specific location to the freestream static pressure, as shown in Equation (48). The contour lines of pressure coefficients are presented in Table 3. It can be observed that, with the same opening configuration, as the slope increased, the pressure stagnation region at the top of the building gradually shifted towards the highest point of the roof, and the leeward-side openings typically exhibited negative *Cp* values. In various building cases, the negative pressure difference at the building's windward outlets was consistently greater than the positive pressure difference at the windward inlets, indicating that the roof slope angle is a crucial parameter for enhancing wind-driven natural cross-ventilation.

Table 3. Pressure coefficient, *Cp*, suitable for different open configurations with gable roofs.

Configuration	Legend	0	9	18	27	36	45
Top–Top	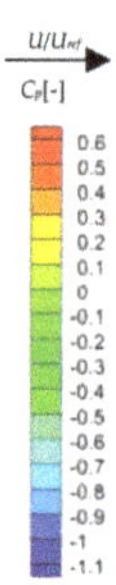		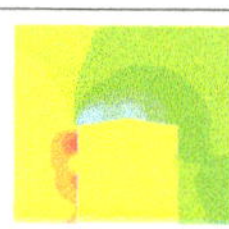	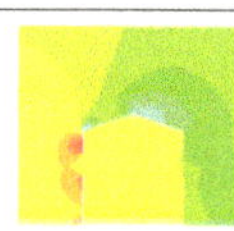	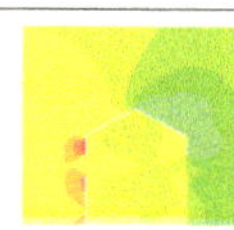	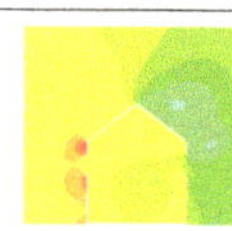	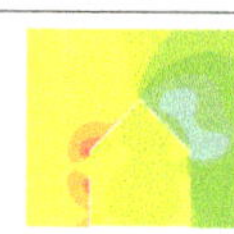
Top–Bottom		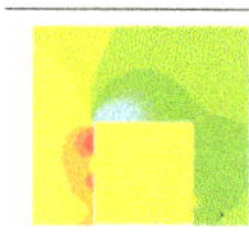	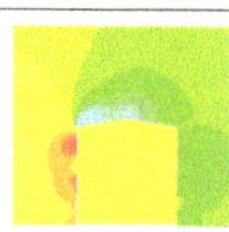	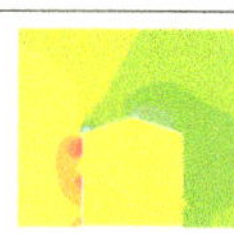	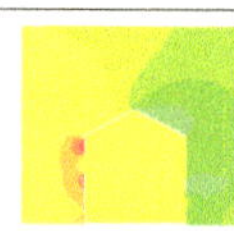	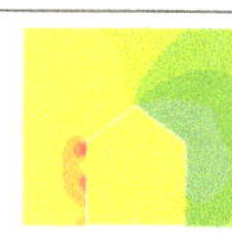	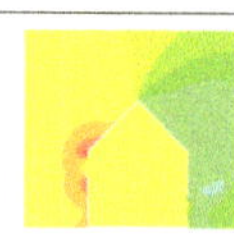
Bottom–Top		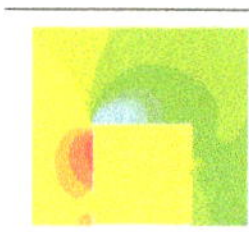		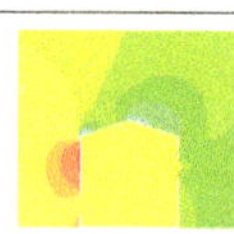		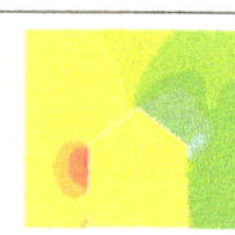	

3.2. Dimensionless U/Uref

The dimensionless quantity $U/Uref$ was obtained by dividing the average streamwise velocity U by the reference velocity $Uref$ (7.4 m/s). The contour lines of the dimensionless quantity $U/Uref$ are depicted in Table 4. From Tables 4 and 5, it can be observed that there was a notable increase in flow velocity in the opening region, a phenomenon evident across all models. Through observation, it was noted that when the opening configuration was top–top or top–bottom, the flow velocity in the exhaust opening area increased with changes in the roof slope, reaching its maximum at 45°, with an acceleration rate 1.2 times that of a flat roof. Conversely, in the case of a bottom–top opening configuration, due to the Coanda effect, the airflow remained attached to the building's bottom, resulting in a lack of sensitivity of flow velocity in the exhaust opening area to changes in roof angle, with flow velocity remaining relatively constant at around 0.6. Overall, the airflow velocity in the exhaust opening area was more sensitive to the vertical opening position than to the slope angle of the gable roof.

Table 4. $U/Uref$ suitable for different open configurations with gable roofs.

Configuration	Legend	0	9	18	27	36	45
Top–Top	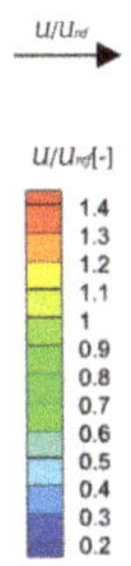	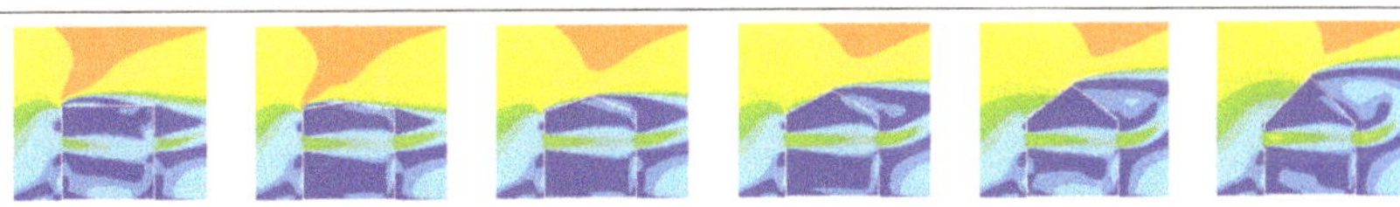					
Top–Bottom		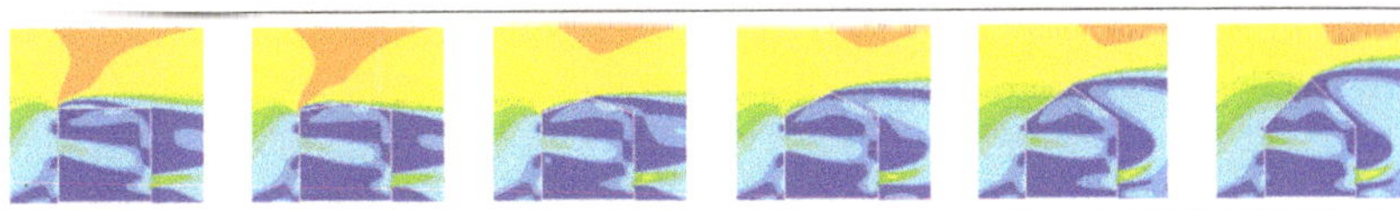					
Bottom–Top		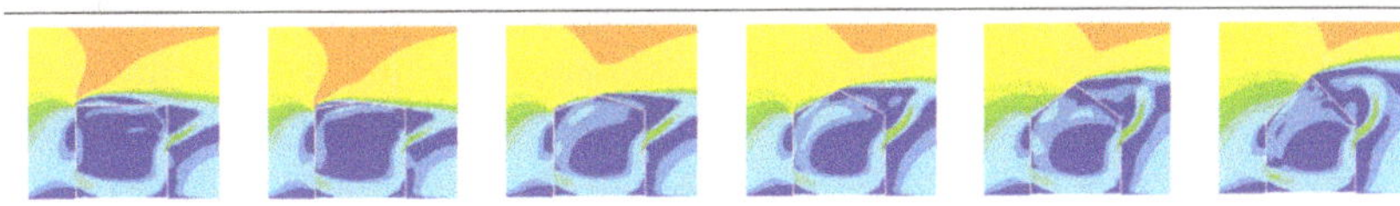					

Table 5. U suitable for different open configurations with gable roofs.

Configuration	Legend	0	9	18	27	36	45
Top–Top	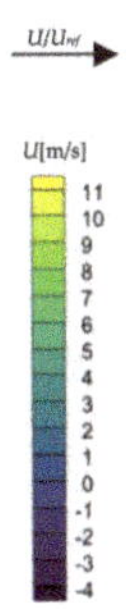	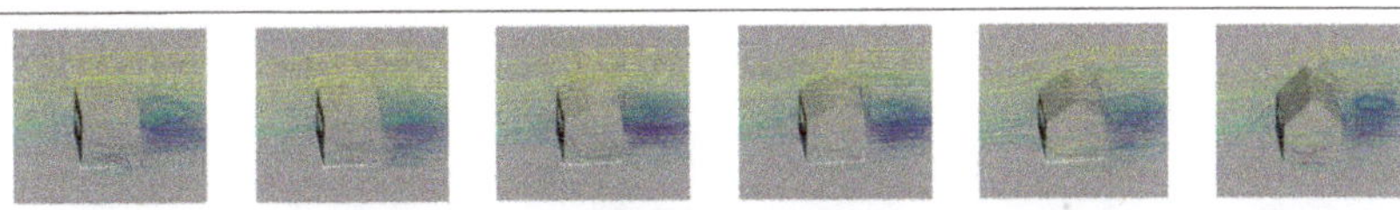					
Top–Bottom		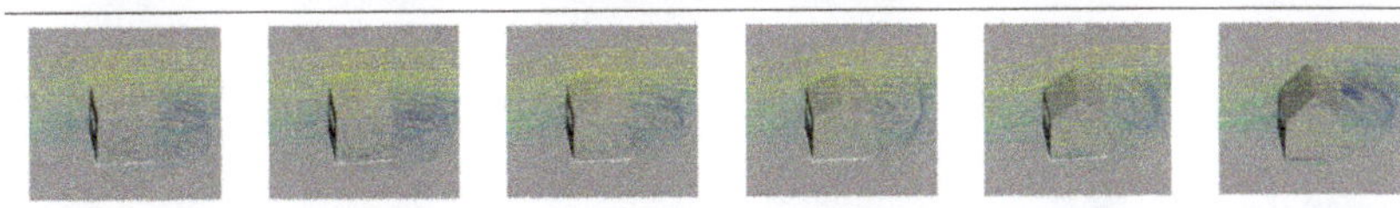					
Bottom–Top		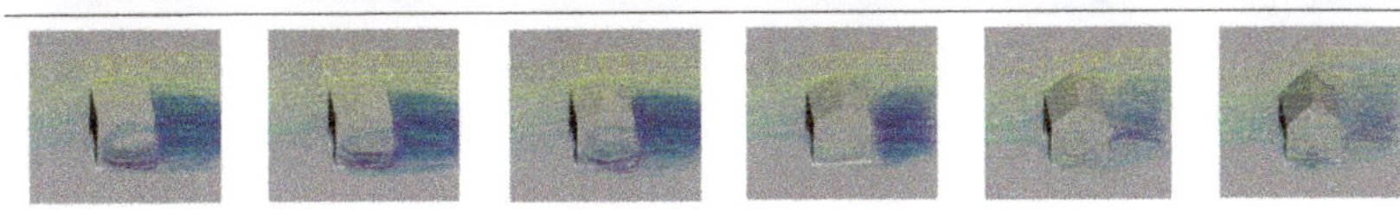					

3.3. Air Exchange Efficiency (AEE)

Air exchange efficiency represents the efficiency of external air flushing the interior of a ventilated building [40]. The AEE is defined by Equation (53), where τ_r represents the AOA at the leeward openings, and τ_{av} is the average volume-averaged AOA within the building.

$$\varepsilon_A = \frac{\bar{\tau}_r}{2\tau_{av}} \times 100[\%] \tag{53}$$

AOA is calculated using the following scalar transport equation [41].

$$\frac{\partial}{\partial t}\rho\varphi + \nabla \cdot (\rho U \varphi) - \nabla \cdot (\Gamma \nabla \varphi) = S_\varphi \tag{54}$$

where φ represents the scalar being solved for, i.e., AOA, with $S_\varphi = 1$ denoting the source term. The diffusion coefficient Γ is determined using a specific formula [41].

$$\Gamma = \rho D + \frac{\mu_t}{S_{c_t}} \tag{55}$$

where $D = 2.88 \times 10^{-5}\mathrm{m}^{-2}/\mathrm{s}$ represents the laminar viscosity of air at a working temperature of 20 °C, μ_t is the local turbulent viscosity, and $S_{ct} = 0.7$ is the turbulent Schmidt number.

Under steady-state conditions, the term $\partial(\rho\varphi)/\partial t$ in Equation (54) equals zero. The transport equation was implemented in FLUENT using user-defined scalars (UDS) to compute AOA within the building. Equation (55) was employed to determine boundary conditions, setting zero values at the inlet and zero gradients at the walls and exit surfaces.

Table 6 presents the AOA distribution for all operating conditions, Figure 8 illustrates the percentage of AEE for all conditions, and Figure 9 displays the average volume AOA within the building. From Table 4, it can be observed that in the top–top opening configuration, at 0°, the AOA between the windward-facing window and the roof was relatively high, gradually decreasing with an increase in roof angle, reaching an optimum at 36°. In the top–bottom configuration, the AOA between the windward-facing window and the roof was lower but higher than the ground as the roof angle increased, with the ground portion of AOA gradually decreasing. In the bottom–top configuration, the formation of eddies within the building led to excessively high AOA in the central region of the structure, and it was insensitive to changes in roof slope.

Table 6. Age of air (AOA) suitable for different open configurations with gable roofs.

Configuration	Legend	0	9	18	27	36	45
Top–Top		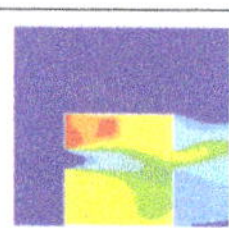	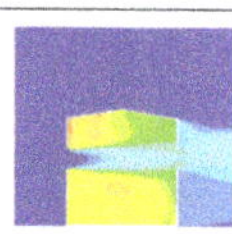	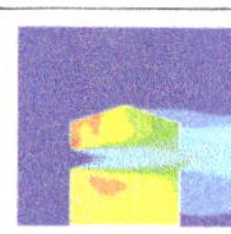	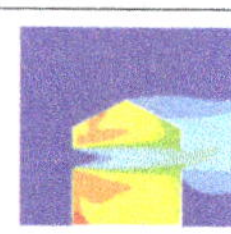	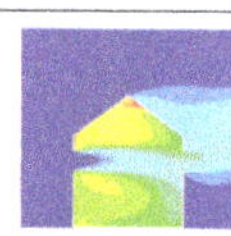	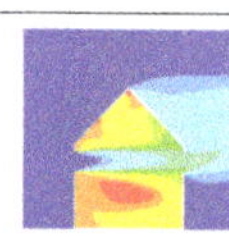
Top–Bottom	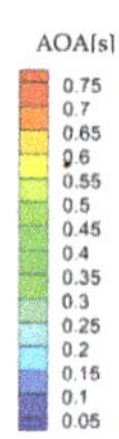	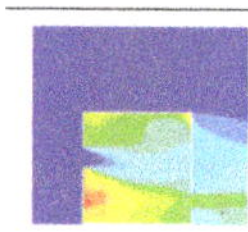	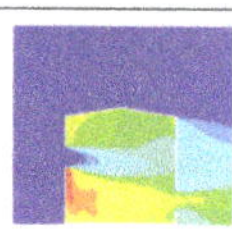	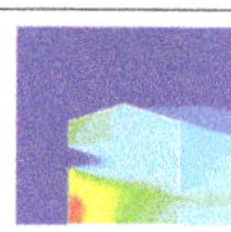	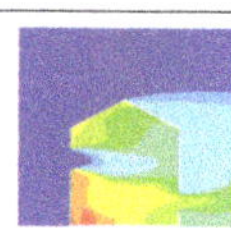	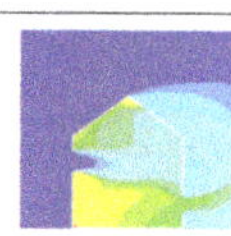	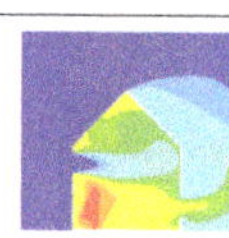
Bottom–Top		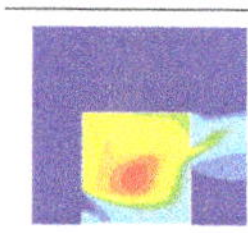	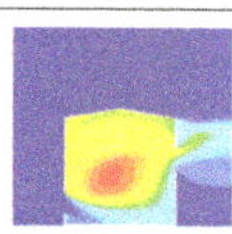	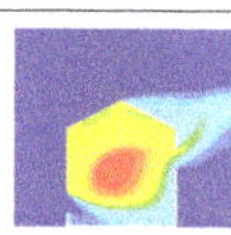	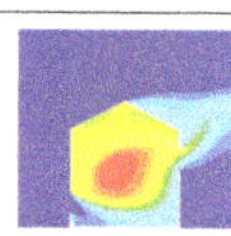	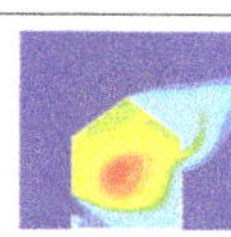	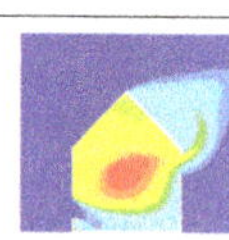

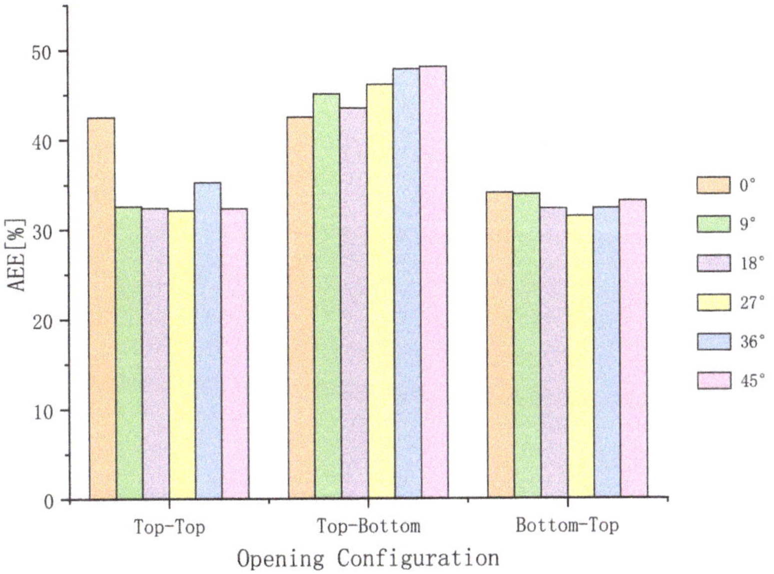

Figure 8. Air exchange efficiency (AEE) with different angles of gable roofs and different opening positions.

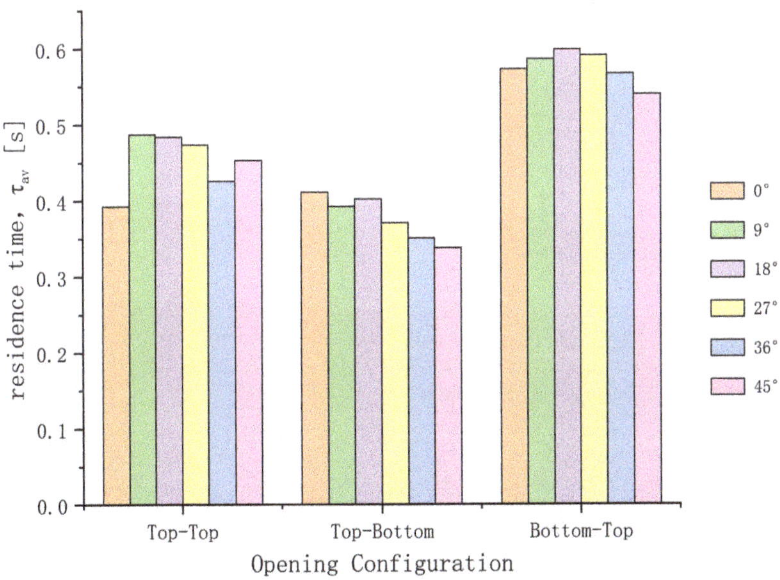

Figure 9. Average volume-averaged age of air (AOA) with different angles of gable roofs and different opening positions.

From Figure 8, it can be observed that the AEE was highest at 45° top–bottom, reaching 48.1%, and lowest at 27° bottom–top, at only 31.4%. In the top–top opening configuration, the AEE for the flat roof was highest at 42.5%, while the AEE for the other five angles was around 32%. This indicates that in this configuration, AEE is not highly sensitive to changes in the gable roof angle. In the top–bottom opening configuration, AEE increased with an increase in roof slope angle, reaching its peak at 45°, at 48.1%. This suggests that in this configuration, the angle of the gable roof effectively promotes indoor airflow, thereby enhancing AEE.

In the bottom–top opening configuration, the AEE for roofs with different slope angles was quite similar, averaging around 32%. This suggests that AEE is not highly sensitive to changes in the gable roof angle in this configuration. Due to the presence of low–high windows, it is easy for vortices to form within the building, leading to a relatively low AEE.

Figure 9 reveals that in the top–bottom opening configuration with a gable roof slope angle of 45°, τav was at its lowest, measuring 0.33 s. In the bottom–top opening configuration with a roof slope angle of 18°, τav was highest, at 0.6 s. When considering Figures 8 and 9 together, it becomes evident that τav maintained a strong negative correlation with AEE. When τav was low, AEE tended to be high, and conversely, when AEE was high, τav was low.

4. Conclusions

The impact of gable roofs with varying slope angles and different vertical opening positions was investigated. The opening configurations included: (1) top–top, (2) top–bottom, and (3) bottom–top, with gable roof angles of 0°, 9°, 18°, 27°, 36°, and 45°, hence a total number of 18 combinations. The design of the computational domain adhered to best practices outlined in the literature and was simulated numerically with the inclusion of an atmospheric boundary layer (ABL). Reference grids underwent grid sensitivity analysis, and numerical simulations were carried out using the SST k-omega model, with results demonstrating good consistency. Subsequently, an analysis of internal velocities, pressure coefficients, AOA, AEE, and other parameters was conducted, contributing to the advancements made in this study.

(1) In all configurations in this paper, the slope angle of the roof and the vertical opening positions did appear to have a significant impact on the wind-pressure coefficients.
(2) In terms of flow velocity, favorable opening positions were more conducive to an acceleration in velocity. With the top–bottom opening configuration, the wind speed at the windward opening was greater than that in the other two opening configurations.
(3) The highest AEE was 48.1%, obtained with the opening configuration of top–bottom and a roof slope angle of 45°, while the lowest AEE was 31.4%, achieved with the opening configuration of bottom–top and a roof slope angle of 27°.
(4) When the opening configuration was top–bottom, the slope angle of gable roofs had a significant impact on AEE, with a 5.5% relative increase in AEE for a 45° gable roof compared to a flat roof. This configuration is encouraged for use in buildings.
(5) When the opening configuration was bottom–top, the slope angle of the gable roof had a very limited impact on the profile wind speed and AEE. Therefore, it is not recommended to use this opening configuration in buildings.

In conclusion, this research has shown that in cross-ventilation, the slope angle of gable roofs and the vertical opening positions play crucial roles in internal airflow, pressure coefficients, AEE, AOA, and other parameters. It is hoped that through this study, buildings' reliance on mechanical cooling can be reduced. Future work should include, but not be limited to, researching wind speed profiles and pressure coefficients under different wind directions or introducing various types of guiding windows and more complex ventilated roofs on the basis of existing openings and studying their ventilation performance.

Author Contributions: Conceptualization, J.S., C.Z. and Y.L.; methodology, J.S.; software, C.Z.; validation, J.S.; formal analysis, J.S., C.Z. and Y.L.; investigation, C.Z.; resources, J.S.; data curation, C.Z.; writing—original draft preparation, J.S.; writing—review and editing, J.S., C.Z. and Y.L.; supervision J.S., C.Z. and Y.L.; project administration, J.S., C.Z. and Y.L. All authors have read and agreed to the published version of the manuscript.

Funding: The authors are grateful for the support of the National Natural Science Foundation of China Youth Fund Project (52208095), the Chongqing Municipal Science and Technology Commission

"Doctoral Direct Train" Project (CSTB2022BSXM-JCX0162), the Chongqing Municipal Education Commission Science and Technology Project (KJQN202100736), and Research on the construction of Genetic Map of Traditional Villages in Chongqing Drieven by Digital Intelligence (23SKGH130).

Data Availability Statement: Not applicable.

Conflicts of Interest: The authors declare no conflict of interest.

References

1. Chenari, B.; Carrilho, J.D.; da Silva, M.G. Towards sustainable, energy-efficient and healthy ventilation strategies in buildings: A review. *Renew. Sustain. Energy Rev.* **2016**, *59*, 1426–1447. [CrossRef]
2. China Building Energy Consumption and Carbon Emissions Research Report 2022. Available online: https://www.cabee.org/site/content/24420.html (accessed on 29 July 2023).
3. Schulze, T.; Eicker, U. Controlled natural ventilation for energy efficient buildings. *Energy Build.* **2013**, *56*, 221–232. [CrossRef]
4. Jiang, Z.; Kobayashi, T.; Yamanaka, T.; Sandberg, M. A literature review of cross ventilation in buildings. *Energy Build.* **2023**, *291*, 113143. [CrossRef]
5. Omrani, S.; Garcia-Hansen, V.; Capra, B.R.; Drogemuller, R. Effect of natural ventilation mode on thermal comfort and ventilation performance: Full-scale measurement. *Energy Build.* **2017**, *156*, 1–16. [CrossRef]
6. Karava, P.; Stathopoulos, T.; Athienitis, A.K. Airflow assessment in cross-ventilated buildings with operable façade elements. *Build. Environ.* **2011**, *46*, 266–279. [CrossRef]
7. Zhang, X.; Weerasuriya, A.U.; Wang, J.; Li, C.Y.; Chen, Z.; Tse, K.T.; Hang, J. Cross-ventilation of a generic building with various configurations of external and internal openings. *Build. Environ.* **2022**, *207*, 108447. [CrossRef]
8. Pierre, L.S.; Kopp, G.A.; Surry, D.; Ho, T.C.E. The UWO contribution to the NIST aerodynamic database for wind loads on low buildings: Part 2. Comparison of data with wind load provisions. *J. Wind. Eng. Ind. Aerodyn.* **2005**, *93*, 31–59. [CrossRef]
9. Quan, Y.; Tamura, Y.; Matsui, M. Mean Wind Pressure Coefficients on Surfaces of Gable-Roofed Low-Rise Buildings. *Adv. Struct. Eng.* **2007**, *10*, 259–271. [CrossRef]
10. Holmes, J.D. Wind pressures on tropical housing. *J. Wind Eng. Ind. Aerodyn.* **1994**, *53*, 105–123. [CrossRef]
11. Holmes, J.D.; Paterson, D.A. Mean wind pressures on arched-roof buildings by computation. *J. Wind. Eng. Ind. Aerodyn.* **1993**, *50*, 235–242. [CrossRef]
12. Stathopoulos, T.; Mohammadian, A.R. Modelling of wind pressures on monoslope roofs. *Eng. Struct.* **1991**, *13*, 281–292. [CrossRef]
13. Cui, B. Wind Effects on Monosloped and Sawtooth Roofs. Ph.D. Thesis, Clemson University, Clemson, SC, USA, 2007.
14. Liu, Z.; Yu, Z.; Chen, X.; Cao, R.; Zhu, F. An investigation on external airflow around low-rise building with various roof types: PIV measurements and LES simulations. *Build. Environ.* **2020**, *169*, 106583. [CrossRef]
15. Prasad, D.; Uliate, T.; Ahmed, M.R. Wind Loads on Low-Rise Building Models with Different Roof Configurations. *Int. J. Fluid Mech. Res.* **2009**, *36*, 231–243. [CrossRef]
16. Vaishnani, Y.; Ali, S.F.; Joshi, A.; Rakshit, D.; Wang, F. Thermal performance analysis of a naturally ventilated system using PMV models for different roof inclinations in composite climatic conditions. *J. Braz. Soc. Mech. Sci. Eng.* **2020**, *42*, 124. [CrossRef]
17. Esfeh, M.K.; Sohankar, A.; Shahsavari, A.R.; Rastan, M.R.; Ghodrat, M.; Nili, M. Experimental and numerical evaluation of wind-driven natural ventilation of a curved roof for various wind angles. *Build. Environ.* **2021**, *205*, 108275. [CrossRef]
18. Kosutova, K.; van Hooff, T.; Vanderwel, C.; Blocken, B.; Hensen, J. Cross-ventilation in a generic isolated building equipped with louvers: Wind-tunnel experiments and CFD simulations. *Build. Environ.* **2019**, *154*, 263–280. [CrossRef]
19. Tai, V.C.; Kai-Seun, J.W.; Mathew, P.R.; Moey, L.K.; Cheng, X.; Baglee, D. Investigation of varying louver angles and positions on cross ventilation in a generic isolated building using CFD simulation. *J. Wind. Eng. Ind. Aerodyn.* **2022**, *229*, 105172. [CrossRef]
20. Perén, J.I.; van Hooff, T.; Leite, B.C.C.; Blocken, B. Impact of eaves on cross-ventilation of a generic isolated leeward sawtooth roof building: Windward eaves, leeward eaves and eaves inclination. *Build. Environ.* **2015**, *92*, 578–590. [CrossRef]
21. Peren, J.I.; van Hooff, T.; Ramponi, R.; Blocken, B.; Leite, B.C.C. Impact of roof geometry of an isolated leeward sawtooth roof building on cross-ventilation: Straight, concave, hybrid or convex? *J. Wind. Eng. Ind. Aerodyn.* **2015**, *145*, 102–114. [CrossRef]
22. Perén, J.I.; van Hooff, T.; Leite, B.C.C.; Blocken, B. CFD analysis of cross-ventilation of a generic isolated building with asymmetric opening positions: Impact of roof angle and opening location. *Build. Environ.* **2015**, *85*, 263–276. [CrossRef]
23. Atmaca, M. Wind tunnel experiments and CFD simulations for gable-roof buildings with different roof slopes. *Acta Phys. Pol. A* **2019**, *135*, 690–693. [CrossRef]
24. Tominaga, Y.; Akabayashi, S.-I.; Kitahara, T.; Arinami, Y. Air flow around isolated gable-roof buildings with different roof pitches: Wind tunnel experiments and CFD simulations. *Build. Environ.* **2015**, *84*, 204–213. [CrossRef]
25. Heidari, A.; Sahebzadeh, S.; Dalvand, Z. Natural Ventilation in Vernacular Architecture of Sistan, Iran; Classification and CFD Study of Compound Rooms. *Sustainability* **2017**, *9*, 1048. [CrossRef]
26. Wu, Y.; Gao, N.; Niu, J.; Zang, J.; Cao, Q. Numerical study on natural ventilation of the wind tower: Effects of combining with different window configurations in a low-rise house. *Build. Environ.* **2021**, *188*, 107450. [CrossRef]
27. Cruz-Salas, M.V.; Castillo, J.A.; Huelsz, G. Experimental study on natural ventilation of a room with a windward window and different windexchangers. *Energy Build.* **2014**, *84*, 458–465. [CrossRef]

28. Horan, J.M.; Finn, D.P. Sensitivity of air change rates in a naturally ventilated atrium space subject to variations in external wind speed and direction. *Energy Build.* **2008**, *40*, 1577–1585. [CrossRef]
29. Asfour, O.S.; Gadi, M.B. Using CFD to investigate ventilation characteristics of vaults as wind-inducing devices in buildings. *Appl. Energy* **2008**, *85*, 1126–1140. [CrossRef]
30. Montazeri, H.; Azizian, R. Experimental study on natural ventilation performance of one-sided wind catcher. *Build. Environ.* **2008**, *43*, 2193–2202. [CrossRef]
31. Alghaseb, M.A.; Hassen, W.; Mesloub, A.; Kolsi, L. Effect of Heat Source Position in Fluid Flow, Heat Transfer and Entropy Generation in a Naturally Ventilated Room. *Mathematics* **2022**, *10*, 178. [CrossRef]
32. Hnaien, N.; Hassen, W.; Kolsi, L.; Mesloub, A.; Alghaseb, M.A.; Elkhayat, K.; Abdelhafez, M.H.H. CFD Analysis of Wind Distribution around Buildings in Low-Density Urban Community. *Mathematics* **2022**, *10*, 1118. [CrossRef]
33. Van Hooff, T.; Blocken, B.; Aanen, L.; Bronsema, B. A venturi-shaped roof for wind-induced natural ventilation of buildings: Wind tunnel and CFD evaluation of different design configurations. *Build. Environ.* **2011**, *46*, 1797–1807. [CrossRef]
34. Blocken, B.; van Hooff, T.; Aanen, L.; Bronsema, B. Computational analysis of the performance of a venturi-shaped roof for natural ventilation: Venturi-effect versus wind-blocking effect. *Comput. Fluids* **2011**, *48*, 202–213. [CrossRef]
35. Available online: http://wind.arch.t-kougei.ac.jp/system/eng/contents/code/tpu (accessed on 29 July 2023).
36. AIJ. *Recommendations for Loads on Buildings*; Architectural Institute of Japan: Tokyo, Japan, 2004.
37. Ramponi, R.; Blocken, B. CFD simulation of cross-ventilation for a generic isolated building: Impact of computational parameters. *Build. Environ.* **2012**, *53*, 34–48. [CrossRef]
38. Yoshie, R.; Mochida, A.; Tominaga, Y.; Kataoka, H.; Harimoto, K.; Nozu, T.; Shirasawa, T. Cooperative project for CFD prediction of pedestrian wind environment in the Architectural Institute of Japan. *J. Wind Eng. Ind. Aerodyn.* **2007**, *95*, 1551–1578. [CrossRef]
39. Wilcox, D.C. *Turbulence Modeling for CFD*; DCW Industries: La Canada, CA, USA, 1998; Volume 2, pp. 103–217.
40. Hang, J.; Li, Y. Age of air and air exchange efficiency in high-rise urban areas and its link to pollutant dilution. *Atmos. Environ.* **2011**, *45*, 5572–5585. [CrossRef]
41. Chanteloup, V.; Mirade, P.-S. Computational fluid dynamics (CFD) modelling of local mean age of air distribution in forced-ventilation food plants. *J. Food Eng.* **2009**, *90*, 90–103. [CrossRef]

buildings

MDPI

Article

Exploring the Implementation Path of Passive Heat-Protection Design Heritage in Lingnan Buildings

Hui Ji [1], Shuqi Wu [1], Bishan Ye [1], Shuxi Wang [2], Yuqing Chen [3] and Ji-Yu Deng [1,*]

[1] School of Architecture and Urban Planning, Guangdong University of Technology, Guangzhou 510080, China; archjihui@gdut.edu.cn (H.J.); archwsq0130@163.com (S.W.); arcybs0413@163.com (B.Y.)
[2] Shuifa Energy Engineering Co., Ltd., Zhuhai 519000, China; wangshuxi@zhsye.com
[3] Guangdong Architectural Design & Research Institute Co., Ltd., Guangzhou 510080, China; cyq@gdadri.com
* Correspondence: jiyudeng@gmail.com

Abstract: To achieve indoor thermal comfort via natural ventilation, traditional buildings in South China's Lingnan region have evolved distinct features tailored to the hot and humid climate conditions, involving site planning, function layout, and construction techniques. This study delves into the influences of these features on aspects such as sun-shading, ventilation, and heat insulation. By analyzing over ten Lingnan buildings in both the traditional and modern forms, several representative standardized models have been developed. Through a hybrid approach of combining qualitative and quantitative methodologies, including simulations, quantifications, and comparisons, several passive heat-protection measures commonly employed in Lingnan buildings were examined and evaluated. The effectiveness of shading, ventilation, and heat insulation in both traditional and modern buildings was assessed, resulting in the compilation of design principles for passive heat protection in buildings located in similar climatic zones. Key findings include (1) Shading: traditional methods reduce sunlight by 54.55%, while modern buildings enhance shading by applying new materials; (2) ventilation: traditional design achieves an outdoor wind speed of 1.5 m/s, improving thermal comfort, while modern Lingnan buildings optimize these principles; (3) insulation: traditional techniques maintain indoor temperatures below 26.0 °C, and modern buildings introduce innovation solutions for improved thermal insulation. In summary, traditional Lingnan design effectively addresses the challenges of the hot and humid climate by employing passive strategies for thermal comfort. Modern Lingnan buildings, in turn, preserve these principles while introducing innovative approaches.

Keywords: Lingnan buildings; thermal comfort; passive design; architectural design; heat protection

1. Introduction

The construction sector plays a pivotal role in global energy consumption, accounting for roughly 30% of total energy usage and more than 50% of electricity demand worldwide [1]. In China, buildings alone contribute to 25% of the country's total energy consumption [2]. While the operational phase of maintaining indoor thermal conditions and basic services constitutes the majority of building energy consumption [3,4], the energy efficiency of a building is predominantly shaped by the design decisions made during the initial planning phase [5]. This implies that designers have the opportunity to select optimal design solutions that can significantly enhance energy efficiency and promote the development of environmentally friendly buildings [6]. To achieve this, extensive efforts are being made to identify both active and passive strategies aimed at improving building energy efficiency [7–11].

However, it is noteworthy that heatwaves have become more frequent, long-lasting, and increasingly intense, especially after the 1990s [12]. By the year 2022, heatwaves in China had doubled in both magnitude and frequency compared to the level of the 1990s, with Southern China experiencing the most significant increases [13,14]. Under a 1.5 °C warming limit, it is projected that the average number of heatwave days and their duration across China will increase by 10.8 days and 3.9 days, respectively [15]. The co-occurrence of drought and heatwave is also on the rise, with a notable increase at a rate of 7–11% per decade from 1990 to 2022. This co–occurrence further intensifies heatwaves [16]. During the summer, the indoor thermal environment is primarily influenced by the absorption and transmission of thermal radiation through the building envelope and the penetration of solar radiation through openings. Simultaneously, the outdoor thermal environment is affected by direct solar radiation and stagnant heat in the absence of wind. Traditional passive energy-saving design techniques for Lingnan buildings aim to physically separate the indoor and outdoor environments, thereby reducing overall energy consumption [17,18]. During the operational phase, significant heat exchange occurs, resulting in the transfer of outdoor heat into the indoor environment, which can lead to what is often referred to as the "oven effect" in hot climates. This effect significantly reduces the thermal comfort of occupants. However, passive energy-efficient design strategies can effectively minimize heat gain into the interior space, leading to lower indoor temperatures and enhanced thermal comfort for occupants. Therefore, the passive thermal design of buildings has a substantial impact on both energy efficiency and indoor environmental quality [19,20].

Currently, cities are confronted with the dual challenges of global warming and the urban heat island (UHI) effect, with buildings playing a significant role as both contributors to and sufferers of urban overheating. Researchers have proposed that the concept of a GB-based UHIM system, or "zero UHI impact building", or "zero-heat building", or "microclimate neutral building", aiming to achieve a zero-heat impact on surrounding environments through reasonably designing and operating buildings, or depending on innovative techniques to eliminate the excessive heat, on the basis of GB's goals [21]. Studies on natural ventilation in building atria have demonstrated its ability to lower indoor temperatures, remove humidity, and reduce air conditioning energy consumption while providing a healthier and more comfortable indoor environment [22]. Dehghani-sanij [23] has proposed the use of a "wind tower," which is a vertical ventilation design leveraging the "chimney effect". This design is particularly effective for natural ventilation under hot and humid conditions, thereby reducing cooling energy consumption. Furthermore, researchers and engineers have recognized the potential of enhancing building thermal insulation by modifying the characteristics of building envelopes. In recent years, optimization models combined with building energy simulation techniques have been developed to assist architects in identifying optimal design solutions [24,25]. For instance, C.K. Cheung et al. [26] demonstrated a remarkable 31.4% energy saving by adding extruded polystyrene insulation to the walls of selected high-rise apartments in Hong Kong. Balaras et al. [27] found that the insulated buildings in Greece consume 20–40% less energy compared to the uninsulated buildings, while low-permeability structures use 20–40% less energy than their uninsulated counterparts.

The warming trend in Chinese cities has shown an accelerating pace, with the average temperature of Chinese cities in the last two decades reaching its highest point in at least one century (CSIRO, 2015). The urban heat island effect compounds this issue, presenting a significant challenge. Wang [28] conducted an analysis of various mitigation strategies and technologies within the natural and built environment. These strategies include urban greenery, green roofs and walls, water-based technologies, cool roofs, and cool pavements. The aim is to assist local governments in mitigating the impact of rising temperatures on their communities and residents. Li [29] proposed a risk assessment framework for urban heat exposure. This framework works to identify the factors influencing the risk of excess or potentially unsafe heat exposure for individuals, locations, and the environment. To adapt to the hot and humid climate of the Lingnan region, traditional Lingnan architecture incorporates various thermal design measures for indoor and outdoor spaces [30,31]. The primary objective is to offer users optimal thermal comfort through effective strategies involving shading, ventilation, and thermal insulation. In recent decades, passive energy-saving technologies, such as external shading systems, natural ventilation, lightweight concrete wall insulation, and green roofs, have been widely implemented in architectural design (Figure 1) [32–35]. The fusion of technology and art within traditional Lingnan architecture has also influenced modern Lingnan building design to a certain extent. Therefore, it is crucial for contemporary architectural design to not only reflect local culture but also to thoroughly consider the unique regional climate characteristics [36].

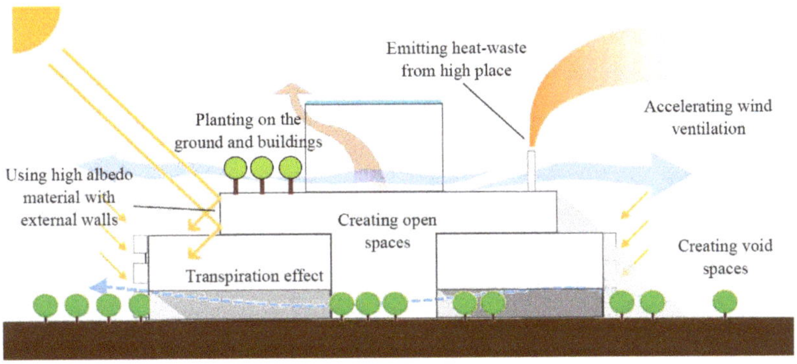

Figure 1. Summary of passive strategies for architectural design.

This paper provides a brief overview of the design characteristics and development trends of Lingnan buildings at the beginning. It then conducts an in-depth analysis of the spatial features of Lingnan buildings, employing simulation and quantitative analysis as research methods to establish a scientific basis for climate-adaptive design approaches. Subsequently, the impact of specific design variables on the thermal environment of buildings is explored, considering factors such as site space, building layout, and construction techniques. The study concludes that shading, ventilation, and insulation are effective design strategies for Lingnan buildings and also summarizes the findings from these three aspects to formulate a comprehensive thermal design strategy tailored specifically to Lingnan buildings.

Furthermore, the study presents the results of simulations and verifications, emphasizing the impact of passive thermal design on both traditional and modern Lingnan buildings. The applied simulation-quantification-comparison analysis method visually demonstrates the scientific rationale of passive technologies for heat protection in Lingnan buildings. The proposed passive thermal design approach, which encompasses shading, ventilation, and insulation, provides valuable insights and technical references for decision makers and architects involved in construction projects within the Lingnan region.

2. Description of the Study Area

2.1. Study Area

This study focuses on the Lingnan region, which is situated to the south of the Five Ridges in South China. The Five Ridges, specifically Yue Cheng Ling, Du Pang Ling, Meng Zhu Ling, Jie Tian Ling, and Dayu Ling, extend from west to east and naturally serve as watersheds between the Yangtze River Basin and the Pearl River Basin. Consequently, the Lingnan region is distinguished by its distinct geographical features when compared to the northern central plains. In a broader context, the Lingnan region, as depicted in Figure 2, encompasses Guangdong, Guangxi, Hainan, Hong Kong, Macao, and the South China Sea islands. It shares borders with Fujian to the east, Yunnan to the west, the South Ridge to the north, and the South China Sea to the south [37]. Guangzhou, a representative city within the Lingnan climate zone, exhibits specific climatic attributes detailed in the accompanying table. All these climatic characteristics play important roles in shaping the features of traditional Lingnan buildings. Figure 3 shows the design features of a typical

Lingnan vernacular dwelling. It is worth noting that all the selected cases discussed in this paper are situated in Guangzhou.

Figure 2. Schematic representation of the extent of the Lingnan region in China.

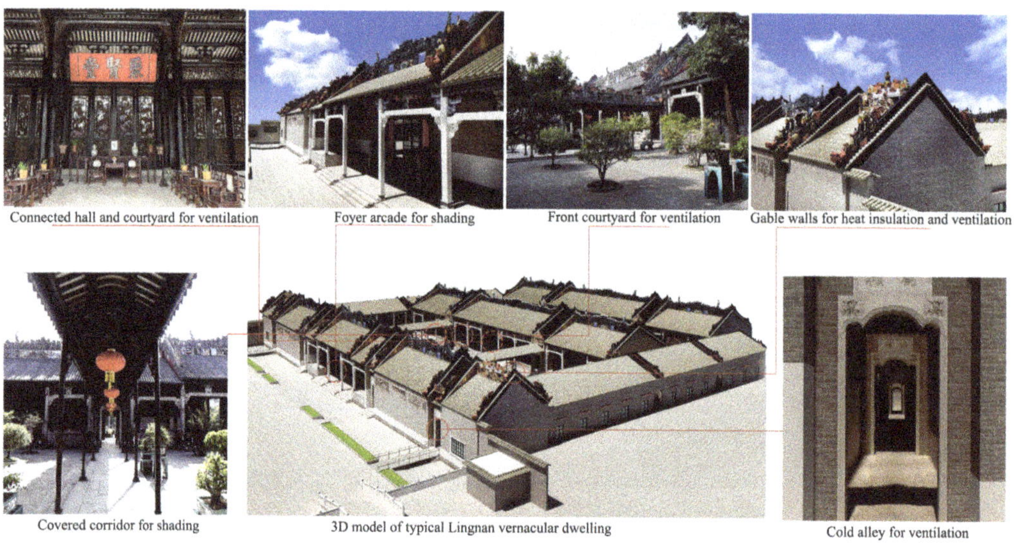

Figure 3. Typical representative of Lingnan architecture: arcade building.

2.2. Climatic Conditions

According to China's Thermal Code for Civil Buildings (GB50176-2016) [38], the Lingnan region falls within the category of hot summer and warm winter climate zones. This climate zone is characterized by a wide temperature range, intense heat radiation, strong solar exposure, and abundant sunlight. The region experiences annual solar radiation hours ranging from 1900 to 2200 h, with an annual solar radiation percentage exceeding 40%. Total annual average solar radiation levels vary from 4500 to 5500 MJ/m^2 [39]. The climate of Guangzhou, a representative city within the Lingnan region, is significantly influenced by its geographical location and natural surroundings. Prevailing winds throughout the year primarily originate from the north and southeast directions, as illustrated in Figure 4. Wind speeds in Guangzhou peak in December, with an average of 3.79 m/s, while September records the lowest wind speeds, averaging at 1.94 m/s. For a more comprehensive understanding, climatic data for Guangzhou City are outlined in Table 1 and illustrated in Figure 5.

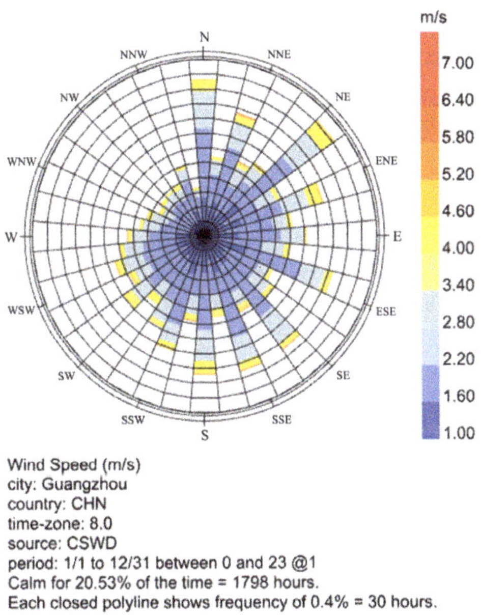

Figure 4. Annual wind rose of Guangzhou.

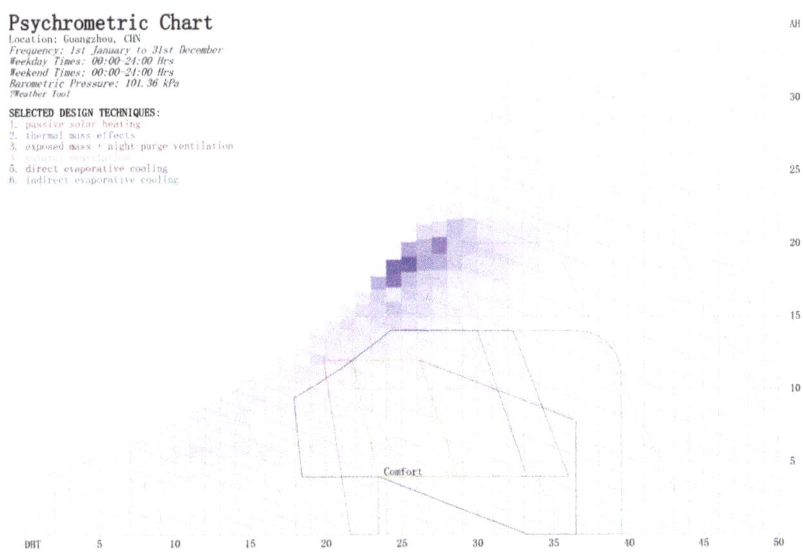

Figure 5. Enthalpy–humidity diagram generated by Weather Tool based on the meteorological data of Guangzhou city.

Table 1. Monthly weather data in Guangzhou city (Source: http://www.cma.gov.cncma.gov.cn, accessed on 23 June 2023).

Months	January	February	March	April	May	June	July	August	September	October	November	December
Average temperature °C	13.8	14.1	18.3	22.3	26.0	27.1	28.8	28.0	27.3	24.3	20.0	15.3
Maximum temperature °C	24.6	22.7	29.7	28.2	32.8	32.9	34.7	34.6	34.9	32.3	28.9	24.2
Minimum temperature °C	6.1	4.7	11.3	17.1	19.4	22.6	22.0	24.0	22.5	18.6	13.0	8.6
Rainfall (mm)	40.9	69.4	84.7	201.2	283.7	276.2	232.5	227.0	166.2	87.3	35.4	31.6
Number of days of rainfall	7.5	11.2	15.0	16.3	18.3	18.2	15.9	16.8	12.5	7.1	5.5	4.9
Average wind speed (m/s)	2.58	3.26	2.54	2.63	2.14	2.67	2.51	2.25	1.94	3.22	2.57	3.79
Average daily solar radiation (kWh/m^2)	3252	2614	2279	2834	4070	3676	4541	4692	4676	5092	4745	4361

Monthly temperature averages in Guangzhou indicate that July and August are the warmest months, with an average monthly peak of 28 °C, while December and January are the coolest, with temperatures averaging around 15 °C. In terms of rainfall, May receives the highest volume, with an average of 283.7 mm, while December sees the lowest, with an average of 31.6 mm. Regarding daily solar radiation duration, October records the highest average daily solar radiation, while March experiences the least, as shown in Figure 6. For a comprehensive overview, Table 1 compiles climatic data, including temperature, rainfall, wind speed, and solar radiation, for Guangzhou.

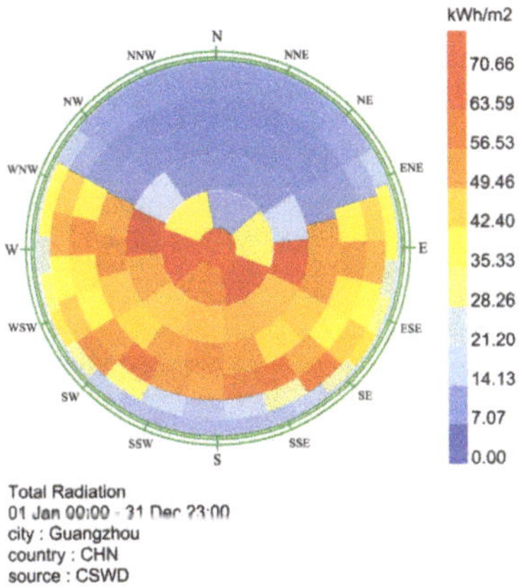

Figure 6. Distribution of solar radiation in Guangzhou.

3. Methodology

3.1. Case Selection

This study focuses on the architectural design in the subtropical region of China, with a specific emphasis on Lingnan architecture. Within the sub-tropical architectural regions of China, the Pearl River Delta area stands out for its distinct climatic characteristics and rapid architectural development. In this context, a series of Lingnan buildings exhibit advanced design concepts and a diverse range of design techniques, showcasing strong typicality and orientation.

Accordingly, a total of 12 typical Lingnan buildings, including 6 traditional and 6 modern Lingnan buildings, were selected to examine their unique passive design features, as shown in Table 2. These cases are geographically centered around Guangzhou, with building completion dates spanning 30 years from 1988 to 2018. These architectural samples are characterized by well-thought-out designs and high technological standards, offering a comprehensive reflection of the development of Lingnan architecture.

Table 2. Selected typical Lingnan buildings.

Ground-Level Elevated for Shading		Overhanging Roof for Shading	
Arcade along Guangzhou Ensi Road	The main entrance of Guangzhou Urban Planning Exhibition Center	The main gate of Xiguan Dawu	The main entrance of Guangzhou Baiyun Airport terminal

Table 2. *Cont.*

Ground-Level Elevated for Shading		**Overhanging Roof for Shading**	
Building complex arrangement for site ventilation			
Site plan of Yu Yin Shan Fang		Site plan of Cantonese Opera Art Museum	
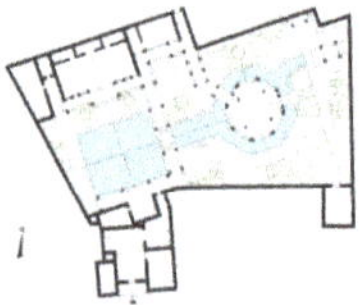			
Climate space for building ventilation			
Courtyard of Bamboo tube house	Patio of Chongdeli residence	Courtyard of vernacular with three rooms and two corridors	Atrium of Library at GDUT

Special wall material for thermal insulation		Special roof structure for thermal insulation	
Oyster shell wall of vernacular house in Siu Chau village, Haizhu district	Masonary wall of Guangzhou Library	Terracotta roof of vernacular house in shawan town, Panyu district	Green roof of Wengyuan meteorological station, Guangzhou

3.2. *Simplified Models*

The essence of this study lies in its utilization of simulation data extracted from simplified models for quantitative analysis, which defines our quantitative research approach. This approach departs from relying on simulation results based on specific building examples and instead utilizes simplified models based on representative traditional and modern Lingnan buildings located in Guangzhou, as illustrated in Table 3. These selected buildings epitomize the unique architectural style of Lingnan, designed to adapt to the local natural climate, and encompass a diverse range of architectural types. Notably, these buildings are of a moderate scale and hold significant research value. Furthermore, the simulations are conducted based on the assumption of ideal experimental conditions. Key climate parameters, such as temperature, wind speed, and solar radiation, are derived from authoritative scientific data published in reputable sources to ensure the accuracy and reliability of the findings.

The development of the simplified model aims to optimize and simplify the form and dimensions of the cases to meet the research requirements. In accordance with the environmental characteristics division of the primary spaces found in Lingnan architecture, this study primarily focuses on three fundamental objects: the external site space, the spatial volume of the building, and the internal building spaces. Accordingly, several representative cases from both traditional Lingnan architecture and modern Lingnan architecture are selected. Consequently, corresponding simplified models are derived. These two types of simplified models are built using professional software, such as Ecotect, Phoenics, and

Ladybug. These software enable researchers to control variables in simulation and generate results for analysis that are scientifically grounded. Subsequently, these analyses facilitate the exploration of how specific thermal protection measures influence the thermal comfort of users.

Table 3. Simplified models of the selected typical Lingnan buildings.

Ground-Level Elevated for Shading		**Overhanging Roof for Shading**	
Arcade along Guangzhou Ensi Road	The main entrance of Guangzhou Urban Planning Exhibition Center	The main gate of Xiguan Dawu	The main entrance of Guangzhou Baiyun Airport terminal
	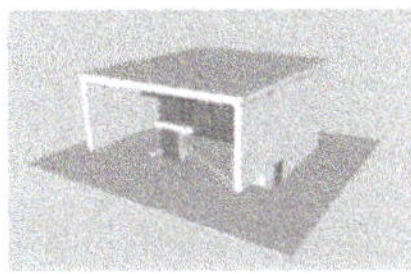	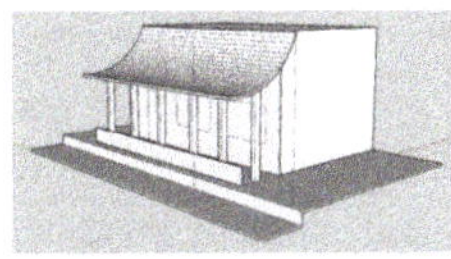	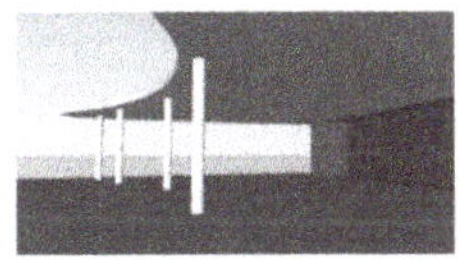
Building complex arrangement for site ventilation			
Site plan of Yu Yin Shan Fang		Site plan of Cantonese Opera Art Museum	
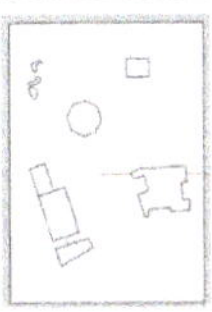	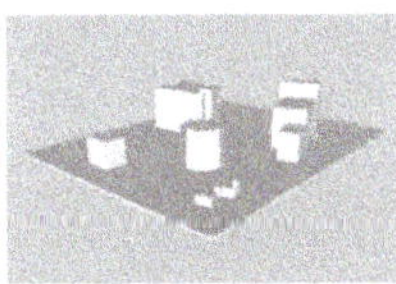		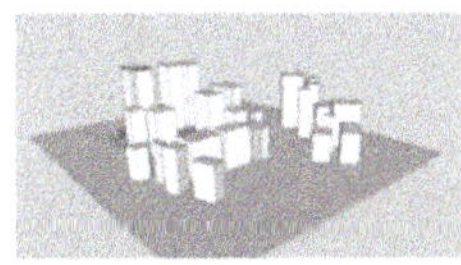
Climate space for building ventilation			
Courtyard of Bamboo tube house	Patio of Chongdeli residence	Courtyard of vernacular with three rooms and two corridors	Atrium of Library at GDUT
	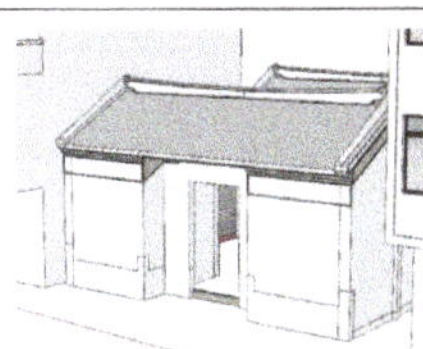	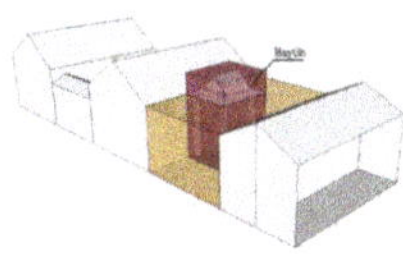	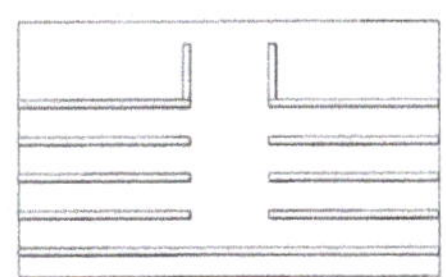
Special wall material for thermal insulation		Special roof structure for thermal insulation	
Oyster shell wall of vernacular house in Xiaozhou village, Haizhu district	Masonry wall of Guangzhou Library	Terracotta roof of vernacular house in Shawan town, Panyu district	Green roof of Wengyuan meteorological station, Guangzhou
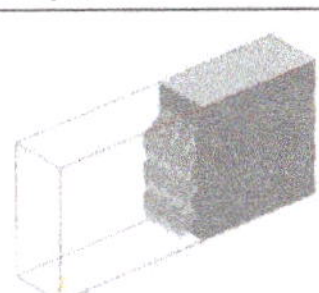			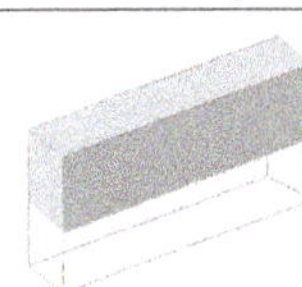

3.3. Combining Quantitative and Qualitative Analysis

The results of the environmental simulations consist of two primary components: the data distribution map and quantitative data. The distribution map provides an overview of the general environmental trends in the study area, while the analysis of sampled quantitative data offers a precise examination of factors such as wind patterns, solar radiation levels, and temperature fluctuations. Building upon this foundational data and considering real-world conditions, a qualitative analysis of individual design variables is conducted to assess whether specific design measures enhance a building's adaptability to hot and humid climates. The study consolidates critical technical aspects related to shading, ventilation, and insulation. Subsequently, it summarizes passive heat-protection design strategies within these three domains to offer practical guidance. By employing a combination of quantitative and qualitative methodologies, the study ensures the objectivity of its conclusions. It also equips architects with a means of rapidly assessing the effectiveness of design measures related to shading, ventilation, and thermal insulation during the early stages of building planning.

3.4. *Simulation-Quantification-Comparison Analysis*

To comprehensively assess passive thermal design in both traditional and contemporary Lingnan buildings, this study introduces a hybrid approach that seamlessly integrates simulation, quantification, and comparative analysis. By combining building simulation analysis with visualization techniques, this method offers a compelling demonstration of the scientific validity behind these design practices. As depicted in Figure 7, this conceptual framework outlines the proposed building simulation visualization method, which consists of three key steps: typical case selection, the development and validation of simplified models, and building simulation analysis. This method Leverages simulation tools such as Ecotect, Phoenics, and the Grasshopper-based Ladybug plug-in to facilitate scenario simulations based on the adaptability of the building and site characteristics. The resulting simulation data are carefully processed to mimic real-world conditions, leading to a detailed analysis. This analysis, in turn, yields specific passive heat-protection design strategies meticulously tailored for Lingnan buildings. Through this approach, the study effectively conveys the effectiveness of heat protection in Lingnan buildings, providing valuable insights for architects and researchers engaged in sustainable design practices.

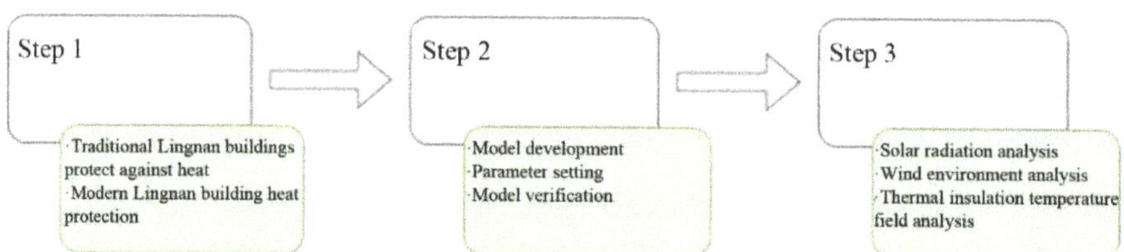

Figure 7. Flow of the proposed building simulation visualization method.

3.5. *Detailed Configurations for Simulations*

3.5.1. Solar Radiation

In this study, the typical meteorological year (TMY) weather data of Guangzhou served as the benchmark weather file. The shading scenario was parameterized using the Rhino platform, and the weather file was imported by utilizing the Ladybug tool (source: https://www.ladybug.tools/epwmap/, accessed on 26 June 2023). for solar radiation simulations. The threshold was set at 300 kWh per unit. The noon time of 12:00 on the summer solstice (22nd June) was selected as the simulation time. The analysis area included the ground-level elevated space and street space, which were divided into a grid with a resolution of 0.2 m × 0.2 m for simulations. The solar radiation simulations were conducted based on four cases: Shops along Ensi Road, Xiguan Dawu, Guangzhou Baiyun Airport Terminal (GBAT), and Guangzhou Urban Planning Exhibition Center (GUPEC).

3.5.2. Outdoor Wind Environment of Building Complexes

The WinAir plug-in from the Ecotect platform was used to simulate the site wind environment of the building complex cases by applying the TMY weather file of Guangzhou. The results were observed at a height of 1.5 m above the ground level. The noon time of 12:00 on the summer solstice (22nd June) was also selected as the simulation time. The site area was divided into a grid with a resolution of 1 m × 1 m for simulation. The wind direction was set to south with a speed of 2.7 m/s. The wind simulation was applied to two cases: Yu Yin Shan Fang and Cantonese Opera Art Museum. To assess the impact of variations in building height, façade dimension, block shape, and orientation of Lingnan buildings on the external wind environment, this study focused on the layout of building groups in the external wind environment. The wind shadow area was determined with a wind speed threshold of 0.2 m/s. Table 4 lists the impact of wind speed on people's work and activities [40].

Table 4. Evaluation of wind speed perception.

Wind Speed (m/s)	Impact on Work and Activities
0~0.25	Imperceptible
0.25~0.5	Pleasant
0.5~1.0	Generally pleasant
1.0~1.5	Unpleasant
1.5~7	Extreme unpleasant

3.5.3. Indoor Wind Environment of Individual Building

The simulations in this group were conducted based on 4 scenarios: bamboo tube house, Chongdeli residence, vernacular with three rooms and two corridors, and the library at Guangdong University of Technology (GDUT). They were simulated by applying the incoming wind with a speed of 1.5 m/s at a reference height of 10 m above the ground level, and 13:00 pm on the summer solstice (22nd June) was set as the simulation time. The wind vector diagram at key sections was analyzed to identify the patterns in wind pressure and thermal pressure ventilation. The simulation results on two sections were selected for indoor wind environment evaluation. One section was within a cold lane, allowing for the observation of horizontal airflow in it, while the other was in a patio, enabling the observation of the updraft in it. Four profiles were selected as follows.

3.5.4. Heat Transfer through Building Envelope

In this section, one-dimensional steady-state heat transfer simulations for various building elements of the selected cases were conducted by using Phoenics. These elements involve the oyster shell wall of a vernacular house, the masonry wall of Guangzhou Library, the terracotta roof of a vernacular house, and the green roof of Wengyuan meteorological station. Some detailed configurations are listed as follows (Table 5).

(1) For convective heat transfer between the air, wall/roof, wind speed, direction, and atmospheric temperatures were specifically set at 0.25 m/s, southeast, and 31 °C;
(2) Temperature differences were considered between the inner and outer surfaces of the wall and roof. The outer surface temperature was set at 31.86 °C, while the inner surface temperature was set at 28.19 °C.
(3) Heat convection effects between the air and the wall/roof were taken into account, with heat radiation between the surfaces being neglected.
(4) The thermal parameters and thickness of the materials used in the simulations are presented in Table 5. The total thickness of the wall was 0.06 m, while the total thickness of the roof was 0.10 m.

Table 5. Thermal properties settings of the materials.

Materials	Thermal Conductivity ($w/m^2 \cdot K$)	Specific Heat Capacity ($J/(g \cdot K)$)	Density (Kg/m^3)	Thickness (m)
Oyster shell	0.12	155	1870	0.6
Concrete	1.74	1680	1900	0.6
Terracotta roof	1.433	920	2100	0.6
Green roof	1.35	740	2000	0.6

4. Passive Shading in Lingnan Buildings

Solar radiation from the sun is the primary driver of temperature increase, introducing heat into both the interior and exterior of a building through transparent and non-transparent surfaces. Consequently, effective shading is crucial in blocking or mitigating the impact of direct solar radiation, with the creation of shadows being a fundamental technique to achieve this objective. Hence, shading stands as a critical element in the heat-protection design of Lingnan buildings. Tables 6 and 7 provide a comprehensive overview of traditional shading techniques employed in Lingnan buildings. These include riding tower shading, roof shading, partition fan shading, and stained-glass shading. These time-tested techniques have demonstrated their efficacy in providing shade and serve as valuable references for contemporary shading practices in Lingnan buildings. The calculation process and results of solar radiation for both traditional and modern Lingnan buildings are shown in Figure 8, Figure 9, and Table 8. The "Xia's sun shading" method has set a precedent for contemporary Lingnan building window and roof shading, making it an essential practice in heat-protection strategies. The in-depth examination of shading techniques presented in this study offers architects valuable insights into addressing solar heat gain in Lingnan buildings, aiding in the design of energy-efficient and sustainable structures.

Table 6. List of shading techniques in traditional Lingnan buildings.

Category	Practice	Technical Description	Scope of Application
Arcade	Partial elevation of the ground floor	The ground floor is occupied by stores on the inner side, sidewalks on the outer side, and residences above.	Vernacular dwelling
Roof	Roof structure shading	With sunshade, rain prevention, lighting and other environmental maintenance comprehensive efficiency.	Vernacular dwelling
Partition	Window and door elements shading	Partition doors and windows, generally a group of four, by the child mullion, frame, wipe head, skirt plate four parts.	Vernacular dwelling
Stained-glass	Glass reflective shading	Inlaid colored glass or stained glass.	Vernacular dwelling

Table 7. List of shading techniques in modern Lingnan buildings.

Category	Practice	Technical Description	Scope of Application
Localized overhead shading	Partial elevation of the ground floor	The ground floor contains the main functional rooms on the inner side and the sunken plaza on the outer side.	Modern public buildings
Metal sunshade elements	Roof shading elements	Shading elements made of metal materials, commonly in the form of eaves shading formed by metal roofs, and façade and roof shading formed by metal frames.	Modern public buildings

Table 7. *Cont.*

Category	Practice	Technical Description	Scope of Application
Electronically controlled color-changing glass sunshade	Glass reflective shading	By laminating the existing glass with dimming film or installing electronically controlled dimming glass, and adjusting the light transmission capacity of the glass through a control system.	Modern public and residential buildings
Vegetative shading	Plant reflective shading	Plant shading is the use of some kind of plant to block the excessive light that we do not need, and at the same time has the functions of heat preservation and heat insulation.	Modern public and residential buildings

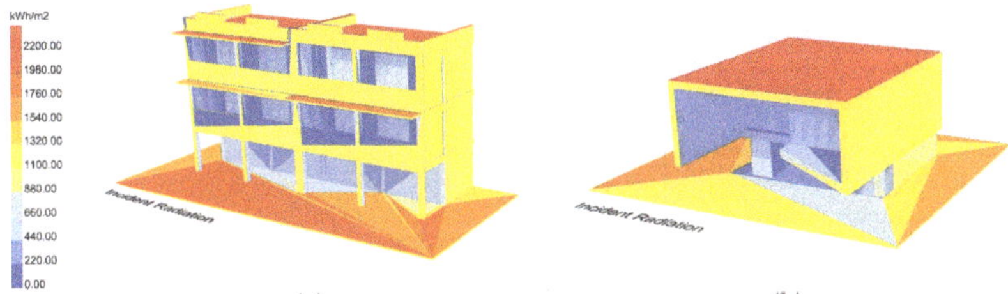

(a) (b)

Figure 8. Comparison of solar radiation results between traditional and modern volume shading. (**a**) Arcade along Ensi Road. (**b**) Main entrance of GUPEC.

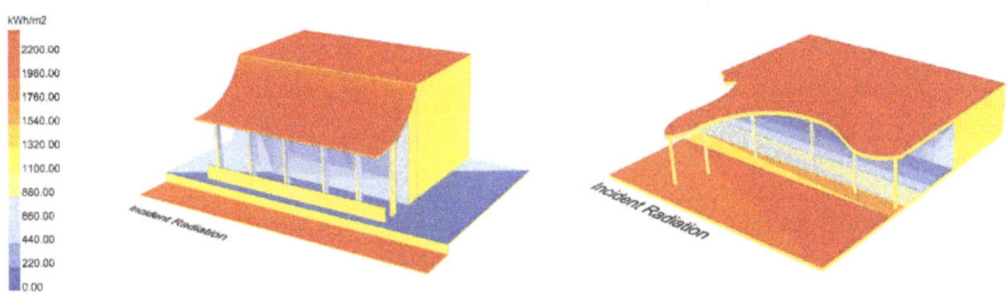

(a) (b)

Figure 9. Comparison of solar radiation results between traditional and modern roof shading. (**a**) Xiguan Dawu. (**b**) Main entrance at GBAT.

Table 8. Comparison of solar radiation between traditional and modern Lingnan buildings.

Form	Average Solar Radiation without Shading (KWh)	Average Solar Radiation with Shading (KWh)	Solar Radiation Reduction Rate (%)
The arcade shaded the sun	1109	504	54.55%
Sunken Plaza Shade	1692	497	70.63%
Floating eaves shade the sun	1208	901	25.41%
Large roof shading	1238	727	41.28%

4.1. Building Volume Shading

The concept of self-shading design in architectural form involves the purposeful utilization of architectural elements such as concavity, convexity, external projections, and the resultant shadows they create [41]. Traditional Lingnan architecture often incorporates partial building elevation as a prevalent form of self-shading, effectively blocking direct solar radiation. A prime example of this self-shading technique can be found in the riding towers commonly seen in Lingnan towns, where transportation, shading, and ventilation seamlessly merge to provide efficient cooling. In this study, the Ensi arcade was selected as a representative case of self-shading in traditional Lingnan buildings. Considering the hot and humid climate, Ladybug Tools software v1.4.0 was employed for radiation simulation analysis and visualization of the cyclorama by applying the hourly weather data from the past decade for Guangzhou City (source: EPW Map (ladybug.tools)). The results, as outlined in Table 8 and Figure 8, highlight that the concave space on the first floor of the cyclorama received 504.64 kWh of solar radiation, while the unshaded street space received 1109.82 kWh, resulting in an average solar radiation reduction of 54.55%. This measure not only provides pedestrians with a comfortable open space and effective sunshade but also offers shelter from rain. It exemplifies a successful integration of climatic adaptation design into the building.

The concept of concave–convex architectural design is a recurring technique in modern Lingnan architecture. It involves creating suspended architectural designs with voids at lower levels and solid structures at upper levels. An excellent example of this design approach can be observed at the main entrance of the GUPEC, drawing inspiration from the space beneath the riding towers in Lingnan. This design seamlessly bridges the gap between indoor and outdoor spaces. At the GUPEC, a three-story suspended volume, combined with a sunken plaza, generates a substantial shaded area at the entrance. Solar radiation analysis and visualization using Ladybug software, as shown in Table 8 and Figure 8, reveal that the sunken space at the main entrance received 497.74 kWh of solar radiation, while the unshaded street space received 1692.61 kWh, resulting in an average solar radiation reduction of 70.63%. This design approach effectively provides shade and mitigates heat, carrying forward traditional Lingnan shading practices while improving the indoor thermal environment. This approach not only prevents the building envelope from overheating but also reduces the indoor heat load caused by secondary radiation and convection, creating a more comfortable and energy-efficient indoor space.

4.2. Shading of Building Elements

Traditional Lingnan buildings are distinguished by their low and expansive structures, which effectively provide shelter from the sun and protection from rain. A notable example is the renowned ancient residential building known as "Xiguan Dawu". Solar radiation analysis and visualization were conducted using Ladybug software. The results, detailed in Figure 9, reveal that Xiguan Dawu received 901.23 kWh of solar radiation, while the unshaded street space received 1208.95 kWh, resulting in an average solar radiation reduction of 25.41%. This ingenious design utilizes eaves and gable corridors to create shaded areas for indoor–outdoor transition spaces. It not only enhances overall environmental efficiency but also ensures structural stability. Furthermore, it introduces a transitional gray space to the traditional building or courtyard gateway, providing effective sunshade benefits. Additionally, traditional Lingnan architecture incorporates waist eaves to compensate for the absence of roof shading. Robust eave roof structures are commonly employed to fulfill the requirements of single large spaces while providing sunshade, rain protection, and natural light. The design of doors and window sunshades is of paramount importance. The concave shape of doors and windows, achieved through wall thickness, offers comprehensive sun-shading benefits. Floating eaves above windows and doors serve as horizontal shades. Furthermore, the distinctive colored door and window glass used in Lingnan architecture boasts a low shading coefficient, efficiently offering sunshade and creating a distinctive indoor color landscape. The combination of roof, windows, and doors effectively prevents solar radiation heat from directly infiltrating the interior.

While modern roofing materials may differ from traditional tiled roofs, contemporary Lingnan architecture still incorporates features like eaves and outer corridor spaces in roof design to mitigate sun exposure. Take the GBAT as an example, where solar radiation analysis and visualization were conducted using Ladybug software. The results, presented in Table 8 and Figure 9, reveal that the Airport terminal in Guangzhou received 727.85 kWh of solar radiation, while the unshaded street space received 1238.14 kWh, resulting in an average solar radiation reduction of 41.28%. This example illustrates how the roof shading technology of traditional Lingnan buildings has evolved into gable porch shading space created by substantial metal roof eaves and sturdy columns in modern Lingnan architecture. This innovative design not only provides shade for pedestrians but also meets requirements for wind and rain protection. Furthermore, it exhibits characteristics such as lightweight construction, thermal insulation, noise reduction, lightning protection, and innovative aesthetics. Modern Lingnan buildings retain the sunshade considerations of traditional Lingnan architecture while adding a distinctive Lingnan touch [42]. In various modern residential areas and public buildings, the influence of "Xia's sunshade" can still be observed, as seen in Building 14 at South China University of Technology and the outpatient clinic building at Sun Yat-sen University of Medical Sciences. Regarding window shading, modern Lingnan buildings take into account the sunlight angle and building orientation in Guangzhou [6]. They incorporate prefabricated sunshade panels that create staggered projection lines. The combination of different materials, structural forms, and shapes in the shading components of modern Lingnan buildings showcases a more integrated and distinctive Lingnan characteristic compared to the past.

5. Passive Ventilation in Lingnan Buildings

The Lingnan region is known for its "hot and humid" climate, emphasizing the need for a robust ventilation strategy in the area [43]. Tables 9 and 10 provide an overview of the traditional ventilation methods employed in Lingnan architecture, encompassing both group layouts and individual climate spaces. From a holistic layout perspective, it is important to note that the prevailing wind direction in the Lingnan region usually comes from the southeast during the summer. Ensuring unobstructed airflow from this direction has become the primary method for achieving natural ventilation in Lingnan buildings. Additionally, the significance of local ventilation systems that create a "micro-environment" with cooling effects should not be underestimated. This underscores the important role of ventilation as a fundamental aspect of Lingnan architecture. It not only addresses the challenges posed by the region's hot and humid climate but also promotes thermal comfort and sustainability.

Table 9. List of ventilation techniques in traditional Lingnan buildings.

Form	Practice	Technical Description	Scope of Application
Outdoor ventilation	Group layout ventilation	The building complex adopts a decentralized overall layout and the space pattern of low south and high north to obtain good natural ventilation to achieve passive cooling of the building.	Landscape architecture in the Lingnan region
Climate space ventilation	Patio ventilation Corridor ventilation Cold-stream ventilation	Typical "climate spaces" of traditional Lingnan buildings include cold alleys, courtyards, patios, corridors, and so on. Various architectural climate spaces do not exist in isolation, but have a mutual linkage effect, which can work together to solve the thermal comfort problem of the indoor environment of the building.	Landscape architecture in the Lingnan region

Table 10. List of ventilation techniques in modern Lingnan buildings.

Form	Practice	Technical Description	Scope of Application
Outdoor ventilation	Group layout ventilation	The complex adopts the spatial pattern of low south and high north to obtain good natural ventilation and forms an organic ventilation system of patios, courtyards, and corridors to enhance the cooling effect.	Modern public and residential buildings
Localized room ventilation	Atrium ventilation	Ventilation in the atrium is mainly vertical ventilation, and the unique "greenhouse effect" and "chimney effect" of the atrium are conducive to inducing ventilation problems in the atrium and even in the whole building.	Modern public and residential buildings

5.1. Decentralized Group Layout

In situations where a building site is congested and building orientation options are limited, adopting a garden-style building layout, as referenced in [44], can serve as an effective solution to address indoor ventilation and lighting challenges. Traditional Lingnan houses have ingeniously utilized a decentralized overall layout and spatial organization to achieve natural ventilation, facilitating passive cooling. This design approach effectively eliminates indoor humidity and stuffiness, thereby providing occupants with a healthy and comfortable indoor environment. Under similar conditions, if buildings can be spaced apart, additional gap spaces can be incorporated, allowing for the infiltration of natural air into the buildings or facilitating wind circulation between them. Moreover, many large-scale Lingnan building complexes exhibit a notable trend of having lower structures in the south and taller structures in the north. The lower southern side welcomes the entry of summer southeast winds, creating favorable conditions for the wind to enter the rooms and enhancing ventilation through open gardens. Conversely, the higher northern side acts as a barrier against winter northwest winds, preventing them from encroaching on the gardens. It is worth noting that in some Southern villages, a comb ventilation layout is employed, aligning the main village lanes with the dominant wind direction during summers, which leads to excellent ventilation [45]. These architectural strategies emphasize the significance of adapting the building layout to local climate conditions to ensure effective passive cooling and maintain a comfortable indoor environment.

Taking the Lingnan garden-style building Yu Yin Shan Fang in Guangzhou as an example, designed to accommodate the dominant summer wind direction from the southeast, it features an overall layout characterized by "two east–courtyards with sun-shading partitions," "south-facing halls with cleverly designed corridors", and "distinct main and secondary areas with shading for secondary spaces". Through analysis using WinAir software v4 and wind environment simulations, it is evident that the comprehensive utilization of wind pressure ventilation and thermal pressure ventilation within the courtyard, patio, and corridor settings results in an average wind speed of 1.25 m/s, creating a cool and comfortable environment (Figure 10a). The concept of having lower structures in the south and higher structures in the north allows for summer winds to easily flow directly into the building area through open courtyards in the south. Furthermore, due to the denser arrangement of buildings and separate blocks, these gaps serve as natural ventilation "corridors", facilitating accelerated airflow. The thermal pressure ventilation effect becomes apparent as the temperature difference between the top and bottom of the buildings, along with the varying air density, contributes to a cool and comfortable environment. The open courtyard at the front of the building receives prolonged sunlight exposure, causing air temperatures to rise rapidly and resulting in upward airflow. In contrast, the shaded area at the back of the house mitigates the influence of solar radiation, leading to relatively lower air temperatures and downward airflow, which complements the front courtyard. Yu Yin Shan Fang exemplifies the typical layout featuring open and spacious courtyards with large water surfaces, referred to as a sparse front and dense back layout. The wind environment simulation analysis validates the significant positive impact of this layout, as detailed in Figure 10.

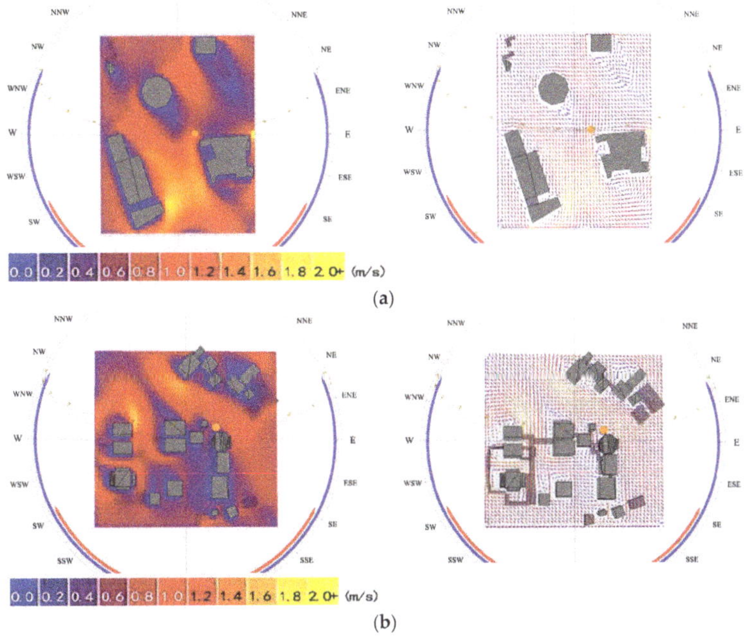

Figure 10. Wind simulation results of Yu Yin Shan Fang and Cantonese Opera Art Museum. (**a**) Wind speed and vector distributions in Yu Yin Shan Fang. (**b**) Wind speed and vector distributions in Cantonese Opera Art Museum.

This spatial arrangement has also been inherited by contemporary architecture, where an organic ventilation system is integrated through the use of patios, courtyards, and porches. These architectural elements vary in scale and serve as both air inlets or outlets, facilitating favorable indoor–outdoor airflow and heat exchange. For instance, the Cantonese Opera Museum is designed to harmonize with the surrounding historic buildings. Inside the museum, a multi-level patio–courtyard–corridor system is implemented. The highest point, Bawo Pavilion, is located in the northwest corner of the garden, while the lower pavilions are staggered on the southeast side and connected by corridors. This results in a general layout of lower structures in the southeast and higher structures in the northwest. Wind environment simulations, conducted using WinAir software, demonstrate effective natural ventilation, with an average wind speed of 1.32 m/s (Figure 10b). Along with wind induction from Liwan Chong and the central water court, the museum successfully creates a favorable overall wind environment in the old urban area. Furthermore, when planning residential areas in the Lingnan region, buildings are ideally oriented to face south. Additionally, architectural designs incorporate a sawtooth arrangement to increase the windward surface area of the buildings [46], resulting in improved ventilation effects.

5.2. Monolithic Building Climate Space

Traditional Lingnan buildings provide ample space designed to regulate the microclimate environment [47]. These buildings incorporate various architectural climate spaces that are interconnected, enhancing indoor and outdoor air convection through techniques like thermal and wind pressure ventilation. This approach aids in heat dissipation and improves thermal comfort. A common feature in traditional Lingnan architecture is the combination of a front courtyard and a back patio. The three-room and two-corridor layout, representing a typical plan form of traditional Lingnan buildings, comprises a main building with three rooms, two corridors, and a front patio connected to three courtyards. This layout exhibits characteristics of external closure and internal openness. The patio serves as a crucial inlet and outlet for air, enabling airflow driven by wind and thermal pressure to enter the rooms. Simulations using Fluent v2021 R1 were conducted to assess the indoor wind environment in the patio. The results, detailed in Figure 11c, indicate that in completely calm outdoor conditions, areas such as building corridors and halls are affected by natural convection ventilation, resulting in airflow rates of approximately 0.6~1.0 m/s. The patio serves a dual purpose by effectively inducing and extracting wind. During periods of strong external wind, the patio functions as a wind pressure vent, while it generates its own thermal pressure ventilation when external wind speeds are low. Consequently, the patio's ventilation efficiency surpasses that of simple air inlets and outlets. Thermal pressure ventilation is particularly significant in this spatial configuration. The relatively small scale of the patio creates cold zones during the day, shielded by surrounding buildings or walls, preventing direct sunlight. This leads to naturally cooler air and lower temperatures. In contrast, the larger courtyard exposed to sunlight becomes a heat source. The interplay between these two sources allows for cold air from the back patio to complement the front courtyard, resulting in cooling through airflow.

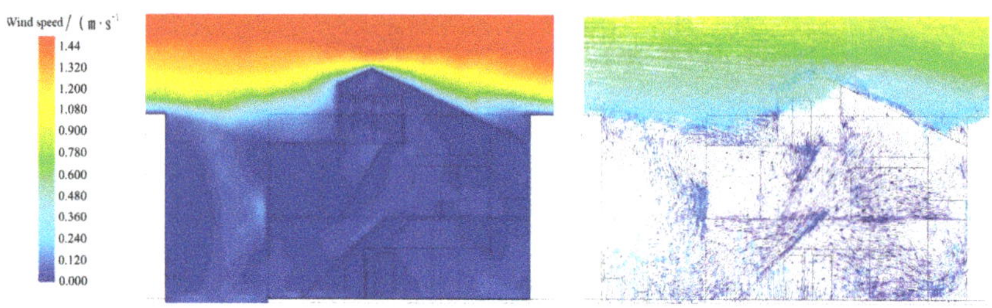

(**a**)

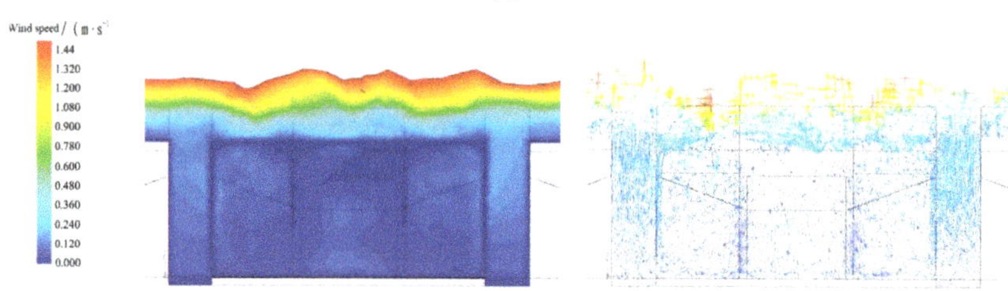

(**b**)

Figure 11. *Cont.*

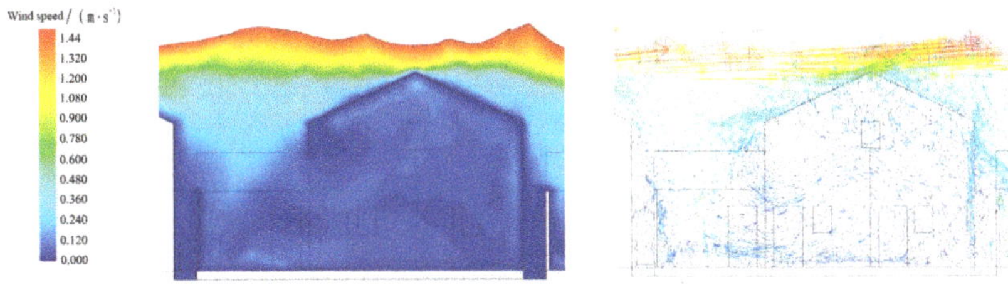

(c)

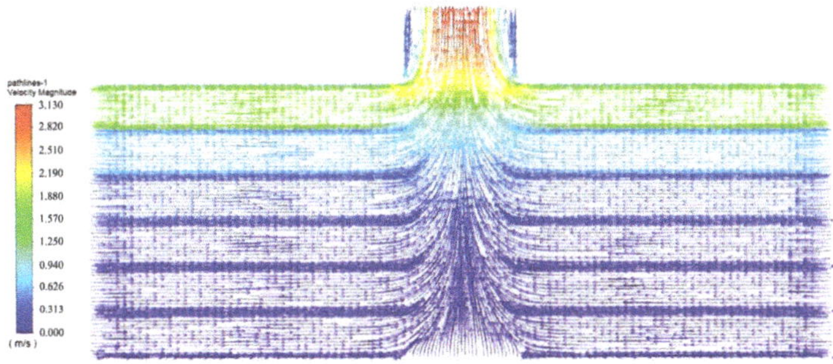

(d)

Figure 11. Wind simulation results in traditional and modern Lingnan buildings. (**a**) Wind velocity and vector distribution in the cold alley of bamboo tube house. (**b**) Wind velocity and vector distribution in the patio of Chongdeli residence. (**c**) Wind velocity and vector distribution in the patio of the vernacular with three rooms and two corridors. (**d**) Wind velocity and vector distribution in the atrium of the library at GDUT.

In addition to the back patio, traditional Lingnan buildings also feature a unique architectural element known as the "cold alley" within the house. Cold alleys play a central role in climate regulation in Lingnan architecture and are particularly prominent in bamboo houses. The Fluent software simulation results reveal that in completely calm outdoor conditions (Figure 11a), areas like building corridors and halls experience natural convection ventilation, resulting in airflows of approximately 0.6~1.0 m/s. Cold alleys serve not only as passageways within the house but also as the principal air ducts for the internal ventilation system. They facilitate the movement of airflow within the house through what is known as the "Venturi effect." When air from open halls or patios enters the confined space of the cold alley, the air mass cannot accumulate significantly, leading to an increase in wind speed within the cold alley. The long, narrow, and tall configuration of the cold alley, oriented in a north–south direction, restricts the penetration of sunlight from the east–west direction. This natural shading effect cools the air within the alley, creating a cold source. Together with other heat sources, such as open and spacious courtyards, this configuration facilitates thermal pressure ventilation. Additionally, the elongated shape of the cold alley promotes wind pressure ventilation, accelerating air movement and enhancing both indoor and outdoor ventilation.

The ventilation elements that are inherent in traditional Lingnan architecture, such as courtyards, patios, and cold alleys, have found a renewed purpose in modern Lingnan architecture. Contemporary Lingnan architectural designs prioritize spatial openness and frequently employ techniques that promote multi-directional openness and excavation, staying true to the inherent spatial permeability of traditional architecture. The integration of cold alleys and patios within modern Lingnan architecture serves as a means to organize and stimulate ventilation, gradually extending a sequence of courtyard spaces from the internal building areas to the external environment. This open layout creates favorable conditions for ventilation. Simulations and analyses conducted by the Chongdeli residence in Guangzhou using Fluent software demonstrate that in the afternoon of midsummer and in completely calm outdoor wind conditions, the air flow rate in building corridors, halls, and other areas is approximately 0.3~0.8 m/s due to natural convection ventilation (Figure 11b). The inclusion of cold alley spaces in these designs establishes connections between outdoor environments and internal open halls, facilitating the introduction of cool air and inducing ventilation. Furthermore, the integration of cold alleys with courtyards and open halls contributes to comfortable natural ventilation in courtyard spaces and blurs the boundaries between indoor and outdoor environments. This results in the creation of pleasant public spaces for relaxation and interaction. The unique spatial configuration of patios remains an irreplaceable component of traditional architecture's sustainable design wisdom. Hence, when incorporating ventilation elements like courtyards and patios into building atriums of varying scales, it is common to position a tall courtyard at the center of the structure. This design approach can be viewed as a modern variation of the traditional deep wells, vertically connecting spaces to create a "big chimney" that spans from the top to the bottom, bridging the indoor and outdoor environments. This "big chimney" functions as a natural wind extractor. Elevating the base of the building allows for the formation of a wind-through hall. For example, the library at GDUT serves as a case study(Figure 11d), where Fluent software is utilized to simulate and analyze the indoor sectional wind environment. The results demonstrate that in the afternoon of midsummer and in completely calm outdoor wind conditions, the air flow rate in building corridors, halls, and other areas is basically 0.8–1.5m/s. The design blends traditional architectural wisdom, creating a modern "climate space system" that reduces the need for artificial climate control, ultimately achieving energy-efficient design. By adhering to Lingnan's natural climate patterns and understanding the appropriate scales for "climate space," a green building space layout strategy tailored to the climatic conditions of the Lingnan region is developed. The building's profile incorporates principles from Lingnan architecture's patio, allowing for vertical ventilation through wind chimneys extending from the first floor to the roof. This organized airflow circulation is achieved through internal interactions, facilitating the smooth exchange of indoor and outdoor airflows. These techniques, reminiscent of the traditional "courtyard-tenjing" in Lingnan houses, can be combined to enhance air circulation.

6. Passive Insulation in Lingnan Buildings

In the Lingnan region, the use of eco-efficient insulation materials and advanced wall masonry techniques in Lingnan buildings serves to facilitate the exchange of heat between the indoor and outdoor environments. The building envelope in Lingnan architecture functions like a living, self-regulating biological skin, adapting to changing environmental conditions. Among the envelope components, the thermal insulation performance plays a pivotal role in determining the overall thermal efficiency of the building. Notably, both the roof and walls are crucial elements of the envelope structure, contributing significantly to thermal insulation and heat preservation. In the context of modern Lingnan buildings, special attention must be given to the arrangement of material layers, as well as the incorporation of double roofs and diverse facades, to maximize the insulation's effectiveness. As elaborated in Tables 11 and 12, various insulation techniques utilized in traditional Lingnan buildings, including roof insulation and façade insulation, provide valuable insights for insulation practices in contemporary Lingnan buildings. These findings offer essential technical guidance for achieving superior insulation performance in modern Lingnan architecture.

Table 11. List of heat insulation techniques in traditional Lingnan building.

Form	Practice	Technical Description	Scope of Application
Roof	Ceramic tile double layer Insulated Roofs	Ceramic tile double-layer heat-insulated roofing is a kind of roofing that takes heat-insulating measures to prevent the sun from directly irradiating the upper surface of the roof. By the upper and lower two layers of roof composition, the lower roof is the main ventilation roof, the upper roof is generally used in the lighter materials, and difference in height of the upper and lower two layers is generally 200mm.	Vernacular dwelling
External wall	Oyster shell wall	Lingnan oyster shell wall is made of local materials, is ecological and environmentally friendly, and its heat transfer coefficient is small, with excellent thermal insulation properties, adapted to the hot and humid climate of Lingnan. The main component of oyster shells is calcium carbonate, which is resistant to erosion and insects, and has the function of wind and moisture protection in the coastal areas of Lingnan.	Vernacular dwelling

Table 12. List of heat insulation techniques in modern Lingnan buildings.

Form	Practice	Technical Description	Scope of Application
Roof	Green roof	Roof greening refers to green planting on the roof, with green plants as the main cover, with nutrient soil layer, water storage layer, etc., together to form a roof system.	Modern public and residential buildings
External wall	Modern insulation	Enhancement of wall thermal insulation and heat preservation performance by using thermal insulation materials	Modern public and residential buildings

6.1. Material Insulation

The walls of traditional Lingnan buildings are often constructed using water-worn and red bricks, which offer excellent insulation and heat preservation properties. In terms of masonry, the gap left in the middle of the brick walls in traditional Lingnan architecture acts as an effective air barrier, enhancing heat preservation and insulation while reducing the need for bricks, mortar, and labor. Modern Lingnan buildings predominantly use wood for doors and windows, which provides outstanding heat preservation capabilities. In particular, black lacquer wood is employed to absorb heat radiation from the environment, contributing to the building's ventilation, heat insulation, and lighting performance. Various styles of window splicing are employed to create partitions between spaces that are both aesthetically pleasing and functionally ventilated for efficient heat dissipation. Furthermore, Guangdong's folk architecture takes a pragmatic approach, emphasizing adaptation to local conditions and the use of locally sourced materials. This approach has resulted in unique craftsmanship, such as the oyster shell walls found in traditional Lingnan architecture. Oyster shells, primarily composed of calcium carbonate, are resistant to erosion and insects, effectively preventing wind and moisture penetration in Lingnan's coastal areas. Through a case analysis of oyster shell houses in Xiaozhou Village, Guangzhou, temperature analysis using Phoenics [48] revealed (Figure 12a) that the outer cavity of the oyster shell wall has a certain heat resistance ability, which reduces the outer boundary temperature of the wall from 45.8 °C to 33.5 °C. The lower the temperature of the oyster shell wall, the higher the temperature drop rate, and 63.4% of the low-temperature field is within the oyster shell wall. Plain white oyster shell walls have higher reflectivity compared to bricks and other ancient building materials. This reduced absorption of solar radiant heat and the irregularly distributed protuberances on the walls minimize direct sunlight exposure. Additionally, the external cavity of oyster shell walls provides a certain degree of heat resistance, reducing the temperature at the outer boundary of the wall. The temperature drop rate increases as one moves towards the interior of the oyster shell wall, creating a larger low-temperature zone within the internal temperature field. This low-temperature zone, similar to an air interlayer due to the presence of air within the oyster shell, exhibits a low coefficient of heat transfer, enhancing its thermal storage capacity and contributing to thermal insulation.

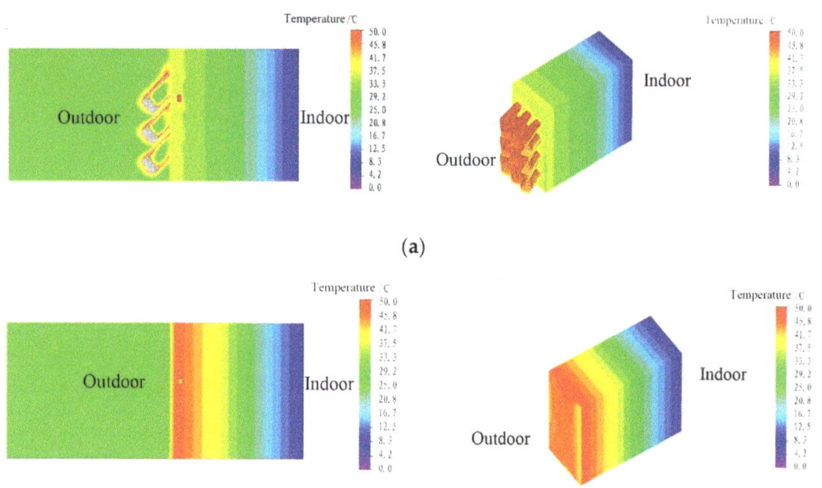

(**a**)

(**b**)

Figure 12. Simulation results for the thermal insulation of different wall materials. (**a**) Temperature distribution across the oyster shell wall. (**b**) Temperature distribution across the masonry wall of Guangzhou Library.

In the past decade, rapid urbanization in China has brought significant advancements in climate-adaptive design in the Lingnan region [49]. The introduction of green building materials, including thermally dimmable glass and heat-insulating wall materials, has revolutionized the heat insulation capabilities of modern Lingnan buildings, surpassing those of traditional materials like oyster shell walls and Manchurian windows. These innovative materials have become essential in the design of contemporary Lingnan buildings. For example, modern Lingnan building facades often incorporate light-colored finishes, reflective coatings, and facade tiles to effectively reduce radiation heat. Vertical greening of building facades has emerged as a crucial ecological compensation measure in green buildings. It not only enhances aesthetics but also contributes to cooling, noise reduction, reduced energy consumption, and decreased carbon emissions. A noteworthy case is the Guangzhou Library, which features Low-E glass windows and doors, providing abundant natural lighting, heat insulation, and comprehensive energy-saving benefits. The library's facade is constructed using light-colored stone with a textured concave and convex design. Temperature field analysis of the library's profile wall using Phoenics reveals (Figure 12b) that the textured facade generates numerous small shadows, significantly reducing solar radiation absorbed by the wall. The temperature of the outer boundary of the wall is reduced from 46.8 °C to 12.8 °C. Moving to the inside of the wall, the temperature drop rate increases, forming a large low-temperature zone in the internal temperature field, effectively achieving a substantial thermal protection effect. This contemporary heat-protection design exhibits similarities to the traditional oyster shell walls of Lingnan architecture. In building design, it is imperative to make thoughtful choices in selecting suitable materials and construction practices to meet green and energy-saving objectives [50].

6.2. Tectonic Insulation

The roofs of traditional Lingnan houses are often adorned with flowers and trees, and sunshade pergolas are constructed to enhance heat insulation. An illustrative example is the use of bamboo trellises on roofs to support climbing plants, which, through transpiration and photosynthesis, dissipate heat and provide effective insulation. Ceramic tiles are a common insulation material used for the roofs of folk residential buildings in Lingnan [51]. Typically, traditional residential architecture features a double-layer tiled roof structure. By conducting temperature field simulations using Phoenics, it becomes evident (Figure 13a) that the combination of bottom and surface tiles forms an active air layer, allowing for ventilation and heat dissipation between the two layers of the roof. This double-layer design offers superior heat insulation compared to a single-layer tiled roof. It not only shields the roof surface from direct sunlight exposure but also utilizes airflow to facilitate heat dissipation. The undulating shadows created by the pitched tiles further contribute to reducing the temperature at the outer boundary of the roof. As can be seen from the temperature field cloud image, when moving to the interior of the roof, the rate of temperature drop increases, which reduces the temperature of the outer boundary of the roof from 48.6 °C to 14.8 °C, leading to the enlargement of the low-temperature zone of the internal temperature field of the roof. This significantly enhances the thermal insulation performance of the roof and effectively serves as a thermal insulation layer.

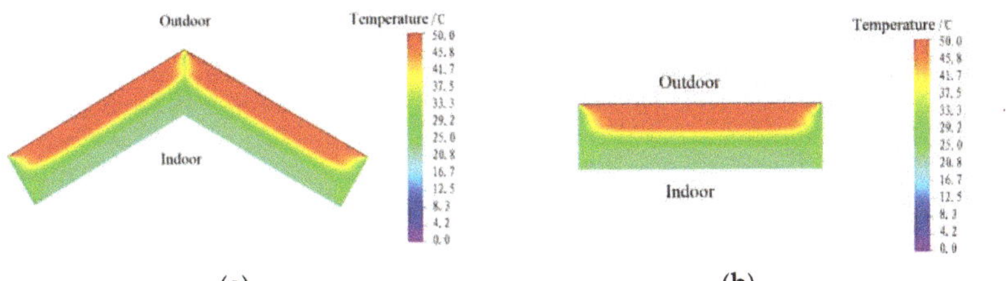

Figure 13. Temperature distribution across different roofs. (**a**) Temperature distribution across the traditional terracotta roof. (**b**) Temperature distribution across the modern green roof.

In modern Lingnan architecture, there is a growing diversity in roof designs, featuring options such as ventilated roofs, water storage roofs, vegetated roofs, and sloped roofs with attic floors [52]. Among these, vegetated roofs with suitable substrate materials play a pivotal role in enhancing the thermal inertia and thermal resistance of the roof, thus delivering effective insulation. Green roofs offer exceptional thermal and heat insulation properties for buildings while also providing added benefits such as structural protection, water storage capacity, and reduced roof runoff, which collectively contribute to an improved microclimate in the vicinity of the building. Water storage roofs leverage the high heat capacity of water and its evaporative properties, effectively absorbing and minimizing heat conduction. This approach prevents the roof panels from overheating under the scorching summer sun. Moreover, the evaporative cooling effect of the water body consumes a significant amount of heat during vaporization, maintaining the water temperature within reasonable limits and reducing the surface temperature of the roof, thereby achieving the desired heat insulation. Passive thermal insulation evaporation roofs, using moisture-absorbing porous materials as heat storage mediums, offer an efficient insulation method [53]. This approach is especially suitable for Lingnan's hot and humid climate. For instance, the

Guangzhou Wengyuan meteorological station incorporates a large green roof with a variety of plantings. The roofing system integrates layers of nutrient soil, water storage, plant root barriers, drainage, and waterproofing, among others. Simulation results show that the green roof effectively delivers thermal insulation properties for the building. As can be seen from the temperature field cloud map (Figure 13b), when moving to the interior of the green roof, the rate of temperature decline increases, reducing the temperature of the outer boundary of the roof from 49.5 °C to 18.6 °C, resulting in the expansion of the low-temperature area of the temperature field inside the roof. This greatly improves the thermal insulation performance of the roof and effectively plays the role of an insulation layer. It also safeguards the building structure, reduces roof runoff, and enhances the microclimate in the vicinity of the building, contributing to the mitigation of the urban heat island effect.

7. Conclusions

In the pursuit of the green, low-carbon concept, modern Lingnan architectural design methods are evolving and adapting. They have, to varying degrees, drawn inspiration from the experiences and practices of passive heat-protection design in traditional Lingnan buildings, building upon this rich heritage. As a result, the energy-efficient design strategies rooted in traditional Lingnan architecture have left a profound imprint on the design of contemporary Lingnan buildings. This paper has presented a comprehensive argument covering three key aspects: shading, ventilation, and heat insulation. It has introduced a qualitative and quantitative method within a case–model–analysis framework (Figure 14), integrating building simulation analysis and visualization. This approach has been illustrated through the selection of representative traditional Lingnan buildings and modern Lingnan structures, which have been analyzed and validated using building simulation and emulation platforms.

(1) Shading and heat protection: Traditional Lingnan architecture's practice of self-shading, including mutual shading from closely arranged building layouts and techniques such as roofing using doors, windows, external porches, and eaves, proves highly effective. These principles can be directly applied in the design of contemporary Lingnan buildings.
(2) Ventilation and heat mitigation: Given the Lingnan region's prolonged hot climate, the incorporation of air ducts in the "southeast to northwest" direction remains a paramount consideration in the design of modern Lingnan buildings. Additionally, innovative interpretations for creating ventilation and cooling effects within the local spaces of modern Lingnan structures are emerging. This includes the flexible use of atriums and elevated floors, especially by exploiting atrium height.
(3) Heat insulation and mitigation: Traditional Lingnan buildings traditionally use air gaps between layers for insulation. Modern Lingnan architecture extends this practice and incorporates new insulation techniques using modern technology. These innovations include green roofs, water storage roofs, and more to enhance heat insulation performance.

Beyond the rich cultural significance of traditional Lingnan architecture, its adaptation to the hot and humid climate through spatial layout, building volume, openings, structural techniques, and greenery is a topic worthy of further exploration. Architects are urged to build upon the demand for heat protection and insulation in Lingnan buildings. Valuable lessons from traditional structures that can be applied and disseminated should be unearthed and passed on, thereby inspiring new architectural creations that harmoniously adapt to the hot and humid climate of the Lingnan region.

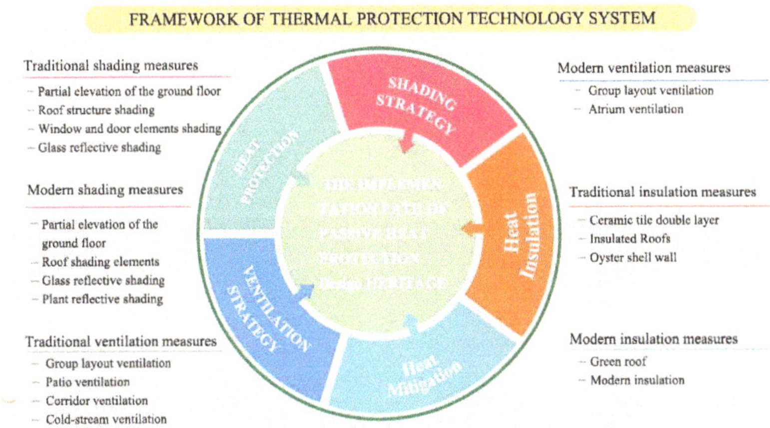

Figure 14. Framework of thermal protection technology system.

Author Contributions: Conceptualization, H.J., J.-Y.D. and Y.C.; methodology, B.Y. and S.W. (Shuqi Wu); investigation, B.Y. and H.J.; resources, S.W. (Shuqi Wu) and S.W. (Shuxi Wang); data curation, B.Y. and S.W. (Shuqi Wu); writing—original draft preparation, H.J. and S.W. (Shuqi Wu); writing—review and editing, S.W. (Shuxi Wang), H.J. and J.-Y.D.; visualization, B.Y. and. S.W. (Shuqi Wu); supervision, H.J. and Y.C. All authors have read and agreed to the published version of the manuscript.

Funding: This research was supported by the National Natural Science Foundation of China (No. 52078140), the Reform of Degree and Postgraduate Education in Guangdong Province (No. 2022JGXM049), and the Undergraduate Teaching Engineering Project of Guangdong University of Technology (No. 2023-51).

Data Availability Statement: The datasets generated during and analyzed during the current study are available from the corresponding author on reasonable request. The data are not publicly available due to privacy.

Acknowledgments: We would like to take this opportunity to express our heartfelt thanks to the Photovoltaic Building Integration Institute of Zhuhai Shuifa Xingye Energy Group and the Green Building Design Research Center of Guangdong Architectural Design and Research Institute for their support and help during the research process.

Conflicts of Interest: Author Shuxi Wang was employed by the company Shuifa Energy Engineering Co., Ltd. and Author Yuqing Chen was employed by the company Guangdong Architectural Design & Research Institute Co., Ltd. The remaining authors declare that the research was conducted in the absence of any commercial or financial relationships that could be construed as a potential conflict of interest.

References

1. Chen, X.; Yang, H.; Zhang, W. Simulation-Based Approach to Optimize Passively Designed Buildings: A Case Study on a Typical Architectural Form in Hot and Humid Climates. *Renew. Sustain. Energy Rev.* **2018**, *82*, 1712–1725. [CrossRef]
2. Ma, G.; Lin, J.; Li, N.; Zhou, J. Cross-Cultural Assessment of the Effectiveness of Eco-Feedback in Building Energy Conservation. *Energy Build.* **2017**, *134*, 329–338. [CrossRef]
3. Delgarm, N.; Sajadi, B.; Delgarm, S. Multi-Objective Optimization of Building Energy Performance and Indoor Thermal Comfort: A New Method Using Artificial Bee Colony (ABC). *Energy Build.* **2016**, *131*, 42–53. [CrossRef]
4. Yang, Z.; Becerik-Gerber, B. A Model Calibration Framework for Simultaneous Multi-Level Building Energy Simulation. *Appl. Energy* **2015**, *149*, 415–431. [CrossRef]
5. Häkkinen, T.; Kuittinen, M.; Ruuska, A.; Jung, N. Reducing Embodied Carbon during the Design Process of Buildings. *J. Build. Eng.* **2015**, *4*, 1–13. [CrossRef]
6. Jalaei, F.; Jrade, A. An Automated BIM Model to Conceptually Design, Analyze, Simulate, and Assess Sustainable Building Projects. *J. Constr. Eng.* **2014**, *2014*, 1–21. [CrossRef]

7. Deng, Q.; Jiang, X.; Zhang, L.; Cui, Q. Making Optimal Investment Decisions for Energy Service Companies under Uncertainty: A Case Study. *Energy* **2015**, *88*, 234–243. [CrossRef]
8. Guo, K.; Zhang, L.; Wang, T. Optimal Scheme in Energy Performance Contracting under Uncertainty: A Real Option Perspective. *J. Clean. Prod.* **2019**, *231*, 240–253. [CrossRef]
9. Li, X.; Shen, C.; Yu, C.W.F. Building Energy Efficiency: Passive Technology or Active Technology? *Indoor Built Environ.* **2017**, *26*, 729–732. [CrossRef]
10. Laws, N.D.; Anderson, K.; DiOrio, N.A.; Li, X.; McLaren, J. Impacts of Valuing Resilience on Cost-Optimal PV and Storage Systems for Commercial Buildings. *Renew. Energy* **2018**, *127*, 896–909. [CrossRef]
11. Zhang, L.; Li, Y.; Stephenson, R.; Ashuri, B. Valuation of Energy Efficient Certificates in Buildings. *Energy Build.* **2018**, *158*, 1226–1240. [CrossRef]
12. Ding, T.; Ke, Z. Characteristics and Changes of Regional Wet and Dry Heat Wave Events in China during 1960–2013. *Theor. Appl. Climatol.* **2015**, *122*, 651–665. [CrossRef]
13. Baldwin, J.W.; Dessy, J.B.; Vecchi, G.A.; Oppenheimer, M. Temporally Compound Heat Wave Events and Global Warming: An Emerging Hazard. *Earth's Future* **2019**, *7*, 411–427. [CrossRef]
14. Wei, J.; Wang, W.; Shao, Q.; Yu, Z.; Chen, Z.; Huang, Y.; Xing, W. Heat Wave Variations Across China Tied to Global SST Modes. *J. Geophys. Res. Atmos.* **2020**, *125*, e2019JD031612. [CrossRef]
15. You, Q.; Jiang, Z.; Kong, L.; Wu, Z.; Bao, Y.; Kang, S.; Pepin, N. A Comparison of Heat Wave Climatologies and Trends in China Based on Multiple Definitions. *Clim. Dyn.* **2017**, *48*, 3975–3989. [CrossRef]
16. Wang, J.; Yan, Z. Rapid Rises in the Magnitude and Risk of Extreme Regional Heat Wave Events in China. *Weather. Clim. Extrem.* **2021**, *34*, 100379. [CrossRef]
17. Koo, C.; Park, S.; Hong, T.; Park, H.S. An Estimation Model for the Heating and Cooling Demand of a Residential Building with a Different Envelope Design Using the Finite Element Method. *Appl. Energy* **2014**, *115*, 205–215. [CrossRef]
18. Junghans, L. Evaluation of the Economic and Environmental Feasibility of Heat Pump Systems in Residential Buildings, with Varying Qualities of the Building Envelope. *Renew. Energy* **2015**, *76*, 699–705. [CrossRef]
19. Méndez Echenagucia, T.; Capozzoli, A.; Cascone, Y.; Sassone, M. The Early Design Stage of a Building Envelope: Multi-Objective Search through Heating, Cooling and Lighting Energy Performance Analysis. *Appl. Energy* **2015**, *154*, 577–591. [CrossRef]
20. De Oliveira Neves, L.; Marques, T.H.T. Building Envelope Energy Performance of High-Rise Office Buildings in Sao Paulo City, Brazil. *Procedia Environ. Sci.* **2017**, *38*, 821–829. [CrossRef]
21. He, B.-J. Towards the next generation of Green Building for Urban Heat Island Mitigation: Zero UHI Impact Building. *Sustain. Cities Soc.* **2019**, *50*, 101647. [CrossRef]
22. Moosavi, L.; Mahyuddin, N.; Ab Ghafar, N.; Azzam Ismail, M. Thermal Performance of Atria: An Overview of Natural Ventilation Effective Designs. *Renew. Sustain. Energy Rev.* **2014**, *34*, 654–670. [CrossRef]
23. Dehghani-sanij, A.R.; Soltani, M.; Raahemifar, K. A New Design of Wind Tower for Passive Ventilation in Buildings to Reduce Energy Consumption in Windy Regions. *Renew. Sustain. Energy Rev.* **2015**, *42*, 182–195. [CrossRef]
24. Li, H.X.; Zhang, L.; Mah, D.; Yu, H. An Integrated Simulation and Optimization Approach for Reducing CO2 Emissions from On-Site Construction Process in Cold Regions. *Energy Build.* **2017**, *138*, 666–675. [CrossRef]
25. Ascione, F.; Bianco, N.; Maria Mauro, G.; Napolitano, D.F. Building Envelope Design: Multi-Objective Optimization to Minimize Energy Consumption, Global Cost and Thermal Discomfort. Application to Different Italian Climatic Zones. *Energy* **2019**, *174*, 359–374. [CrossRef]
26. Cheung, C.K.; Fuller, R.J.; Luther, M.B. Energy-Efficient Envelope Design for High-Rise Apartments. *Energy Build.* **2005**, *37*, 37–48. [CrossRef]
27. Balaras, C.A.; Droutsa, K.; Argiriou, A.A.; Asimakopoulos, D.N. Potential for Energy Conservation in Apartment Buildings. *Energy Build.* **2000**, *31*, 143–154. [CrossRef]
28. Wang, Y.; Berardi, U.; Akbari, H. Comparing the Effects of Urban Heat Island Mitigation Strategies for Toronto, Canada. *Energy Build.* **2016**, *114*, 2–19. [CrossRef]
29. Li, D.; Bou-Zeid, E.; Oppenheimer, M. The Effectiveness of Cool and Green Roofs as Urban Heat Island Mitigation Strategies. *Environ. Res. Lett.* **2014**, *9*, 055002. [CrossRef]
30. Cao, J.; Zheng, L.; Guo, Y. Research on Adaptive Application of Traditional Lingnan Building Materials—Taking Macau as an Example. In Proceedings of the E3S Web of Conferences 2023, Changsha, China, 22–24 December 2023; p. 371. [CrossRef]
31. Yan, L.; Chen, Y. Study of Roof Tiles in Lingnan Traditional Buildings and Roof Drainage Technologies. *IOP Conf. Ser. Earth Environ. Sci.* **2021**, *768*, 012151. [CrossRef]
32. Harkouss, F.; Fardoun, F.; Biwole, P.H. Passive Design Optimization of Low Energy Buildings in Different Climates. *Energy* **2018**, *165*, 591–613. [CrossRef]
33. Alhuwayil, W.K.; Abdul Mujeebu, M.; Algarny, A.M.M. Impact of External Shading Strategy on Energy Performance of Multi-Story Hotel Building in Hot-Humid Climate. *Energy* **2019**, *169*, 1166–1174. [CrossRef]
34. Roberz, F.; Loonen, R.C.G.M.; Hoes, P.; Hensen, J.L.M. Ultra-Lightweight Concrete: Energy and Comfort Performance Evaluation in Relation to Buildings with Low and High Thermal Mass. *Energy Build.* **2017**, *138*, 432–442. [CrossRef]
35. Liu, M. (Max) Probabilistic Prediction of Green Roof Energy Performance under Parameter Uncertainty. *Energy* **2014**, *77*, 667–674. [CrossRef]

36. Li, W.; Zhong, F. Current characteristics and inheritance and development of traditional architecture in Duxiong Lingnan. *Urban Build.* **2019**, *16*, 96–97. (In Chinese)
37. Yu, W. Study on Han Dynasty painted pottery in Lingnan area. *J. Natl. Mus. China* **2021**, *9*, 67–81. (In Chinese)
38. *GB50176-2016*; Code for Thermal Design of Civil Buildings. Ministry of Housing and Urban-Rural Development of the People's Republic of China: Beijing, China, 2016. (In Chinese)
39. Wang, X.; Chen, W. Huang Yuping's research on regional adaptability creation strategy of Lingnan high-rise office building—taking Shunde Rural Commercial Bank back-office service center scheme design as an example. *Cent. China Build.* **2021**, *39*, 28, 32. (In Chinese)
40. Song, D. (Ed.) *Energy-Saving Building Design and Technology*; Tongji University Press: Shanghai, China, 2003. (In Chinese)
41. Zhuo, J. "Barrier" and "shelter": Research on Climate Adaptive Design Strategy of Buildings in Lingnan Region. Master's Thesis, South China University of Technology, Guangzhou, China, 2018. (In Chinese). [CrossRef]
42. Tang, G. "Xia's shade" and Lingnan building heat protection. *New Build.* **2005**, *6*, 17–20. (In Chinese) [CrossRef]
43. Liu, J. Way of cooling buildings. *China Rep.* **2010**, *8*, 32–35. (In Chinese)
44. Deng, Q. Garden and environmental protection. *Guangdong Gard.* **1981**, *3*, 1–4. (In Chinese)
45. Lu, Y. Ventilation and heat protection of traditional buildings in the southern region. *J. Archit.* **1978**, *4*, 36–41+63–64. (In Chinese)
46. Xia, G. Study on the Adaptability of Lingnan Architecture Based on Modernity Concept. Ph.D. Thesis, South China University of Technology, Guangzhou, China, 2010. (In Chinese).
47. Xiao, Y.; Liu, S. Research on the scale of climate space in Lingnan traditional architecture. *Dynamic (Eco-City Green Build.)* **2015**, *2*, 73–79. (In Chinese)
48. Xu, M. Study on Thermal and Moisture Characteristics of Traditional Oyster-Shell Wall in the South of Xu Minling. Master's Thesis, Guangzhou University, Guangzhou, China, 2021. (In Chinese).
49. Wang, R.; Guo, W.; Dou, J.; Xie, H. Study on green construction wisdom of Lingnan traditional buildings adapted to hot and humid climate—A case study of three houses and two corridors. *Archit. Cult.* **2021**, *4*, 257–259. (In Chinese) [CrossRef]
50. Jiang, X. Research on climate adaptive design strategy of contemporary Lingnan architecture. *House* **2018**, *25*, 112–113. (In Chinese)
51. Chen, X. Inherits regional culture creation with The Times—Practice and thinking of contemporary Lingnan architectural design. *Contemp. Archit.* **2020**, *1*, 26–28. (In Chinese)
52. Lin, Q. On the improvement of residential environment and building energy saving in Lingnan. *J. South China Univ. Technol. (Nat. Sci. Ed.)* **1997**, *1*, 48–52. (In Chinese)
53. Liu, Z. Research on Passive Energy Saving Design in the Renovation of Existing Buildings in Lingnan University. Master's Thesis, South China University of Technology, Guangzhou, China, 2017. (In Chinese).

buildings

MDPI

Article

Daylighting Performance of CdTe Semi-Transparent Photovoltaic Skylights with Different Shapes for University Gymnasium Buildings

Yanpeng Wu *, Shaoxiong Li, Xin Gao and Huifang Fan

School of Civil and Resource Engineering, University of Science and Technology Beijing, Beijing 100083, China; m202110070@xs.ustb.edu.cn (S.L.); s20200070@xs.ustb.edu.cn (X.G.); fanhuifang@ustb.edu.cn (H.F.)
* Correspondence: wuyanpeng@ustb.edu.cn

Abstract: The daylighting environment in university gymnasiums affects daily teaching and sports training. However, direct sunlight, glare, and indoor overheating in summer are common problems. Semi-transparent photovoltaic glass can solve these issues by replacing shading facilities, blocking solar radiation, and generating electricity. This study examines the influence of different types of CdTe semi-transparent film photovoltaic glass on the daylighting environment of six typical university gymnasium skylights. The optimal types of CdTe semi-transparent film photovoltaic glass are determined by dynamic daylighting performance metrics DA, DAcon, DAmax, and UDI. The results show that, for instance, centralized rectangular skylights benefit from the 50–60% transmittance type, while centralized X-shaped skylights require the 70–80% transmittance type to enhance indoor daylighting. The research results offer specific recommendations based on skylight shapes and photovoltaic glass types and can provide a reference for the daylighting design of university gymnasium buildings with different forms of photovoltaic skylights in the future.

Keywords: university gymnasium; daylighting; dynamic daylighting performance; skylights; BIPV

1. Introduction

According to statistics, by 2021, there will be a total of 3012 higher education institutions in China, including 2756 universities (1270 universities, 1486 junior colleges) and 256 adult education institutions. To meet the daily physical education and training needs, most colleges and universities build gymnasiums according to national standards. The daylighting environment in the gymnasium is an important factor influencing people's activities [1]. Some studies show that 80% of the information obtained by the sports crowd comes from the visual information caused by the daylighting environment [2]. At the same time, the university gymnasiums not only provide the venue for physical education activities in colleges and universities but also sometimes host some international competitions. For example, some of the venues for the 2008 Beijing Olympic Games are set in several university gymnasiums in Beijing, which has higher demands on the quality of the daylighting environment in the gymnasium. The university gymnasium is a tall space building with a large depth. In order to meet the daylighting requirements, it is necessary to set up openings for daylighting. The main daylighting methods include side window daylighting, skylight daylighting, and side window skylight co-daylighting [3,4]. Although the window structure improves the indoor illumination level, it will bring problems such as direct sunlight and glare. Also, it can cause indoor overheating in summer and increase building energy consumption [5–7].

Most of the studies on the light environment aspects of gymnasium buildings mainly use a single window form and mainly consider the impact on the indoor environmental objectives by changing parameters such as building shape, shading parameters, window sizes and materials, and lack of studies on the light environment aspects of gymnasiums with different shapes of skylights [1,8–10], and there is a lack of research on different shapes of skylight gymnasiums in terms of the light environment.

Flat skylight is the most commonly used skylight in university gymnasiums, with high lighting efficiency, uniform illumination distribution, flexible layout, and low glare probability. The layout is mainly divided into centralized and uniform distributed skylights; centralized skylights include centralized rectangular skylights, centralized ribbon skylights, and centralized X-shaped skylights; uniform distributed skylights include distributed rectangular skylights, distributed vertical strip skylights, and distributed horizontal strip skylights, as shown in Figure 1 [3]. For the above different types of skylights, it is necessary to study their impact on the lighting performance of gymnasium buildings so as to make up for the deficiency in the study of the gymnasium light environment.

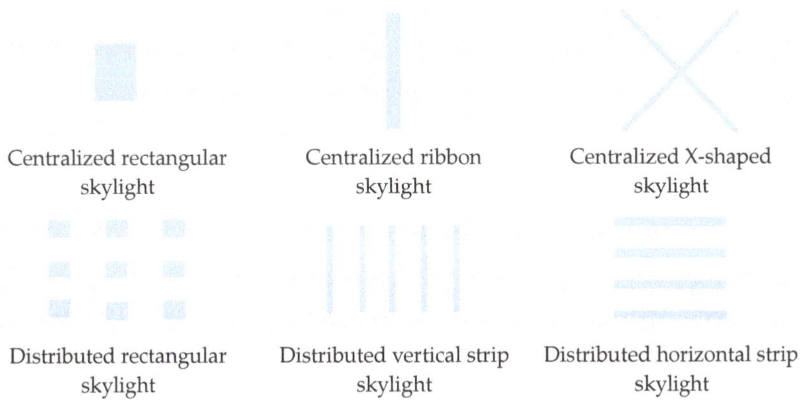

Figure 1. Skylight layout.

Building integrated photovoltaic (BIPV) is a technology that integrates solar power generation products into buildings [11–13]. As a part of the external structure of the building, BIPV not only has the function of generating electricity, but also has the function of building components and materials. It can also improve the beauty of the building and form a perfect unity with the building. As a form of BIPV, solar photovoltaic glass can replace sunshade facilities to solve the problem of indoor glare caused by direct sunlight. Its semi-transparency also blocks a certain amount of solar radiation into the room, alleviates the phenomenon of indoor overheating in summer, and reduces summer cooling power consumption; the electricity generated can be used for the building itself or the use of the grid, with great energy saving potential. For instance, in the China National Stadium, one of the three main venues for the 2008 Beijing Olympics, 24 solar photovoltaic glass panels are installed on its roof, with a rated output of 100 kW per day, reducing CO_2 emissions by about 94 tons per year. With the progress of photovoltaic technology and the promotion of building energy conservation, solar photovoltaic glass is very likely to be applied to the skylight and glass curtain wall structure of the university gymnasium. However, the existing research on photovoltaic glass modules is mainly focused on energy conservation [14–22], and the research on indoor daylighting environments of buildings rarely involves university gymnasiums [23–27]. It is not known whether the actual illumination level of solar photovoltaic skylights applied in university gymnasiums can meet indoor daylighting requirements.

In the past 20 years, research has mainly focused on glare and illuminance to establish reliable indicators for assessing visual comfort [28–31]. Currently, the daylight factor (DF) static index is widely used in China's architectural daylighting design field to evaluate indoor daylighting under ideal CIE cloudy conditions [32]. However, this type of lighting assessment presents static characteristics, merely reflecting the illumination levels of a specific day. It fails to capture the variations occurring at different times and seasons, thereby limiting the evaluation of building illumination [33]. Based on the development of Climate-Based Daylight Modeling (CBDM) [34], dynamic daylight metrics have been introduced to assess the annual daylighting performance and are widely adopted worldwide. Daylight Autonomy (DA) and Useful Daylight Illuminance (UDI) have been proposed as dynamic metrics to quantify the amount of daylight [29]. However, due to issues such as insufficient lighting and excessive glare, two dynamic daylight metrics, Continuous Daylight Autonomy (DAcon) and Maximum Daylight Autonomy (DAmax), are also introduced for comprehensive analysis of indoor lighting conditions.

DA measures the proportion of time that a point in the building meets the minimum illumination requirement throughout the year. DAcon quantifies the degree of insufficient lighting below the minimum requirement. DAmax describes the possibility of glare generation with 10 times the minimum daylighting illumination as a reference value [35]. UDI measures the proportion of time that a point in the building is in the effective utilization range during the operating period of the year. Nabil [35] studied building daylighting environments and categorized indoor illuminance levels into three intervals using 100 lx and 2000 lx as thresholds. Illumination below 100 lx is considered too low for visual activities, while illumination between 100 lx and 2000 lx is considered moderate. Illumination exceeding 2000 lx may cause visual discomfort.

Based on the above four dynamic daylighting evaluation indexes, this paper conducts dynamic daylighting simulation for six skylight forms common in university gymnasiums, explores the influence of different types of CdTe thin film photovoltaic glass on building daylighting, and makes the optimal selection. The noteworthy contributions of this study extend beyond the elucidation of optimal photovoltaic glass types. By placing emphasis on skylight shapes and their specific requirements, our research provides architects and researchers with concrete recommendations for informing the daylighting design of university gymnasium structures.

2. Methodology

Firstly, the typical university gymnasium and skylight with different shapes are modeled, and the material and boundary conditions of the envelope structure are set. Secondly, the simulation software is used to simulate the dynamic daylighting of CdTe thin film photovoltaic glass of different transmittance types, and the indoor daylighting environment is analyzed by obtaining the dynamic daylighting index value. Finally, the most optimal type of CdTe thin film photovoltaic glass for skylights with six shapes is summarized. The methodology is presented in Figure 2.

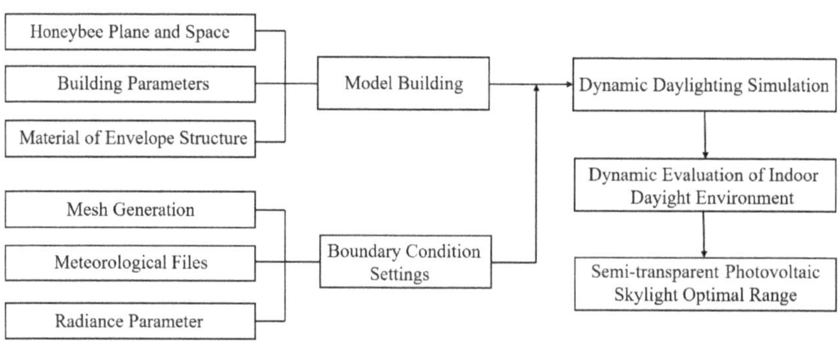

Figure 2. Steps involved in the methodology.

The basic building is modeled using Rhinoceros 3D software (version 7.4) [36] and imported to Ladybug and Honeybee plug-in [37] for Grasshopper for dynamic daylighting simulation and assessment for the actual design. Parameters such as the geometry, size, and location of the skylight are taken into account in the gymnasium model building, and accurate material properties are set in the model to ensure that the simulated results match the actual environment. The Grasshopper parametric tool [38] has been used to model the skylight and provide multiple variations for the skylight shape.

After completing the creation of the parametric model, we use the Honeybee plug-in to perform detailed model property settings, which include the precise definition of the materials for the building envelope and skylight. In this process, we pay special attention to the optical properties, reflectance, and glazing transmittance of the building envelope and skylight materials. We further proceed with the detailed setup of the simulation boundary conditions. This step covers the generation of the simulation mesh, the setting of the simulation parameters, and the use of the meteorological file. Ladybug reads imported meteorological files, correlates Radiance with a parameterized platform through the Honeybee interface plug-in, generates the simulation mesh and inputs set parameters into simulation analysis software, and gets visualization data results after calculation.

Dynamic daylighting simulations for different scenarios are carried out using Ladybug and Honeybee plug-ins. The aim of this phase of the simulation is to evaluate the dynamic daylighting performance of CdTe thin film photovoltaic glass in different skylight forms with different transmittance types and to analyze its impact on the indoor daylighting environment. Through this process, we obtain key data, such as the dynamic daylighting indexes, which are used to quantitatively evaluate the practical effectiveness of different transmittance types of CdTe thin film photovoltaic glass in a typical university gymnasium. This provides an important basis for our subsequent summary of the optimal CdTe thin film photovoltaic glass types.

Finally, a comprehensive assessment of the indoor daylighting environment is carried out using dynamic daylighting metrics. This dynamic assessment involves comparing the performance of CdTe thin-film photovoltaic glass under six different skylight forms to determine the optimal type of CdTe thin-film photovoltaic glass.

2.1. Software Introduction

In this study, the building was modeled parametrically in the Rhino and Grasshopper. Grasshopper is a visual programming plugin and one of the leading software tools in the field of parametric design. We carried out the dynamic daylighting simulation of the university gymnasium by the Ladybug and Honeybee component, a free and open-source simulation plugin based on Grasshopper. The core of daylight simulation in Honeybee is primarily based on the Radiance. Radiance is a validated daylighting simulation engine that employs a backward raytracing algorithm created by Greg Ward at Lawrence Berkeley National Laboratory [39]. The accuracy of building performance simulation software has been verified in detail by predecessors [40].

2.2. Photovoltaic Glass Properties

Solar photovoltaic glass mainly includes crystalline silicon photovoltaic glass [41] and thin film photovoltaic glass. Crystalline silicon photovoltaic glass is composed of photovoltaic cells and tempered glass; the internal photovoltaic cell is not transparent; by controlling the cell gap and edge gap between the double-sided glass to control the sunlight transmittance, crystalline silicon photovoltaic glass is easy to cause indoor illumination uneven. Thin film photovoltaic glass includes silicon thin film type [42], compound semiconductor thin film type [43], new material thin film type battery [44], etc. A nano Cadmium telluride solar cell (CdTe) is a thin-film solar cell based on the heterogeneous combination of P-type CdTe and N-type CdS. Compared with traditional solar products, CdTe has better low-light performance and higher conversion efficiency. Products with different light transmittance can be made by laser etching, and the structure is shown in Figure 3.

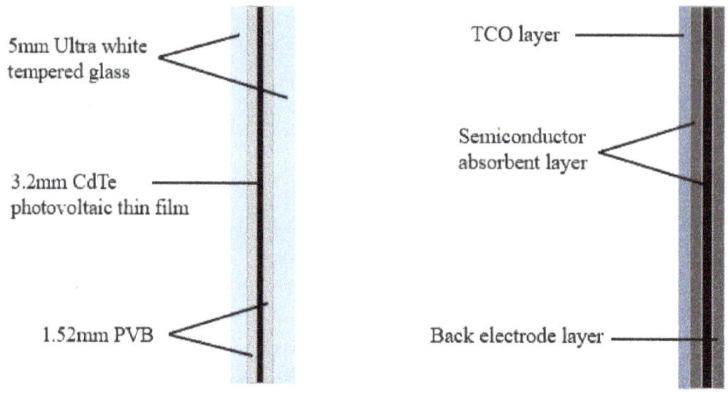

Figure 3. Structure of CdTe power generation glass and CdTe power generation film.

In this paper, CdTe thin film photovoltaic glass is taken as the research object to explore the influence of different types of thin film photovoltaic glass on indoor daylighting in the university gymnasium. The CdTe thin film photovoltaic glass of 10–80% transmittance type is set, and the step size is 10%, corresponding to a 90–20% power generation thin film coverage area ratio, respectively. The minimum coverage ratio of power generation film is set at 20% because the photovoltaic glass with lower film coverage has too low revenue. In the following content, 10–80% CT is used to represent the CdTe thin film photovoltaic glass of 10–80% transmittance type, respectively. Based on the hypothesis of Miyazaki [45] on thin film photovoltaic cells, the transmittance, thickness, and other parameters of front and rear panel glass are input into the software Window 7.7, and the photothermal properties parameters of CdTe thin film photovoltaic glass of different transmittance types are obtained, as shown in Table 1.

Table 1. Photothermal properties parameters.

Photovoltaic Glass Type	Subsequent Name	Visible Transmittance	Shading Coefficient	Solar Heat Gain Coefficient
10% transmittance type	10% CT	0.091	0.418	0.363
20% transmittance type	20% CT	0.181	0.48	0.417
30% transmittance type	30% CT	0.272	0.541	0.471
40% transmittance type	40% CT	0.363	0.603	0.524
50% transmittance type	50% CT	0.454	0.664	0.577
60% transmittance type	60% CT	0.546	0.725	0.631
70% transmittance type	70% CT	0.637	0.786	0.684
80% transmittance type	80% CT	0.729	0.848	0.738

2.3. Model Parameter

According to the field survey and data analysis of 29 university gymnasiums in China by predecessors [3] and the Gymnasium Design Standard (JGJ31-2003) [46], the simulated building, oriented with its long side facing east-west and short side facing north-south, has a plane size of a 46 m × 70 m rectangle. The competition area within this rectangle measures 38 m × 44 m, and the building has a height of 18m with fixed stands on both sides. No side windows were considered, and a gymnasium with an integrated skylight was modeled to accurately represent the daylighting conditions. The simulation location is set in Beijing, located in the Chinese daylighting climate zone III. Table 2 shows the natural light utilization hours and design illuminance in different daylighting climate zones.

According to China National Standard for Daylighting Design of Buildings (GB50033-2013) [47], for daylighting in class III climate zones and buildings with daylighting grade IV, the skylight ratio of glazing to floor area is set at 1/13, to calculate the area covered by the glazing, this ratio is multiplied by the floor area. In our case, the floor area is 3220 square meters, and the calculated glass area for each of the six skylight shapes is approximately 248 m^2. The indoor interface reflectance is also set according to the real utilization condition, as shown in Table 3.

Table 2. Hours of natural light utilization in different climatic zones.

Daylighting Climate Zones	Number of Stations (lx)	Annual Average Total Illuminance (lx)	Design Illuminance of Exterior Daylight (lx)	The Number of Hours of Natural Light Utilization for Design Illuminance (h)	Critical Illuminance of Exterior Daylight (lx)	The Number of Hours of Natural Light Utilization for Critical Illuminance (h)
I	29	48,781	18,000	3356	6000	3975
II	40	42,279	16,500	3234	5500	3921
III	71	37,427	15,000	3154	5000	3909
IV	102	32,886	13,500	3055	4500	3857
V	31	27,138	12,000	2791	4000	3689

Table 3. Indoor interface reflectance.

Type	Material	Reflectance
Ceiling	Concrete and steel construction	0.2
Wall	White paint	0.75
Floor	Light color wood floors	0.58

2.4. Simulation Parameter Settings

Based on the China National Standard for Daylighting Design of Buildings (GB50033-2013), the minimum illuminance level of the gymnasium in the form of top lighting is set at 150 lx. The height of the calculating plane is set as 1.5 m above the ground to be closer to the real illumination of the human eye, and a 1 m × 1 m grid is set on the calculating plane, resulting in a total of 1748 grids, the 3D image of the reference plane grid is shown in Figure 4. Other relative simulation parameters for RADIANCE are set in Table 4. Ambient bounces represent the number of reflections between the surfaces. Ambient divisions and super-samples set the number of samples sent. Ambient resolution deals with the maximum error, scene dimension, and the sampling cutoff point. The ambient accuracy (-aa) is usually set to 0.1 to 1, and lower numbers result in better accuracy. The daylight simulation quality defined in Honeybee is set to a value of 1, which corresponds to medium quality.

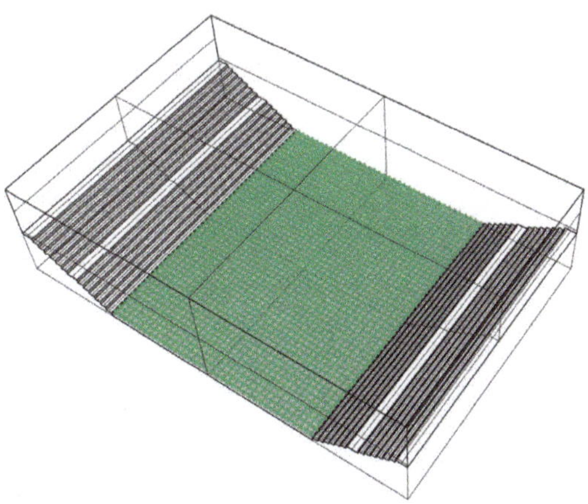

Figure 4. Plane grid diagram (green area represents the plane grid).

Table 4. Environment parameter settings.

Parameter	Setting Value
Quality	1
Ambient bounces	5
Ambient divisions	2048
Ambient super-samples	512
Ambient resolution	512
Ambient accuracy	0.08

Table 5 presents the primary analysis options for the simulations. Based on the above-mentioned preset simulation conditions, a comprehensive simulation analysis study was conducted on the indoor lighting environment with different application scenarios of CdTe thin film photovoltaic glass.

Table 5. Analysis settings in RADIANCE.

Parameter	Setting Value
Occupancy schedule	8:00–18:00
Minimum illumination	150 lx
Grid size	1 m × 1 m
Number of grids	1748

3. Results and Discussions

3.1. DA Analysis of Six Shapes of Skylights

The average DA variation curves and varying ranges of DA of six shapes of skylights under the condition of 10–80% CT photovoltaic glass are shown in Figure 5. The average DA is obtained by calculating the average of all measuring points on the calculating plane. A higher average DA in the upper part of the range suggests that there are more measuring points with adequate daylighting, while a lower average DA in the lower part of the range indicates that there are more measuring points with inadequate daylighting. Moreover, the average DA of the six skylight shapes increases as the transmittance of the CT photovoltaic glass increases.

During the working period from 8 am to 6 pm, when the measured DA is less than 25%, daylighting is considered very poor; when the measured DA is between 25% and 55%, daylighting is considered insufficient; when the measured DA is between 55% and 75%, daylighting is considered acceptable; when the measured DA is above 75%, daylighting is considered ideal [48].

From Figure 5, it can be seen that for the centralized rectangular skylight, distributed horizontal strip skylight, and distributed vertical strip skylight, when the photovoltaic glass is 80% CT, all of the measured DA value is above 75%; for the distributed rectangular skylight and centralized ribbon skylight, when the photovoltaic glass is 70–80% CT, all of the measured DA value is above 75%; for the centralized X-shaped skylight, there is no such situation where the DA value of all the measuring points is above 75%. Compared with distributed skylights, the DA value of centralized skylights has a wider range. The reason is that some areas are far from directly under the skylight, and the lighting quality in these areas differs greatly from that directly under the skylight.

To better analyze the indoor daylighting situation of the gymnasium, the detailed DA values of the six kinds of skylights are shown in Tables 6–11. DA $\geq$ 75%, 55% $\leq$ DA < 75%, 25% $\leq$ DA < 55%, and DA < 25% represent the proportion of measuring points that meet the requirements of their respective DA to all measuring points.

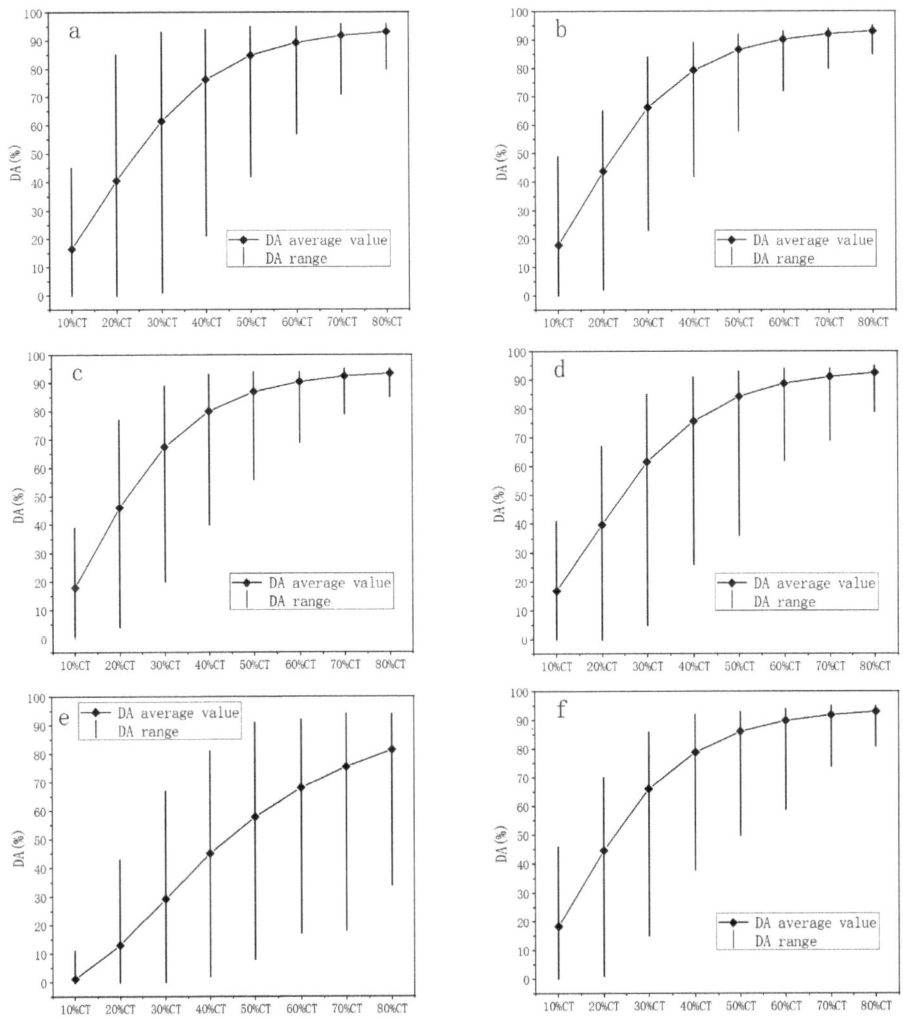

Figure 5. DA curves of six kinds of skylight: (**a**) centralized rectangular skylight; (**b**) distributed rectangular skylight; (**c**) centralized ribbon skylight; (**d**) distributed vertical strip skylight; (**e**) centralized X-shaped skylight; (**f**) distributed horizontal strip skylight.

Table 6. Detailed data of DA for a centralized rectangular skylight.

Type	Centralized Rectangular Skylight			
	DA ≥ 75%	55% ≤ DA < 75%	25% ≤ DA < 55%	DA < 25%
10% CT	0%	0%	30.55%	69.45%
20% CT	9.55%	21.4%	42.91%	26.14%
30% CT	34.5%	31.35%	25.57%	8.58%
40% CT	61.04%	27.8%	10.87%	0.29%
50% CT	83.06%	14.82%	2.12%	0%
60% CT	95.54%	4.46%	0%	0%
70% CT	99.6%	0.4%	0%	0%
80% CT	100%	0%	0%	0%

Table 7. Detailed data of DA for a distributed rectangular skylight.

Type	Distributed Rectangular Skylight			
	DA ≥ 75%	55% ≤ DA < 75%	25% ≤ DA < 55%	DA < 25%
10% CT	0%	0%	28.66%	71.34%
20% CT	0%	25.34%	64.87%	9.97%
30% CT	32.44%	49.03%	18.53%	0%
40% CT	73.68%	24.54%	1.78%	0%
50% CT	94.45%	5.55%	0%	0%
60% CT	99.66%	0.34%	0%	0%
70% CT	100%	0%	0%	0%
80% CT	100%	0%	0%	0%

Table 8. Detailed data of DA for a centralized ribbon skylight.

Type	Centralized Ribbon Skylight			
	DA ≥ 75%	55% ≤ DA < 75%	25% ≤ DA < 55%	DA < 25%
10% CT	0%	0%	30.43%	69.57%
20% CT	2.29%	34.1%	51.37%	12.24%
30% CT	41.93%	33.3%	24.43%	0.34%
40% CT	69.5%	26.6%	3.9%	0%
50% CT	89.47%	10.53%	0%	0%
60% CT	99.03%	0.97%	0%	0%
70% CT	100%	0%	0%	0%
80% CT	100%	0%	0%	0%

Table 9. Detailed data of DA for a distributed vertical strip skylight.

Type	Distributed Vertical Strip Skylight			
	DA ≥ 75%	55% ≤ DA < 75%	25% ≤ DA < 55%	DA < 25%
10% CT	0%	0%	32.44%	67.56%
20% CT	0%	23.51%	55.21%	21.28%
30% CT	26.95%	45.54%	24.03%	3.48%
40% CT	63.5%	28.49%	8.01%	0%
50% CT	87.87%	10.98%	1.15%	0%
60% CT	97.71%	2.29%	0%	0%
70% CT	99.54%	0.46%	0%	0%
80% CT	100%	0%	0%	0%

Table 10. Detailed data of DA for a centralized X-shaped skylight.

Type	Centralized X-Shaped Skylight			
	DA ≥ 75%	55% ≤ DA < 75%	25% ≤ DA < 55%	DA < 25%
10% CT	0%	0%	0%	100%
20% CT	0%	0%	16.65%	83.35%
30% CT	0%	4.46%	56.58%	38.96%
40% CT	3.66%	29.12%	52.06%	15.16%
50% CT	22.2%	37.64%	36.27%	3.89%
60% CT	43.25%	33.81%	22.48%	0.46%
70% CT	60.07%	30.38%	9.44%	0.11%
80% CT	76.66%	19.22%	4.12%	0%

Table 11. Detailed data of DA for a distributed horizontal strip skylight.

Type	Distributed Horizontal Strip Skylight			
	DA ≥ 75%	55% ≤ DA < 75%	25% ≤ DA < 55%	DA < 25%
10% CT	0%	0%	30.43%	69.57%
20% CT	0%	29.4%	60.93%	9.67%
30% CT	35.58%	41.82%	22.31%	0.29%
40% CT	69.62%	26.72%	3.66%	0%
50% CT	91.65%	8.06%	0.29%	0%
60% CT	98.34%	1.66%	0%	0%
70% CT	100%	0%	0%	0%
80% CT	100%	0%	0%	0%

3.1.1. DA Analysis for Centralized Rectangular Skylight

Detailed data on the DA of centralized rectangular skylights are shown in Table 6. For 10–30% CT photovoltaic glass, the measured DA is mostly less than 75%, indicating a poor indoor daylighting environment; for 40–80% CT photovoltaic glass, the DA of more than half of the measurement points is greater than 75%, indicating that the daylighting environment of most of the space is better in the gymnasium. Although the DA of more than half of the 40% CT photovoltaic glass measurement points is greater than 75%, the DA of more than 10% of the measurement points is less than 55%. To create a better daylighting environment, 50–80% CT photovoltaic glass is a better choice.

3.1.2. DA Analysis for Distributed Rectangular Skylight

Detailed data of DA of distributed rectangular skylights are shown in Table 7. For 10–20% CT photovoltaic glass, the DA of all measurement points is below 75%, implying a very poor indoor daylighting environment; for the 40–80% CT photovoltaic glass, the DA of more than half of the measuring points is greater than 75%, indicating that the daylighting environment of most of the space is better in the gymnasium. The DA of 30% CT photovoltaic glass measuring points over 80% is greater than 55%, which belongs to the acceptable daylighting environment, but the proportion of measuring points with DA less than 55% is more than 10%. To create a better daylighting environment, 40–80% CT photovoltaic glass is a better choice.

3.1.3. DA Analysis for Centralized Ribbon Skylight

Detailed data on the DA of centralized ribbon skylights are shown in Table 8. For 10–30% CT photovoltaic glass, the measured DA is mostly less than 75% and shows a poor daylighting level; for the 40–80% CT photovoltaic glass, the DA of more than half of the measuring points is greater than 75%, indicating that the daylighting environment of most of the space is better in the gymnasium. The DA of more than half of the 40% CT photovoltaic glass measuring points is greater than 75%, and the proportion of measuring points with DA less than 55% is very small. Therefore, 40–80% CT photovoltaic glass is a better choice for a centralized ribbon skylight.

3.1.4. DA Analysis for Distributed Vertical Strip Skylight

Detailed data on the DA of distributed vertical strip skylights are shown in Table 9. For 10–20% CT photovoltaic glass, the DA of all measurement points is less than 75%, implying a very poor indoor daylighting environment; for 30% CT photovoltaic glass, the DA of more than 70% measurement points is greater than 55%, but the DA of less than 55% is relatively high; for the 40–80% CT photovoltaic glass, the DA of more than half of the measuring points is greater than 75%, indicating that the daylighting environment of most of the space is better in the gymnasium. Therefore, 40–80% CT photovoltaic glass is a better choice for a distributed vertical strip skylight.

3.1.5. DA Analysis for Centralized X-Shaped Skylight

Detailed data on the DA of centralized X-shaped skylights are shown in Table 10. For 10–30% CT photovoltaic glass, the DA of all measurement points is below 75%, indicating a very poor indoor daylighting environment; for 30–60% CT photovoltaic glass, only a few measuring points of the DA is greater than 75%, the daylighting environment is still not ideal; for the 70–80% CT photovoltaic glass, the DA of more than half of the measuring points is greater than 75%, implying that most of the indoor space has ideal daylighting. Therefore, 70–80% CT photovoltaic glass is a better choice for a centralized X-shaped skylight.

3.1.6. DA Analysis for Distributed Horizontal Strip Skylight

Detailed data on the DA of distributed horizontal strip skylights are shown in Table 11. For 10–20% CT photovoltaic glass, the DA of all measurement points is less than 75%, implying a very poor indoor daylighting environment; for 30% CT photovoltaic glass, the DA of more than 70% measurement points is greater than 55%, but the DA of less than 55% is relatively high. Therefore, 40–80% CT photovoltaic glass is a better choice for a distributed horizontal strip skylight.

3.2. DAcon and DAmax Analysis of Six Shapes of Skylights

Compared with DA, DAcon is a more comprehensive evaluation index for the daylighting of buildings. When the DAcon of indoor measuring points is greater than 80%, the daylighting environment can be considered acceptable even if the indoor illuminance cannot meet the design requirements. When the proportion of measuring points greater than 5% of DAmax is relatively high, indoor glare is more likely to occur.

The variation curves of DAcon and DAmax of six shapes of skylights under the condition of 10–80% CT photovoltaic glass are shown in Figure 6. It can be observed that the six types of skylights exhibit a similar trend in terms of the variations of the DAmax and DAcon. According to the variation curves of DAcon and DAmax, the daylighting of the six shapes of skylights is best as follows: centralized rectangular skylight: 50–60% CT; distributed rectangular skylight: 40–60% CT; centralized ribbon skylight: 50–60% CT; distributed vertical strip skylight: 40–50% CT; centralized X-shaped skylight: 70–80% CT; distributed horizontal strip skylight: 40–50% CT. However, all of them have the potential to cause glare.

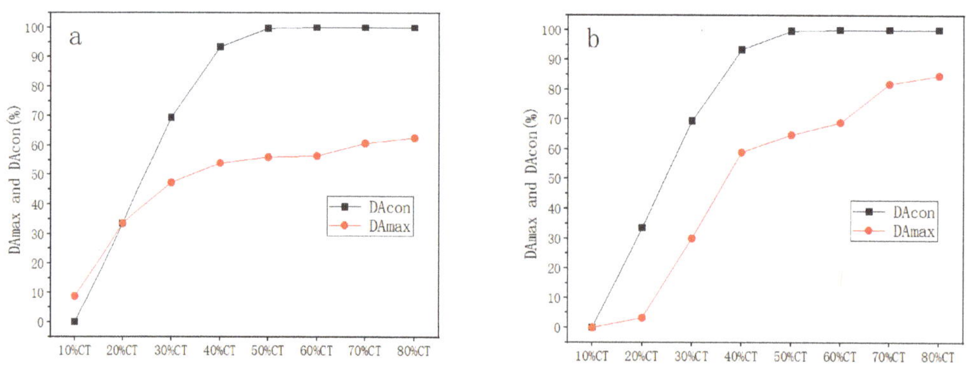

Figure 6. *Cont.*

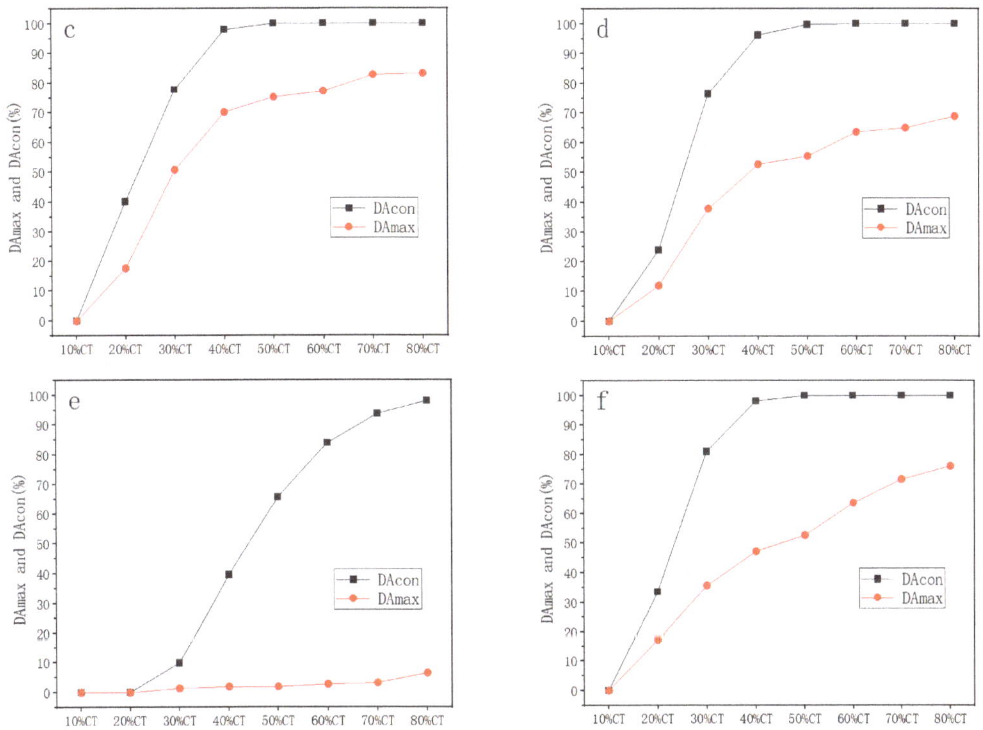

Figure 6. DAcon and DAmax curves of six kinds of skylight: (**a**) centralized rectangular skylight; (**b**) distributed rectangular skylight; (**c**) centralized ribbon skylight; (**d**) distributed vertical strip skylight; (**e**) centralized X-shaped skylight; (**f**) distributed horizontal strip skylight.

3.3. UDI Analysis of Six Shapes of Skylights

The variation curves of UDI of six shapes of skylights under the condition of 10–80% CT photovoltaic glass are shown in Figure 7. It can be seen that with the increase of the transmittance of CT photovoltaic glass, the proportion of UDI < 100 lx decreases rapidly; the proportion of 100 lx < UDI < 2000 lx first increased and then decreased except for the centralized X-shaped skylight, the reason is that the daylighting of the centralized X-shaped skylight is poor, and the illumination of few indoor measuring points exceeds 2000 lx; the proportion of UDI > 2000 lx increased slowly.

For centralized rectangular skylight, the best indoor daylighting environment is 50–80% CT; distributed rectangular skylight: 40–80% CT; Centralized ribbon skylight: 40–80% CT; distributed vertical strip skylight: 50–80% CT; Centralized X-shaped skylight: 70–80% CT; distributed horizontal strip: 40–70% CT.

From the simulated results of various daylighting metrics above, it can be observed that under different skylight forms, there are variations in the degree of change in dynamic daylighting metrics with the increase in the transmittance of photovoltaic glass. However, overall, they exhibit similar changing trends. According to a similar study [27], this trend is also noticeable under different photovoltaic glass conditions in different climatic regions. Given the diverse skylight forms in this study, when compared to the aforementioned similar study, variations in the numerical values and degrees of change in each daylighting metric differ due to differences in building types, skylight glass, and other relevant design parameters.

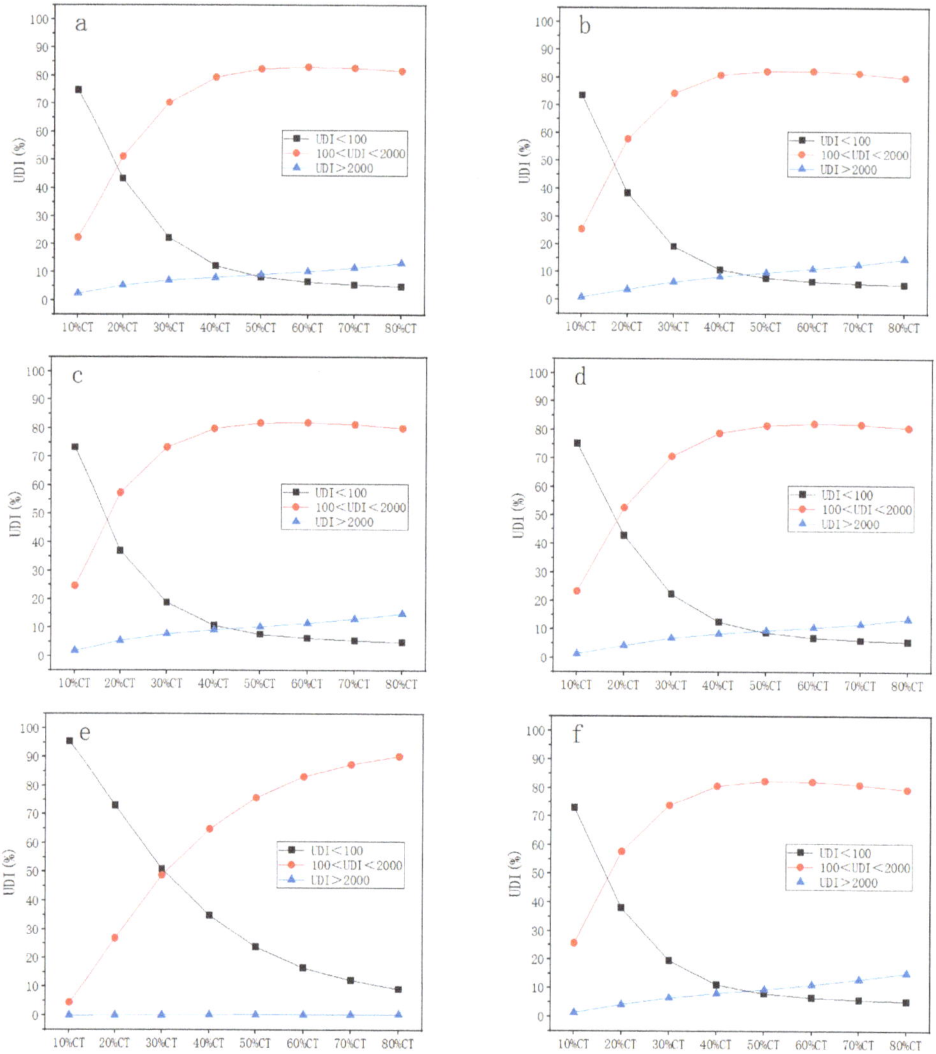

Figure 7. UDI curves of six kinds of skylight: (**a**) centralized rectangular skylight; (**b**) distributed rectangular skylight; (**c**) centralized ribbon skylight; (**d**) distributed vertical strip skylight; (**e**) centralized X-shaped skylight; (**f**) distributed horizontal strip skylight.

In terms of research on different skylight distributions, a similar study [49] is also present. Focusing on railway stations, this study selected two cities and two different skylight distribution patterns. From this, it can be observed that each city, under different skylight distribution patterns, exhibits similar forms of variation in daylighting metrics. This finding aligns with the discoveries made in our study.

3.4. Optimal Type of Photovoltaic Glass for Six Kinds of Skylight

The actual layout of skylights in university gymnasiums is varied, and the area is also different. This study simulates six typical shapes of skylight in a typical university gymnasium. The results cannot be applied to all university gymnasiums. DA, DAcon, DAmax, and UDI are used to analyze indoor daylighting and select the optimal scheme.

The optimal CT photovoltaic glass types of centralized rectangular skylight are shown in Table 12. It can be seen from Table 12 that 50–60% CT photovoltaic glass is the optimal type to create a good indoor daylighting environment, which can ensure that the indoor illumination level meets the minimum requirements for a long time. However, there is the possibility of glare, and certain shading facilities are needed in the period of intense sunlight.

Table 12. Optimal type of CT photovoltaic glass for a centralized rectangular skylight.

Daylighting Index	Evaluation Methodology	Optimal Type
DA	Proportion of measuring points with value $\geq$ 55%	50–80% CT
DAcon	Proportion of measuring points with value > 80%	50–80% CT
DAmax	Proportion of measuring points with value > 5%	10–60% CT
UDI	Average value within 100–2000 lx	50–80% CT

The optimal CT photovoltaic glass types of distributed rectangular skylights are shown in Table 13. It can be seen from Table 13 that 40–60% CT photovoltaic glass is the optimal type to create a good indoor daylighting environment. Shading facilities are needed in the period of intense sunlight.

Table 13. Optimal type of CT photovoltaic glass for a distributed rectangular skylight.

Daylighting Index	Evaluation Methodology	Optimal Type
DA	Proportion of measuring points with value $\geq$ 55%	40–80% CT
DAcon	Proportion of measuring points with value > 80%	40–80% CT
DAmax	Proportion of measuring points with value > 5%	10–60% CT
UDI	Average value within 100–2000 lx	40–80% CT

The optimal CT photovoltaic glass types of centralized ribbon skylight are shown in Table 14. It can be seen from Table 14 that 50–60% CT photovoltaic glass is the optimal type to create a good indoor daylighting environment. Shading facilities are needed in the period of intense sunlight.

Table 14. Optimal type of CT photovoltaic glass for a centralized ribbon skylight.

Daylighting Index	Evaluation Methodology	Optimal Type
DA	Proportion of measuring points with value $\geq$ 55%	40–80% CT
DAcon	Proportion of measuring points with value > 80%	40–80% CT
DAmax	Proportion of measuring points with value > 5%	10–60% CT
UDI	Average value within 100–2000 lx	50–80% CT

The optimal CT photovoltaic glass types of distributed vertical strip skylights are shown in Table 15. It can be seen from Table 15 that 40–50% CT photovoltaic glass is the optimal type to create a better indoor daylighting environment. Shading facilities are needed in the period of intense sunlight.

Table 15. Optimal type of CT photovoltaic glass for a distributed vertical strip skylight.

Daylighting Index	Evaluation Methodology	Optimal Type
DA	Proportion of measuring points with value ≥ 55%	40–80% CT
DAcon	Proportion of measuring points with value > 80%	40–80% CT
DAmax	Proportion of measuring points with value > 5%	10–50% CT
UDI	Average value within 100–2000 lx	50–80% CT

The optimal CT photovoltaic glass types of centralized X-shaped skylight are shown in Table 16. It can be seen from Table 16 that 70–80% CT photovoltaic glass is the optimal type to create a good indoor daylighting environment. However, the possibility of glare from the centralized X-shaped skylight is very small.

Table 16. Optimal type of CT photovoltaic glass for a centralized X-shaped skylight.

Daylighting Index	Evaluation Methodology	Optimal Type
DA	Proportion of measuring points with value ≥ 55%	70–80% CT
DAcon	Proportion of measuring points with value > 80%	70–80% CT
DAmax	Proportion of measuring points with value > 5%	10–80% CT
UDI	Average value within 100–2000 lx	70–80% CT

The optimal CT photovoltaic glass types of distributed horizontal strip skylights are shown in Table 17. It can be seen from Table 17 that 40–50% CT photovoltaic glass is the optimal type to create a good indoor daylighting environment. Shading facilities are needed in the period of intense sunlight.

Table 17. Optimal type of CT photovoltaic glass for a distributed horizontal strip skylight.

Daylighting Index	Evaluation Methodology	Optimal Type
DA	Proportion of measuring points with value ≥ 55%	40–80% CT
DAcon	Proportion of measuring points with value > 80%	40–80% CT
DAmax	Proportion of measuring points with value > 5%	10–50% CT
UDI	Average value within 100–2000 lx	40–70% CT

3.5. Verification

In order to verify the accuracy of the results, the gymnasium model is established in Rhino according to the original scale and compared with the measured results. The gymnasium is Zhuoer Gymnasium of Wuhan University. The skylight shape is a centralized ribbon skylight, the ceiling is whitewashed, the wall material is gray sound-absorbing board, and the floor is wooden. The reflection ratio of the enclosure structure is set according to the China National Standard for Daylighting Design of Buildings (GB50033-2013) [48]. The pollution reduction coefficient of the window glass is set as 0.6, and the light-blocking reduction coefficient of the interior component is set as 0.65. The location distribution of measuring points is shown in Figure 8.

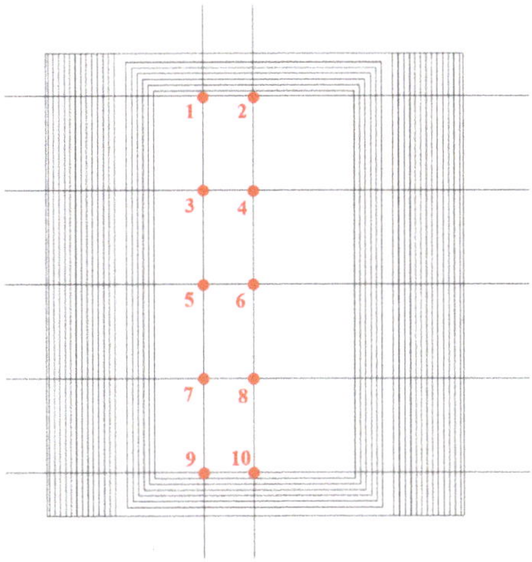

Figure 8. Distribution of measuring points for indoor illuminance (Points 1–10).

The measured data are from the literature [3]. The sky condition is rainy, and the outdoor illumination is 5384 lx. The comparison between measured illumination and simulated illumination is shown in Figure 9. It can be seen that the measured value is almost consistent with the simulated value, and the simulated illuminance is slightly larger than the measured illuminance because of the plane simplification of the seat in the modeling process.

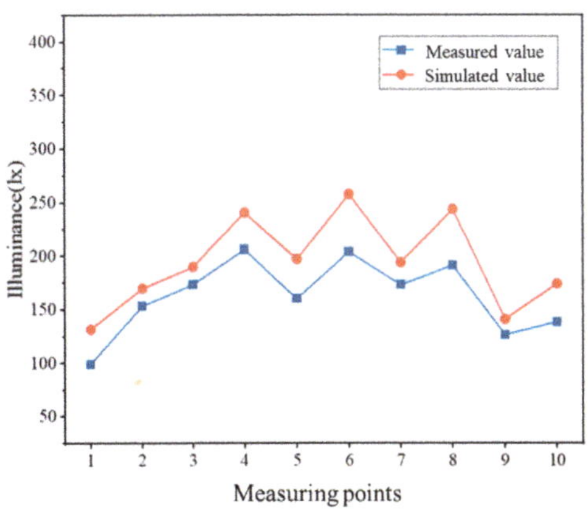

Figure 9. Comparison of illuminance measurement values with simulation.

3.6. Summary

The optimal types of six different shapes of skylights with CT photovoltaic glass for the gymnasium are summarized in Table 18. It can be seen that the daylighting environment of a centralized X-shaped skylight is relatively poor, and CT photovoltaic glass with high transmittance is needed to ensure indoor daylighting. Different shapes of skylights have a certain impact on the daylighting of the gymnasium. The optimal type of CT photovoltaic glass varies with the skylight shape.

Table 18. Optimal type of CT photovoltaic glass for shapes of skylights.

Skylight Shape	Optimal Type
Centralized rectangular skylight	50–60% CT
Distributed rectangular skylight	40–60% CT
Centralized ribbon skylight	50–60% CT
Distributed vertical strip skylight	40–50% CT
Centralized X-shaped skylight	70–80% CT
Distributed horizontal strip skylight	40–50% CT

4. Conclusions

In this paper, Rhino software (version 7.4) is used to model a typical university gymnasium to explore the impact of six different shapes of CdTe thin film photovoltaic skylights on the indoor daylighting environment of the university gymnasium. The indoor daylighting environment of the university gymnasium varies greatly due to the different shapes of the skylight. Based on the dynamic daylighting evaluation indexes DA, DAcon, DAmax, and UDI, the optimal types of CT photovoltaic glass suitable for skylights with different shapes are summarized.

The mean DA of the six shapes of skylight increases with the increase of transmittance of CT photovoltaic glass. For the centralized rectangular skylight, distributed horizontal strip skylight, and distributed vertical strip skylight, when the photovoltaic glass is 80% CT, DA of the competition area is all above 75%; for the distributed rectangular skylight and centralized ribbon skylight, when the photovoltaic glass is 70–80% CT, the DA of the competition area is above 75%; for the centralized X-shaped skylight, all types of CT photovoltaic glass contain areas with DA below 75%. Compared with distributed skylights, the DA of centralized skylights has a wider distribution range.

With the increase of transmittance of CT photovoltaic glass, DAcon and DAmax gradually increased. All skylights are prone to glare except the centralized X-shaped skylights. Therefore, certain shading facilities should be set up during periods of strong sunlight. The results show that for the centralized rectangular skylight, The optimal type of CT photovoltaic glass is 50–60% CT, the optimal type of distributed rectangular skylight is 40–60% CT, the optimal type of centralized ribbon skylight is 50–60% CT, the optimal type of distributed vertical strip skylight is 40–50% CT, the best type of centralized X-shaped skylight is 70–80% CT, and the optimal type of distributed horizontal strip skylight is 40–50% CT.

At present, there are still few studies on the influence of semi-transparent photovoltaic skylights on university gymnasium daylighting. The results of this work focus on the coupled consideration of six different skylight shapes and different types of CT photovoltaic glass only for the city of Beijing, which is located in the Chinese III lighting climate zone, and the conclusions drawn may not be applicable to the rest of the lighting climate zones locations, while the skylight shapes and glass types considered are limited. In the future, the influence of skylight area, geographical location, building orientation, and other factors on university gymnasium daylighting should be studied.

Author Contributions: Conceptualization, methodology, Y.W.; software, X.G.; validation, X.G. and S.L.; formal analysis, X.G.; investigation, Y.W.; resources, Y.W.; writing—original draft preparation, X.G. and S.L.; writing—review and editing, S.L.; visualization, X.G.; supervision, H.F.; project administration, H.F.; funding acquisition, Y.W. All authors have read and agreed to the published version of the manuscript.

Funding: This research was funded by Beijing Natural Science Foundation (8202034) and Central Guidance for Local Scientific and Technological Development Funding Project (236Z5202G). The

APC was funded by Beijing Natural Science Foundation (8202034) and Central Guidance for Local Scientific and Technological Development Funding Project (236Z5202G).

Data Availability Statement: The data presented in this study are available on request from the corresponding author. The data are not publicly available due to privacy requirements of testing workshop.

Conflicts of Interest: The authors declare no conflict of interest.

References

1. Fan, Z.; Liu, M.; Tang, S. A multi-objective optimization design method for gymnasium facade shading ratio integrating energy load and daylight comfort. *Build. Environ.* **2022**, *207*, 108527. [CrossRef]
2. Li, J. Research on Energy-Saving Design of Daylighting and Natural Ventilation of University Gymnasium Based on System Optimization. Ph.D. Thesis, Harbin Institute of Technology, Harbin, China, 2010.
3. Fan, Q. Research on Daylighting Design of College University Gymnasium Based on Parameterized Platform. Master's Thesis, Tianjin University, Tianjin, China, 2020.
4. Lee, B.; Pyo, Y.; Kim, M.; Kim, J. Evaluation of the natural lighting performance of rooftop daylight installations for multi-purpose sports hall in seoul. *KIEAE J.* **2022**, *22*, 21–34. [CrossRef]
5. Webb, A.R. Considerations for lighting in the built environment: Non-visual effects of light. *Energy Build.* **2006**, *38*, 721–727. [CrossRef]
6. Ochoa, C.E.; Aries, M.B.C.; van Loenen, E.J.; Hensen, J.L.M. Considerations on design optimization criteria for windows providing low energy consumption and high visual comfort. *Appl. Energy* **2012**, *95*, 238–245. [CrossRef]
7. Tzempelikos, A. Advances on daylighting and visual comfort research. *Build. Environ.* **2017**, *113*, 1–4. [CrossRef]
8. Li, L. Daylighting environment in gymnasinms competition space. *Harbin Univ. Arch. Eng.* **2002**, *35*, 64–67.
9. Zhao, Y.; Mei, H. Dynamic simulation and analysis of daylighting factors for gymnasiums in mid-latitude China. *Build. Environ.* **2013**, *63*, 56–68. [CrossRef]
10. Acosta, I.; Navarro, J.; Sendra, J.J. Daylighting design with lightscoop skylights: Towards an optimization of shape under overcast sky conditions. *Energy Build.* **2013**, *60*, 232–238. [CrossRef]
11. Shukla, A.K.; Sudhakar, K.; Baredar, P. Recent advancement in BIPV product technologies: A review. *Energy Build.* **2017**, *140*, 188–195. [CrossRef]
12. Martín-Chivelet, N.; Kapsis, K.; Wilson, H.R.; Delisle, V.; Yang, R.; Olivieri, L.; Polo, J.; Eisenlohr, J.; Roy, B.; Maturi, L.; et al. Building-integrated photovoltaic (BIPV) products and systems: A review of energy-related behavior. *Energy Build.* **2022**, *262*, 111998. [CrossRef]
13. Yu, G.; Yang, H.; Luo, D.; Cheng, X.; Ansah, M.K. A review on developments and researches of building integrated photovoltaic (BIPV) windows and shading blinds. *Renew. Sustain. Energy Rev.* **2021**, *149*, 111355. [CrossRef]
14. Chen, L.; Zheng, X.; Yang, J.; Yoon, J.H. Impact of BIPV windows on building energy consumption in street canyons: Model development and validation. *Energy Build.* **2021**, *249*, 111207. [CrossRef]
15. Alrashidi, H.; Issa, W.; Sellami, N.; Sundaram, S.; Mallick, T. Thermal performance evaluation and energy saving potential of semi-transparent CdTe in façade BIPV. *Sol. Energy* **2022**, *232*, 84–91. [CrossRef]
16. Wang, C.; Peng, J.; Li, N. Study of overall energy performance of amorphous silicon photovoltaic window based on variable transmittances. *Acta Energiae Solaris Sin.* **2019**, *40*, 1607–1615.
17. Uddin, M.M.; Wang, C.; Zhang, C.; Ji, J. Investigating the energy-saving performance of a CdTe-based semi-transparent photovoltaic combined hybrid vacuum glazing window system. *Energy* **2022**, *253*, 124019. [CrossRef]
18. Alrashidi, H.; Issa, W.; Sellami, N.; Ghosh, A.; Mallick, T.K.; Sundaram, S. Performance assessment of cadmium telluride-based semi-transparent glazing for power saving in façade buildings. *Energy Build.* **2020**, *215*, 109585. [CrossRef]
19. Barman, S.; Chowdhury, A.; Mathur, S.; Mathur, J. Assessment of the efficiency of window integrated CdTe based semi-transparent photovoltaic module. *Sustain. Cities Soc.* **2018**, *37*, 250–262. [CrossRef]
20. Zhang, W.; Lu, L. Overall energy assessment of semi-transparent photovoltaic insulated glass units for building integration under different climate conditions. *Renew. Energy* **2019**, *134*, 818–827. [CrossRef]
21. Wu, Z.; Zhang, L.; Wu, J.; Liu, Z. Experimental and numerical study on the annual performance of semi-transparent photovoltaic glazing in different climate zones. *Energy* **2022**, *240*, 122473. [CrossRef]
22. Alrashidi, H.; Ghosh, A.; Issa, W.; Sellami, N.; Mallick, T.K.; Sundaram, S. Thermal performance of semitransparent CdTe BIPV window at temperate climate. *Sol. Energy* **2020**, *195*, 536–543. [CrossRef]
23. Qiu, C.; Yang, H. Daylighting and overall energy performance of a novel semi-transparent photovoltaic vacuum glazing in different climate zones. *Appl. Energy* **2020**, *276*, 115414. [CrossRef]
24. Cheng, Y.; Gao, M.; Dong, J.; Jia, J.; Zhao, X.; Li, G. Investigation on the daylight and overall energy performance of semi-transparent photovoltaic facades in cold climatic regions of China. *Appl. Energy* **2018**, *232*, 517–526. [CrossRef]
25. Gao, J.; Peng, J.; Wang, T. Daylighting analysis of semi-transparent photovoltaic windows with different cell widths. *Acta Energiae Solaris Sin.* **2022**, *43*, 223–230. [CrossRef]

26. Kapsis, K.; Dermardiros, V.; Athienitis, A.K. Daylight performance of perimeter office façades utilizing semi-transparent photovoltaic windows: A simulation study. *Energy Procedia* **2015**, *78*, 334–339. [CrossRef]
27. Fan, Z.; Yang, Z.; Yang, L. Daylight performance assessment of atrium skylight with integrated semi-transparent photovoltaic for different climate zones in China. *Build. Environ.* **2021**, *190*, 107299. [CrossRef]
28. Suk, J.Y.; Schiler, M.; Kensek, K. Development of new daylight glare analysis methodology using absolute glare factor and relative glare factor. *Energy Build.* **2013**, *64*, 113–122. [CrossRef]
29. Carlucci, S.; Causone, F.; Rosa, F.D.; Pagliano, L. A review of indices for assessing visual comfort with a view to their use in optimization processes to support building integrated design. *Renew. Sustain. Energy Rev.* **2015**, *47*, 1016–1033. [CrossRef]
30. Konstantzos, I.; Tzempelikos, A. Daylight glare evaluation with the sun in the field of view through window shades. *Build. Environ.* **2017**, *113*, 65–77. [CrossRef]
31. Song, Y. Research on Anti Glare Design of Skylight for Complex Shape Roof of Gymnasium. Master's Thesis, Harbin Institute of Technology, Harbin, China, 2022.
32. Leslie, R.P.; Radetsky, L.C.; Smith, A.M. Conceptual design metrics for daylighting. *Light. Res. Technol.* **2012**, *44*, 277–290. [CrossRef]
33. Zhang, W. Study on Energy Saving of Office Buildings with the Influence of Daylighting. Ph.D. Thesis, Tianjin University, Tianjin, China, 2005.
34. Mardaljevic, J. Examples of climate-based daylight modelling. In Proceedings of the CIBSE National Conference 2006: Engineering the Future, London, UK, 2 May 2006.
35. Nabil, A.; Mardaljevic, J. Useful daylight illuminances: A replacement for daylight factors. *Energy Build.* **2006**, *38*, 905–913. [CrossRef]
36. Eltaweel, A.; SU, Y. Parametric design and daylighting: A literature review. *Renew. Sustain. Energy Rev.* **2017**, *73*, 1086–1103. [CrossRef]
37. Roudsari, M.S.; Pak, M.; Smith, A. Ladybug: A parametric environmental plugin for grasshopper to help designers create an environmentally-conscious design. In Proceedings of the 13th International IBPSA Conference, Lyon, France, 25–28 August 2013; pp. 3128–3135.
38. Touloupaki, E.; Theodosiou, T. Performance simulation integrated in parametric 3D modeling as a method for early stage design optimization—A review. *Energies* **2017**, *10*, 637. [CrossRef]
39. Ward, G.J. The RADIANCE lighting simulation and rendering system. In Proceedings of the 21st Annual Conference on Computer Graphics and Interactive Techniques; SIGGRAPH '94. Association for Computing Machinery: New York, NY, USA, 1994; pp. 459–472. [CrossRef]
40. Zhou, B. Research on Multi-Objective Optimization of Multilayer Office Building Space in Cold Region Based on Daylight Performance. Master's Thesis, Harbin Institute of Technology, Harbin, China, 2017.
41. Ghosh, A.; Sundaram, S.; Mallick, T.K. Investigation of thermal and electrical performances of a combined semi-transparent PV-vacuum glazing. *Appl. Energy* **2018**, *228*, 1591–1600. [CrossRef]
42. Wang, M.; Peng, J.; Li, N.; Yang, H.; Wang, C.; Li, X.; Lu, T. Comparison of energy performance between PV double skin facades and PV insulating glass units. *Appl. Energy* **2017**, *194*, 148–160. [CrossRef]
43. Selvaraj, P.; Ghosh, A.; Mallick, T.K.; Sundaram, S. Investigation of semi-transparent dye-sensitized solar cells for fenestration integration. *Renew. Energy* **2019**, *141*, 516–525. [CrossRef]
44. Hamed, M.S.G.; Oseni, S.O.; Kumar, A.; Sharma, G.; Mola, G.T. Nickel sulphide nano-composite assisted hole transport in thin film polymer solar cells. *Sol. Energy* **2020**, *195*, 310–317. [CrossRef]
45. Miyazaki, T.; Akisawa, A.; Kashiwagi, T. Energy savings of office buildings by the use of semi-transparent solar cells for windows. *Renew. Energy* **2005**, *30*, 281–304. [CrossRef]
46. *JGJ 31-2003*; Design Code for Sports Building. China Building Industry Press: Beijing, China, 2003.
47. *GB50033-2013*; Standard for Daylighting Design of Buildings. China Building Industry Press: Beijing, China, 2012.
48. Li, Z.; Wang, L.; Zhang, H. Research on the lighting environment using photovoltaic glass in office space—A case study in Tianjin. *China Illum. Eng. J.* **2015**, *26*, 23–28.
49. Zhao, N.; Fan, Z.; Liu, J. Daylight oriented optimization of photovoltaic integrated skylights for railway station waiting hall represented large space buildings in China. *Energy Build.* **2023**, *285*, 112777. [CrossRef]

buildings

Article

Trade-Off Judgement for Daylighting and Energy Consumption in the High and Large Space of the University Gymnasium in Beijing

Yanpeng Wu [1,*], Meitong Jin [1] and Tianhu Zhang [2]

[1] School of Civil and Resource Engineering, University of Science and Technology Beijing, Beijing 100083, China; m202310072@xs.ustb.edu.cn

[2] School of Civil Engineering and Mechanics, Yanshan University, Qinhuangdao 066004, China; thzhang@ysu.edu.cn

* Correspondence: wuyanpeng@ustb.edu.cn

Abstract: Taking the high and large space of the University of Science and Technology Beijing Gymnasium as this research object, this paper analyzes the influence of different window positions, window-to-wall ratio (WWR), solar heat gain coefficient (SHGC), heat transfer coefficient (K), and visible light transmittance (VT) on the total indoor energy consumption in winter and summer and obtains the relationship between the daylight factor and VT formed when the window is opened per unit area. Through energy consumption simulation, the variation law and calculation formula for indoor total energy consumption are obtained. The results show that the SHGC and K of the exterior window have a significant influence on the total energy consumption. By using the energy consumption simulation of different types of exterior windows, it is concluded that the SHGC of the south-facing window is negatively correlated with the variation of air conditioning energy consumption per unit area $\Delta e_{1,w}$, while the others are positively correlated. Moreover, the SHGC and K of the skylight have the most significant influence on the $\Delta e_{1,w}$. The total energy consumption decreases and then increases with the increase in the window area, and there is a lowest point, so the right side of the lowest point is less than or equal to 105% of the lowest total energy consumption as a reasonable window area zone. Finally, a progressive optimization method for weighing daylighting and energy consumption in university gymnasiums in Beijing is proposed.

Keywords: university gymnasium; high and large space; trade-off judgement; daylighting; energy consumption

1. Introduction

In recent years, with the rapid development of China's higher education, gymnasiums have occupied an increasing proportion of the public buildings of universities [1], and many newly built campuses have built gymnasiums. As large public buildings, gymnasiums generally have large spans and large spaces [2], which greatly increase the energy consumption of office lighting, air conditioning systems, and other energy-using equipment. Zheng Yan et al. [3] conducted a survey on the University of Science and Technology Beijing (USTB) Gymnasium and concluded that the main energy resources of the gymnasium were office lighting and the air conditioning system, which together accounted for 84% of the total energy consumption.

At present, more and more transparent envelope structures, such as exterior windows, skylights, glass curtain walls, and other transparent materials, are widely used in the construction of university gymnasiums. Daylighting entering the room through transparent envelope structures can enhance the indoor visual effect, make people feel comfortable, and reduce lighting energy consumption [4–7]. Taking an office building as the simulation object, Peng Peng [8] concluded that the energy-saving rate when the lighting density decreased by

1 W/m^2 was equal to the energy-saving rate when the COP of the air conditioning system was increased by 0.3 or the indoor air conditioning temperature was increased by 1 °C. Hence, daylighting has great potential for energy savings [9]. However, the transparent envelope structures have poor shading capacity and a poor thermal insulation effect. If the window-to-wall ratio (WWR) of a building is too large, the energy consumption of air conditioning and heating will inevitably increase by a large margin [10–12].

Therefore, more and more scholars use various methods to find the optimal window design scheme based on the contradiction between daylighting and building energy consumption. Taking the lowest energy consumption of air conditioning, heating, and lighting in the classroom as the starting point, Futrell et al. [13] studied the impact of windows with different orientations on energy consumption and lighting. They adopted the Genopt clustering optimization algorithm to optimize them, concluding that the north window contributed the least to the indoor lighting effect and received the least solar thermal radiation. Konis et al. [14] developed a Passive Performance Optimization Framework (PPOF) and corresponding generative modeling workflow to inform early-stage design by optimizing building geometry, orientation, fenestration, and shading device geometry configurations in response to annual climate-based thermal comfort and daylighting performance outcomes. Motamedi et al. [15] raised an algorithm for the optimal skylight design of office buildings that combined lighting and energy consumption and compared different skylight ratios according to lighting energy consumption and air conditioning energy consumption to obtain the optimal skylight design. Yuan Fang et al. [16] put forward a building performance optimization process that uses parametric design, building simulation, and genetic algorithms to optimize design. They took a small office building as this research subject to verify the effectiveness of the optimization process. Lakhdari et al. [17] showed how the genetic algorithms used to search for high-performing design solutions could be applied to optimize the thermal, lighting, and energy performance of a middle school classroom in a hot and dry climate. Using a parametric approach and evolutionary multi-objective computation via the Octopus plug-in for Grasshopper, various windows-to-wall ratios (WWRs), wall materials, glass types, and shading devices were combined to derive potential solutions that achieve a good balance between daylight provision and thermal comfort while ensuring low energy consumption. Yan Li [18] used theoretical analysis and numerical simulation to obtain the variation trends of energy consumption, the calculation formulas of total building energy consumption under the different areas of transparent envelopes, and the evaluation method of the rationality of building windows.

Existing studies mainly focus on the relationship between lighting and energy consumption in small functional spaces, such as office spaces and classrooms, or in high and large spaces, such as railway stations. In contrast, few studies have been conducted on the balance between daylighting and energy consumption in high and large spaces, such as university gymnasiums. University gymnasiums and railway stations are different in space scale, flow density, equipment operation, interior space, and interior environment design requirements. Railway stations have a high occupant density, a long operating time, and a high frequency of equipment use. And their interior space structure is compact and open [19]. Compared with railway stations, university gymnasiums have a larger span, centralized activity space, and intermittent operation. In addition, the indoor environmental temperature requirements vary for different types of venues and competition events. Gymnasiums usually require a large amount of space to accommodate spectators, competition venues, and other facilities, making the indoor space relatively closed [20].

Aiming at the contradiction between daylighting and energy consumption in university gymnasiums in Beijing, this paper takes the high and large space of the University of Science and Technology Beijing (USTB) Gymnasium as this research object, simulates it by Ecotect software version 2010 to obtain the relationship between the daylight factor and visible light transmittance (VT) formed by skylights and side windows per unit area, and analyzes the variation law of total energy consumption with different transparent envelope structures by EnergyPlus software version 22.2. Ultimately, by analyzing the impact of daylighting, artificial lighting, and air conditioning energy consumption, a progressive optimization calculation process for weighing daylighting and energy consumption in university gymnasiums in Beijing is proposed. Use this calculation process to determine whether the specific window design scheme of university gymnasiums in Beijing is reasonable. If it is unreasonable, adjust the window design scheme and reevaluate until it is reasonable, which provides a theoretical basis for the window design scheme of university gymnasiums in Beijing.

2. Methodology

2.1. Model Establishment and Parameter Setting

Figure 1 shows the interior of the main stadium of the University of Science and Technology Beijing Gymnasium. The gymnasium model is shown in Figure 2. Furthermore, the thermal parameters of the envelope structure and internal disturbance parameters are set according to the code [21]. The illumination control and standard values of indoor daylighting for side and top lighting are set according to the standard [22]. The parameters for the base case are shown in Table 1.

Figure 1. Interior of the main stadium of USTB Gymnasium.

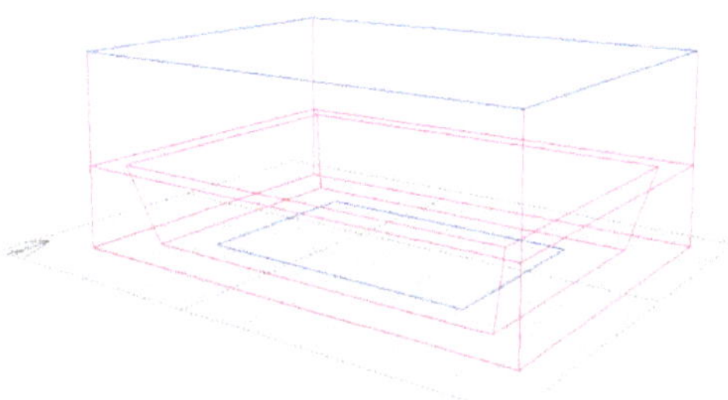

Figure 2. Gymnasium model.

Table 1. Parameters for the base case.

Parameter	Description
Location	Beijing
Gymnasium model	Simplified as 75.1 m × 54.6 m × 23.75 m cuboid
Main competition zone	60 m × 40 m rectangle
Lighting zone	45 m × 30 m rectangle
All around	Trapezoidal stands
Stand height	10 m

Table 1. *Cont.*

Parameter		Description
Roof		Horizontal roof
Reflectance of an opaque envelope structure	Wall	0.750
	Roof	0.200
	Floor	0.580
K of opaque envelope structure	Wall	$0.25\ W/(m^2 \cdot K)$
	Roof	$0.25\ W/(m^2 \cdot K)$
Exterior window	Type	Double-glazed aluminum framed glass
	K	$1.50\ W/(m^2 \cdot K)$
	SHGC	0.81
	VT	0.639
Critical illuminance of outdoor daylighting		5000 lux
Standard value of indoor daylighting	Side	300 lx
	Top	150 lx
Summer indoor air conditioning temperature		24 °C
Winter indoor air conditioning temperature		18 °C
Cooling operation period		6.1–9.30
Heating operation period		11.15–3.15
Artificial lighting power		$20\ W/m^2$
Occupant density		$0.32\ person/m^2$
Equipment power		$10\ W/m^2$
Operating time		08:00–22:00
Air change rate		0.5 per hour

2.2. *Composition of Total Energy Consumption*

Due to the previous research, it was found that the main energy sources of the University of Science and Technology Beijing (USTB) Gymnasium were office lighting and air conditioning systems. Therefore, this paper believes that the total annual energy consumption of the gymnasium is composed of lighting energy consumption and air conditioning energy consumption in summer and winter. This paper only discusses the impact of daylighting on the three types of energy consumption mentioned above, which can be divided into three parts: (1) the impact of window heating on indoor energy consumption; (2) changes in lighting energy consumption caused by daylighting; and (3) changes in energy consumption of air conditioning and heating caused by changes in lighting energy consumption. Figure 3 demonstrates the interaction between daylighting and indoor energy consumption.

As the window area increases, the solar heat gain into the room through the transparent envelope structure increases, and the heat transfer caused by the temperature difference between inside and outside through the glass also increases correspondingly. At the same time, the amount of daylight entering the interior also increases accordingly. Furthermore, the indoor daylight factor increases and the lighting energy consumption decreases accordingly, which leads to the reduction of heat gain from lighting, thus reducing the air conditioning energy consumption in summer but increasing air conditioning energy consumption (heating energy consumption) in winter.

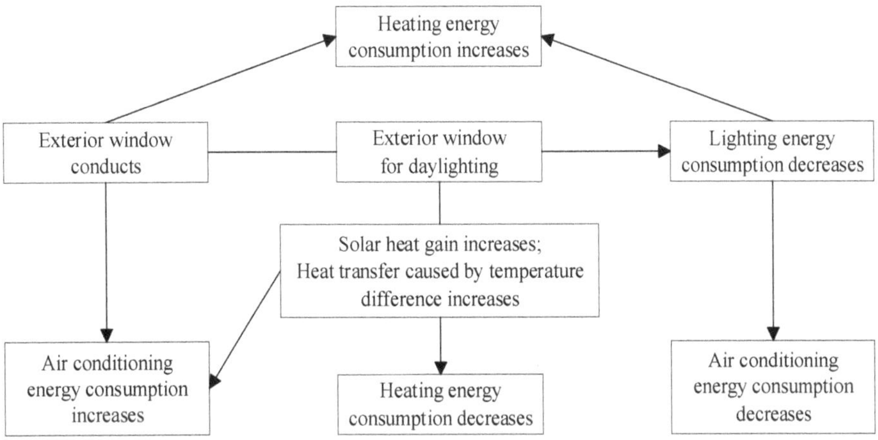

Figure 3. Interaction between daylighting and indoor energy consumption.

The total energy consumption is analyzed as follows:

When the windows are closing, the total energy consumption E_0 is the sum of the air conditioning energy consumption E_1 and lighting energy consumption E_2, that is

$$E_0 = E_1 + E_2 \tag{1}$$

When the windows are opening, the total energy consumption E' is the sum of the total energy consumption E_0 when the window is closed and the total energy consumption variation ΔE caused by the window opening.

$$E' = E_0 + \Delta E \tag{2}$$

The variation in total energy consumption ΔE caused by window opening is the sum of the variation in air conditioning energy consumption ΔE_1 and the variation in lighting energy consumption ΔE_2, that is

$$\Delta E = \Delta E_1 + \Delta E_2 \tag{3}$$

The variation in air conditioning energy consumption ΔE_1 includes two parts: (1) the variation in air conditioning energy consumption $\Delta E_{1,W}$ caused by the change in heat transfer performance of the envelope structure (indoor and outdoor temperature difference heat transfer and solar radiation heat gain) after opening the window; and (2) the variation in air conditioning energy consumption $\Delta E_{1,L}$ caused by the reduction of lighting energy consumption after opening the window:

$$\Delta E_1 = \Delta E_{1,W} + \Delta E_{1,L} \tag{4}$$

Substitute Equation (4) into Equation (3) to obtain

$$\Delta E = \Delta E_{1,W} + \Delta E_{1,L} + \Delta E_2 \tag{5}$$

Since $\Delta E_{1,L}$ and ΔE_2 are only related to the indoor light environment, which $\Delta E_{1,L} + \Delta E_2$ is denoted as ΔE_L, then Equation (5) is simplified as

$$\Delta E = \Delta E_{1,W} + \Delta E_L \tag{6}$$

The total energy consumption E' after opening the window is:

$$E' = E_1 + E_2 + \Delta E_{1,W} + \Delta E_L \tag{7}$$

$\Delta E_{1,W}$ can be obtained by changing the window area without changing the indoor lighting conditions (no lighting control) and calculating the variation in total energy consumption under different window design conditions; $E_1 + E_2 + \Delta E_L$ can be obtained by setting the amount of lighting that is required to be added under different window design conditions in the model without windows and calculating the total energy consumption. Both of these variables can be calculated by EnergyPlus.

$E_1 + E_2 + \Delta E_L$ is denoted as $E_{0,L}$, then the total energy consumption E' after opening the window is

$$E' = \Delta E_{1,W} + E_{0,L} \tag{8}$$

Through the calculation of EnergyPlus, the variation law of $\Delta E_{1,W}$ and $E_{0,L}$ with the window area is obtained, and the total energy consumption E' under different window design conditions can be obtained by superposing $\Delta E_{1,W}$ and $E_{0,L}$, as shown in Figure 4.

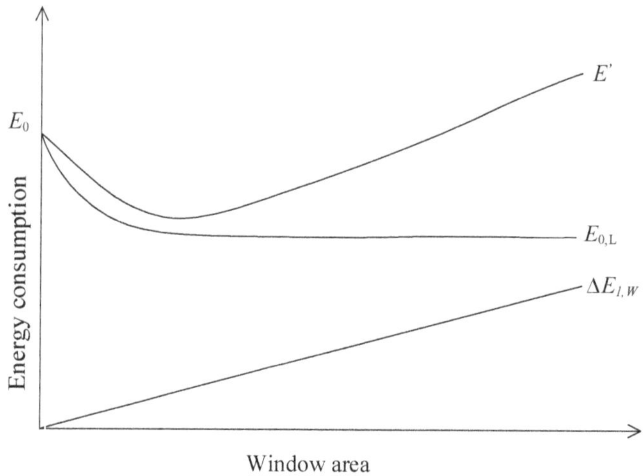

Figure 4. Influence of window area on energy consumption.

3. Results and Discussion

3.1. Daylighting Simulation

The relationship between the daylight factor and VT was analyzed by Ecotect software version 2010. The WWR was set at 1% in the simulation condition, and the simulation results of daylight factors formed by the reference plane of the indoor lighting area under different VTs were obtained when the skylight and side window were opened per unit area, respectively, as shown in Figure 5.

The straight lines formed by the skylight and side window are fitted, respectively, and the fitting relation between the daylight factor (c) and VT formed when the window is opened per unit area is obtained.

$$\text{Skylight: } c = 1.92667(VT) + 0.09556 \tag{9}$$

$$\text{Side window: } c = 0.365(VT) + 0.14417 \tag{10}$$

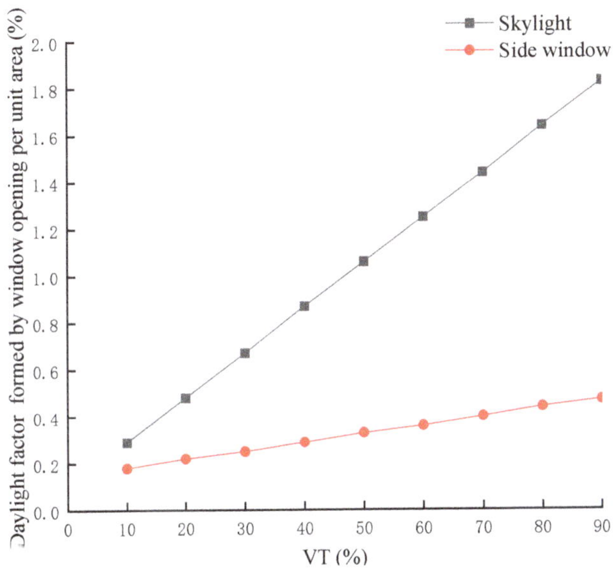

Figure 5. Variation law of daylight factor with VT.

3.2. Total Energy Consumption Simulation

This paper uses EnergyPlus to analyze the influence of different skylight ratios, solar heat gain coefficients (SHGC), K, and VT on total energy consumption (including lighting energy consumption and air conditioning energy consumption in summer and winter) under lighting control, as shown in Figures 6–9. It can be seen from the figure that the SHGC and K of the exterior window have a significant influence on the total energy consumption. All material parameters are selected according to the code [21] and software material libraries.

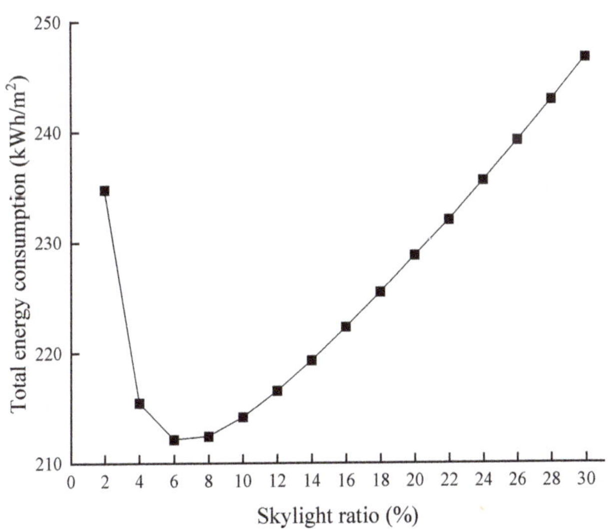

Figure 6. Impact of skylight ratio on total energy consumption.

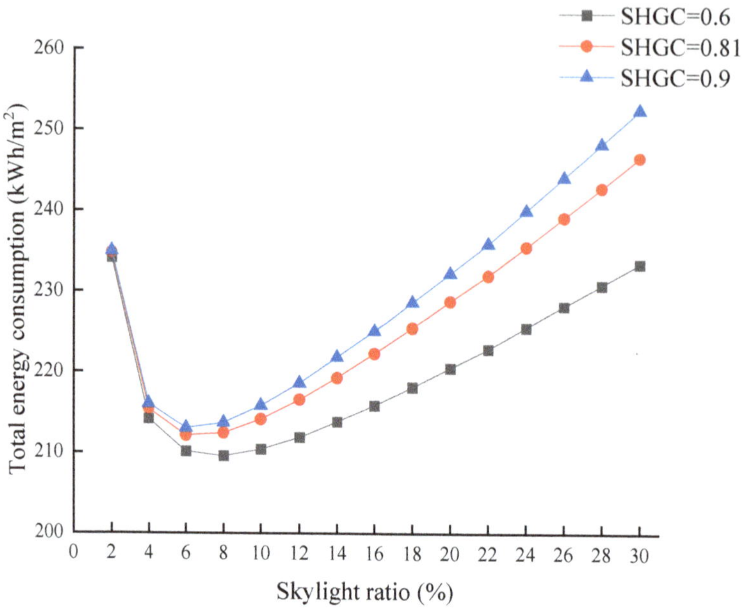

Figure 7. Influence of SHGC on total energy consumption.

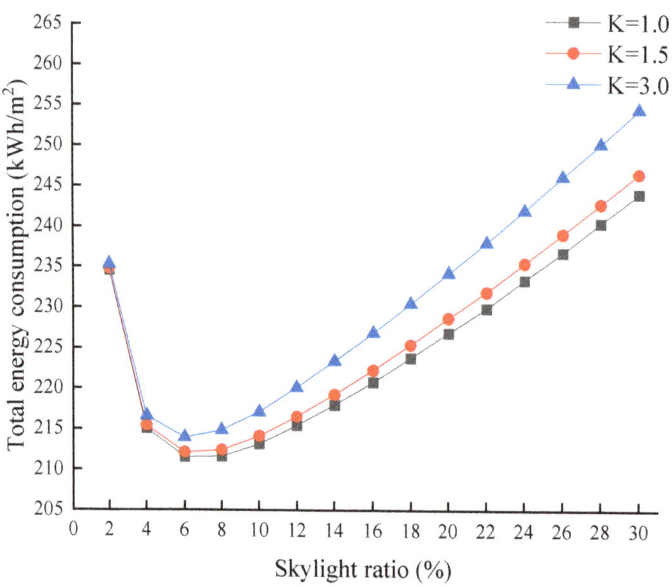

Figure 8. Influence of K on total energy consumption.

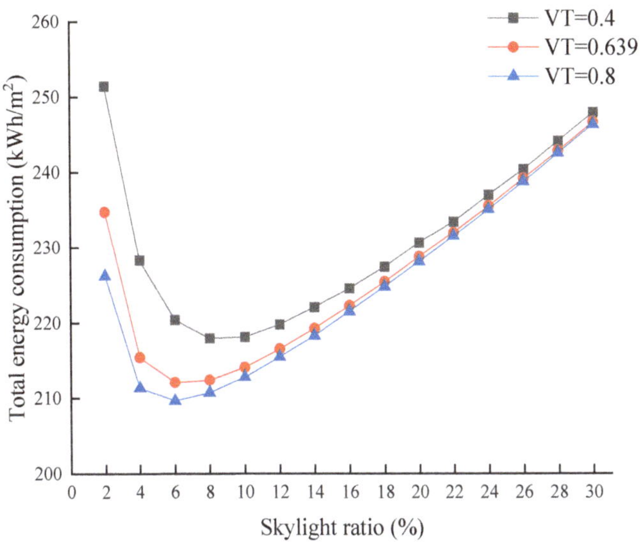

Figure 9. Influence of VT on total energy consumption.

3.3. Summer Air Conditioning Energy Consumption Simulation

This section analyzes the influence of different skylight ratios (SHGC, K, and VT) on air conditioning energy consumption in summer under lighting control, as shown in Figures 10–13.

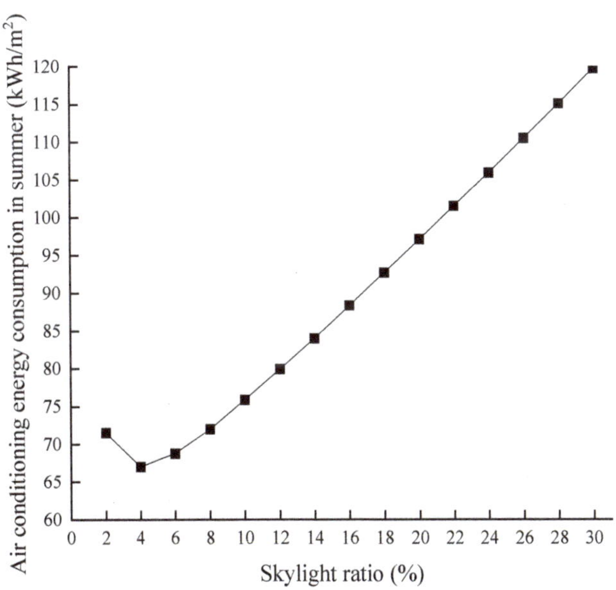

Figure 10. Impact of skylight ratio on air conditioning energy consumption in summer.

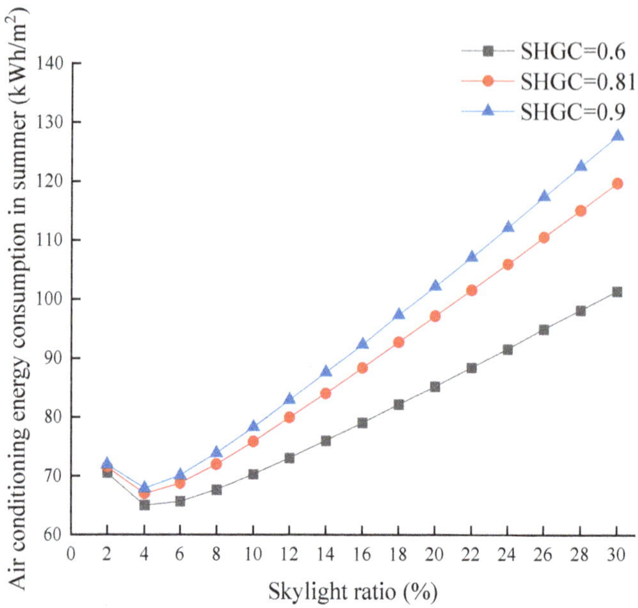

Figure 11. Influence of SHGC on air conditioning energy consumption in summer.

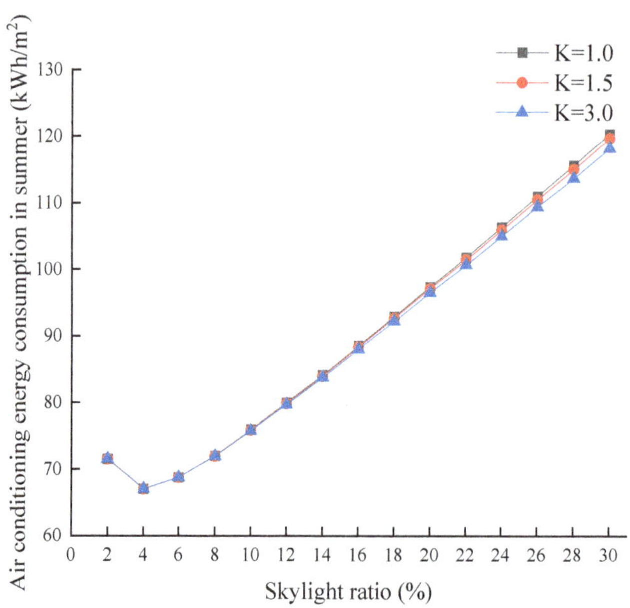

Figure 12. Influence of K on air conditioning energy consumption in summer.

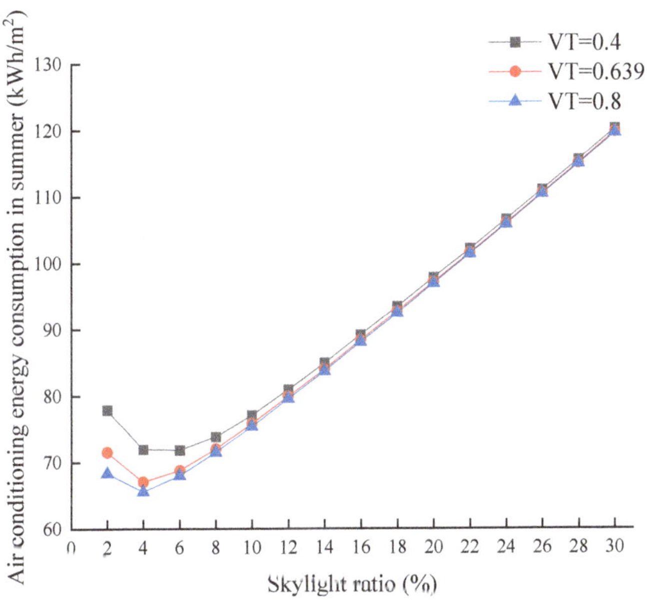

Figure 13. Influence of VT on air conditioning energy consumption in summer.

From Figure 10, it can be concluded that the air conditioning energy consumption in summer decreases and then increases with the increase in the skylight ratio, reaching its lowest point at 4%.

The reasons why air conditioning energy consumption in summer decreases first and then increases are:

(1) When the skylight ratio is small, the heat gain from lighting is the dominant factor. As the skylight ratio increases, the energy consumption and heat gain from lighting both decrease rapidly. The decrease in heat gain from lighting is beneficial for air conditioning energy consumption in the summer. In order to maintain indoor temperatures in the summer, air conditioning energy consumption decreases.

(2) As the skylight ratio continues to increase, the solar heat gain and heat transfer caused by the temperature difference are the dominant factors. At this time, although the reduction in lighting energy consumption gradually slows down, the solar heat gain and heat transfer caused by the temperature difference increase, which are not conducive to air conditioning energy consumption in the summer. To maintain indoor temperatures in the summer, air conditioning energy consumption increases rapidly.

(1) When the skylight ratio is the same, the air conditioning energy consumption in summer increases with a larger SHGC.

(2) The air conditioning energy consumption in summer under three different SHGCs decreases first and then increases with an increase in the skylight ratio.

The reason why air conditioning energy consumption in summer increases with a larger SHGC when the skylight ratio is the same as:

Under the same skylight ratio, the reduction in lighting energy consumption leads to the same reduction in air conditioning energy consumption in the summer. However, the solar heat gain entering the room increases with a larger SHGC, which is not conducive to air conditioning energy consumption in the summer.

(1) When the skylight ratio is the same, the air conditioning energy consumption in summer decreases with a larger K.

(2) The air conditioning energy consumption in summer under three different Ks decreases first and then increases with the increase in skylight ratio.

In summer, the daytime temperature is higher, and the air conditioning energy consumption increases. However, at night, the temperature of the outdoor air is lower than that of the indoor air. The larger the K is, the better the heat dissipation effect will be. When the decrease in daytime air conditioning energy consumption is smaller than the increase in nighttime air conditioning energy consumption, the air conditioning energy consumption in summer decreases with a larger K.

(1) When the skylight ratio is less than 10%, the air conditioning energy consumption in summer decreases with a larger VT under the same skylight ratio; when the skylight ratio is greater than 10%, there is not much difference in air conditioning energy consumption in summer among the three VTs.

(2) The air conditioning energy consumption in summer under three different VTs decreases first and then increases with the increase in skylight ratio.

The reason why the air conditioning energy consumption in summer decreases with a larger VT when the skylight ratio is the same as:

The reduction in lighting energy consumption caused by high-transmittance glass is greater than that caused by low-transmittance glass. The reduction in lighting heat caused by the reduction in lighting energy consumption will be much reduced, which is beneficial for air conditioning energy consumption in the summer.

3.4. Winter Air Conditioning Energy Consumption Simulation

This section analyzes the influence of different skylight ratios (SHGC, K, and VT) on air conditioning energy consumption in winter under lighting control, as shown in Figures 14–17.

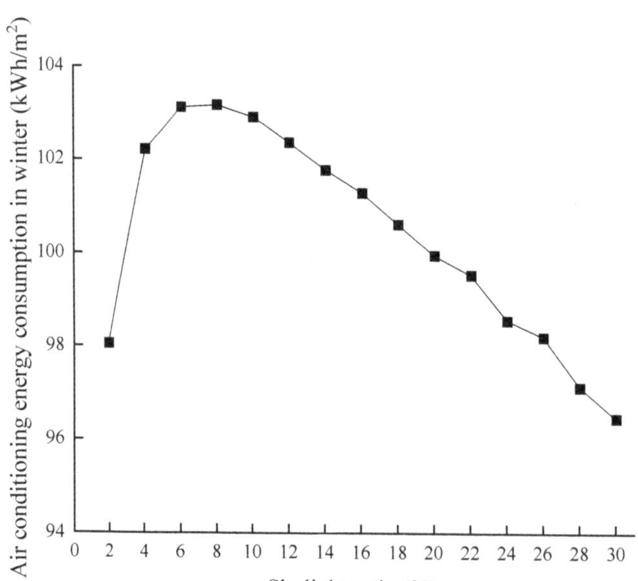

Figure 14. Impact of skylight ratio on air conditioning energy consumption in winter.

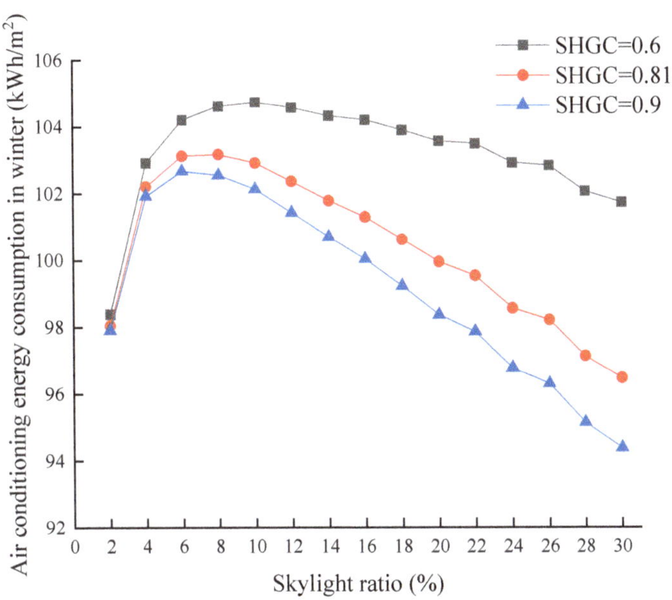

Figure 15. Influence of SHGC on air conditioning energy consumption in winter.

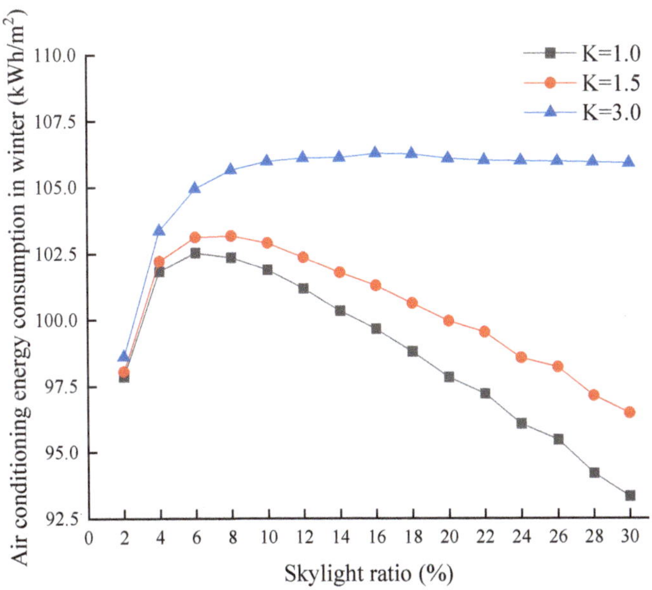

Figure 16. Influence of K on air conditioning energy consumption in winter.

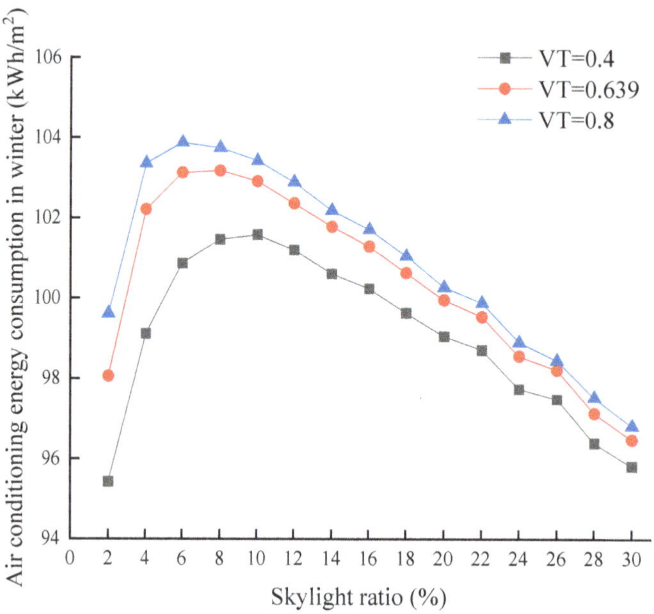

Figure 17. Influence of VT on air conditioning energy consumption in winter.

From Figure 14, it can be concluded that the air conditioning energy consumption in winter increases first and then decreases with the increase in the skylight ratio, which is opposite to the reason why the air conditioning energy consumption in summer decreases first and then increases.

(1) When the skylight ratio is the same, the air conditioning energy consumption in winter decreases with a larger SHGC, which is opposite to summer.

(2) The air conditioning energy consumption in winter under three different SHGCs increases first and then decreases with the increase in the skylight ratio.

(1) When the ratio of skylights is the same, the air conditioning energy consumption in winter increases with a larger K.

(2) The air conditioning energy consumption in winter under three different Ks increases first and then decreases with the increase in skylight ratio.

(1) When the skylight ratio is the same, the air conditioning energy consumption in winter increases with a larger VT, which is opposite to summer.

(2) The air conditioning energy consumption in winter under three different Ks increases first and then decreases with the increase in skylight ratio.

3.5. Validation

Liu Yao [23] analyzed the influence of the skylight ratio, shading coefficient, heat transfer coefficient, and light transmission rate on the energy consumption under lighting control. The results showed that after changing different parameters, the total energy consumption and air conditioning energy consumption in summer first decreased with the increase in the skylight ratio and then increased with the increase in the skylight ratio after reaching the lowest point. In addition, the air conditioning energy consumption in winter increased first and then decreased with the increase in the skylight ratio. The energy consumption variation law simulated in this paper is the same as the above law. The objects of both studies are the large spaces of public buildings, and the influence of transparent envelope parameters on energy consumption is studied under lighting control, so this reference has certain comparative verification significance.

3.6. Determination of $\Delta E_{1,W}$

$\Delta E_{1,W}$ can be regarded as the variation in total energy consumption with the change of window area under no lighting control in the room. Therefore, it is necessary to separately simulate and analyze the variation laws of the total energy consumption of side windows and skylights in different orientations with the window-wall ratio under no lighting control, as shown in Figures 18 and 19.

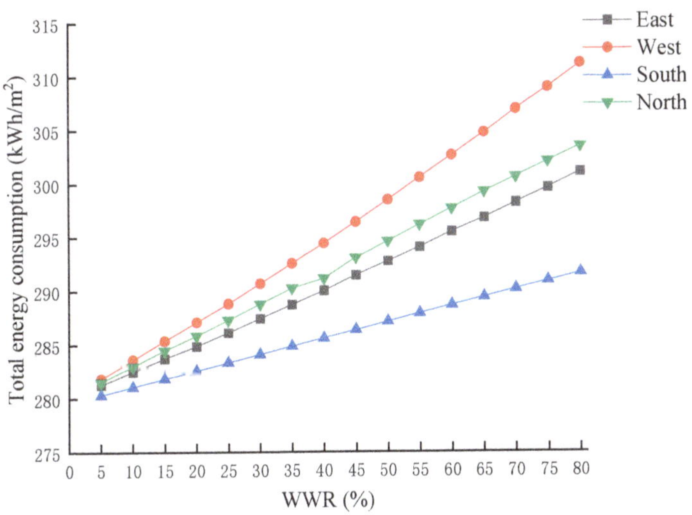

Figure 18. Influence of WWR on total energy consumption.

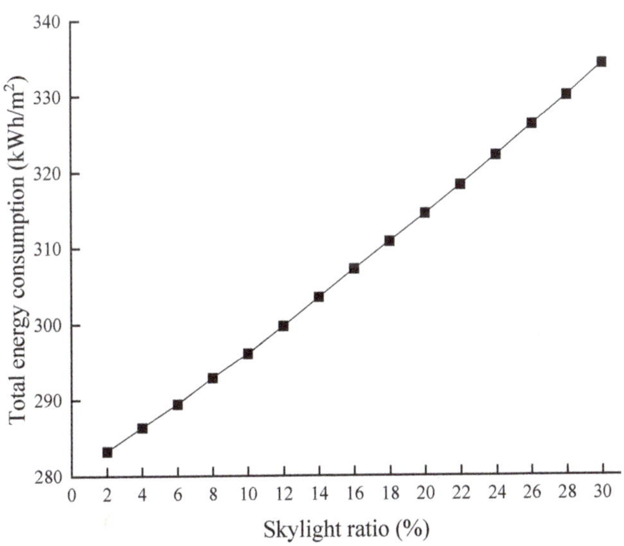

Figure 19. Impact of skylight ratio on total energy consumption.

Due to the different areas for each orientation, the window area represented by the same WWR for each orientation is also different, and the impact of opening the window on the variation of air conditioning energy consumption per unit area $\Delta e_{1,W}$ cannot be directly compared according to the WWR for each orientation. Therefore, 1% of the roof area is taken as the unit area, and the $\Delta E_{1,W}$ for each orientation is converted into the unit window area for comparison. By increasing the window area (1% of the roof area) by one unit, the variation in air conditioning energy consumption per unit area $\Delta e_{1,W}$ for each orientation indicates that the total energy consumption of skylights increases by 1.83 kWh/m^2, the east-facing energy consumption increases by 0.26 kWh/m^2, the west-facing energy consumption increases by 0.39 kWh/m^2, the south-facing energy consumption increases by 0.15 kWh/m^2, and the north-facing energy consumption increases by 0.29 kWh/m^2.

The significant influencing factors of the $\Delta e_{1,W}$ are SHGC and K of the glass. Therefore, by selecting different types of exterior windows from software material libraries, EnergyPlus was used for energy consumption simulation to gain the variation of $\Delta e_{1,W}$ for each orientation under different SHGCs and Ks of the exterior. The results are shown in Table 2.

Table 2. Variation in air conditioning energy consumption per unit area $\Delta e_{1,W}$.

Orientation	Exterior Window Type	SHGC	K [W/(m^2·K)]	$\Delta e_{1,W}$ (kWh/m^2)
Crown	White single-layer glass window	0.72	5.39	2.120
	White hollow glass window	0.64	3.60	1.660
	White single-layer glass window	0.62	4.74	1.815
	White double-layer glass window	0.56	2.67	1.360
	White hollow glass window	0.55	2.68	1.365
	Low-E-coated hollow glass window	0.52	1.77	1.180
	Low-E-coated hollow glass window	0.42	1.60	1.015
East	White single-layer glass window	0.72	5.39	0.404
	White hollow glass window	0.64	3.60	0.322
	White single-layer glass window	0.62	4.74	0.362
	White double-layer glass window	0.56	2.67	0.250
	White hollow glass window	0.55	2.68	0.248
	Low-E-coated hollow glass window	0.52	1.77	0.198
	Low-E-coated hollow glass window	0.42	1.60	0.174
West	White single-layer glass window	0.72	5.39	0.520
	White hollow glass window	0.64	3.60	0.420
	White single-layer glass window	0.62	4.74	0.472
	White double-layer glass window	0.56	2.67	0.342
	White hollow glass window	0.55	2.68	0.336
	Low-E-coated hollow glass window	0.52	1.77	0.284
	Low-E-coated hollow glass window	0.42	1.60	0.240
South	White single-layer glass window	0.72	5.39	0.274
	White hollow glass window	0.64	3.60	0.188
	White single-layer glass window	0.62	4.74	0.252
	White double-layer glass window	0.56	2.67	0.146
	White hollow glass window	0.55	2.68	0.148
	Low-E-coated hollow glass window	0.52	1.77	0.096
	Low-E-coated hollow glass window	0.42	1.60	0.096

Table 2. *Cont.*

Orientation	Exterior Window Type	SHGC	K [W/(m^2·K)]	$\Delta e_{1,W}$ (kWh/m^2)
North	White single-layer glass window	0.72	5.39	0.514
	White hollow glass window	0.64	3.60	0.410
	White single-layer glass window	0.62	4.74	0.466
	White double-layer glass window	0.56	2.67	0.332
	White hollow glass window	0.55	2.68	0.330
	Low-E-coated hollow glass window	0.52	1.77	0.252
	Low-E-coated hollow glass window	0.42	1.60	0.218

Establish quadratic linear regression models of the SHGC and K to the $\Delta e_{1,W}$ for different orientations, as shown in Equations (11)–(15).

$$\text{Crown}: \ \Delta e_{1,W} = 1.4264SHGC + 0.17624K + 0.11571 \tag{11}$$

$$\text{East}: \ \Delta e_{1,W} = 0.21212SHGC + 0.04588K + 0.01045 \tag{12}$$

$$\text{West}: \ \Delta e_{1,W} = 0.31572SHGC + 0.05037K + 0.03011 \tag{13}$$

$$\text{South}: \ \Delta e_{1,W} = -0.08436SHGC + 0.05397K + 0.04692 \tag{14}$$

$$\text{North}: \ \Delta e_{1,W} = 0.30082SHGC + 0.05638K + 0.00628 \tag{15}$$

Standardizing the window area for each orientation to be the same as the skylight area, the equivalent skylight and the original window design scheme have the same $\Delta E_{1,W}$. The total energy consumption variation $\Delta e_{1,W}{}'$ caused by the equivalent window per unit area represents the heat transfer characteristics of the glass. Since the window area and $\Delta E_{1,W}$ are linearly related, $\Delta e_{1,W}{}'$ can represent the weighted average of $\Delta e_{1,W}$ for each orientation, that is

$$\Delta e_{1,W}{}' = \frac{A_0}{A'}\left(\frac{A_e}{A_0}\Delta e_{1,W}(e) + \frac{A_w}{A_0}\Delta e_{1,W}(w) + \frac{A_s}{A_0}\Delta e_{1,W}(s) + \frac{A_n}{A_0}\Delta e_{1,W}(n) + \frac{A_t}{A_0}\Delta e_{1,W}(t)\right) \tag{16}$$

In the formula, e, w, s, n, and t respectively represent east, west, south, north, and crown; A' is the total area of the equivalent skylight, $A' = A_e + A_w + A_s + A_n + A_t$; A_0 is the unit area (1% of the roof area).

3.7. Determination of $E_{0,L}$

The variation in $E_{0,L}$ is ΔE_L, and ΔE_L is only related to the indoor light environment. Therefore, $E_{0,L}$ is expressed as the function of the daylight factor (C), $E_{0,L} = f(C)$, and the value of the daylight factor (C) represents the lighting characteristics of glass. With the increase in the window area, the daylight factor (C) increases linearly. The daylight factor c' formed by the unit equivalent skylight can also be written as the weighted average of the daylight factor c formed by the window per unit area for each orientation, that is

$$c' = \frac{A_0}{A'}\left(\frac{A_e}{A_0}c_e + \frac{A_w}{A_0}c_w + \frac{A_s}{A_0}c_s + \frac{A_n}{A_0}c_n + \frac{A_t}{A_0}c_t\right) \tag{17}$$

The curve of $E_{0,L}$ can be considered as to be solely related to the local meteorological conditions. The daylight factor is expressed in units of 2% to adjust the indoor lighting energy consumption according to 2%, 4%, 6%, ..., and 30%. The variation curve of $E_{0,L}$ is shown in Figure 20.

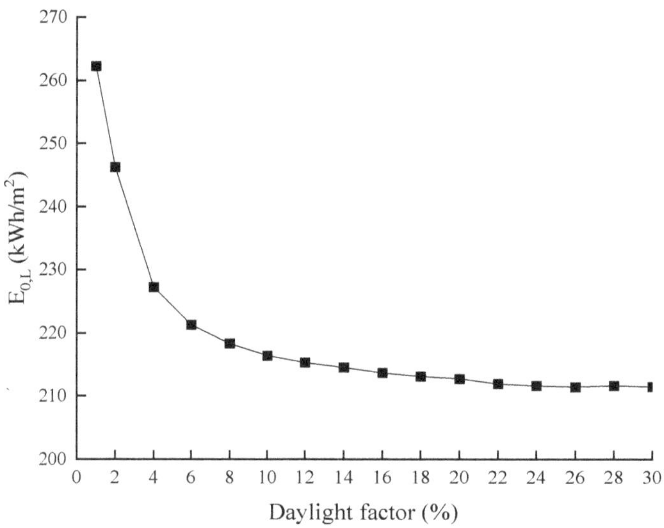

Figure 20. Variation law of $E_{0,L}$ with the daylight factor.

As can be seen from Figure 20, $E_{0,L}$ decreases rapidly as the daylight factor increases, between approximately 2–20%. After that, its rate of decrease gradually slows down, eventually reaching a constant value. For different daylight factor intervals, the variation of $E_{0,L}$ is denoted as $\Delta E_{0,L}$, and its values are shown in Table 3. If the curves in each interval are approximated as straight lines, then $\Delta E_{0,L}$ is the average slope of the $E_{0,L}$ curves in different daylight factor intervals.

Table 3. Of each daylight factor interval.

Daylight Factor Interval	$\Delta E_{0,L}$ (kWh/m^2)
2~4%	−18.99
4~6%	−5.95
6~8%	−2.96
8~10%	−1.91
10~12%	−1.11
12~14%	−0.76
14~16%	−0.73
16~18%	−0.66
18~20%	−0.40
20~30%	>−0.40

3.8. Trade-Off Judgement for Daylighting and Energy Consumption

The trade-off judgment for daylighting and energy consumption proposed in this paper is to find the window design condition with the lowest total energy consumption under daylighting conditions and judge the rationality of the window design scheme by the degree of actual total energy consumption deviating from the lowest one. Therefore, it is not necessary to calculate the entire curve, but only to calculate the actual total energy consumption $E'(A)$ and the lowest total energy consumption $E'(a)$, as shown in Figure 21.

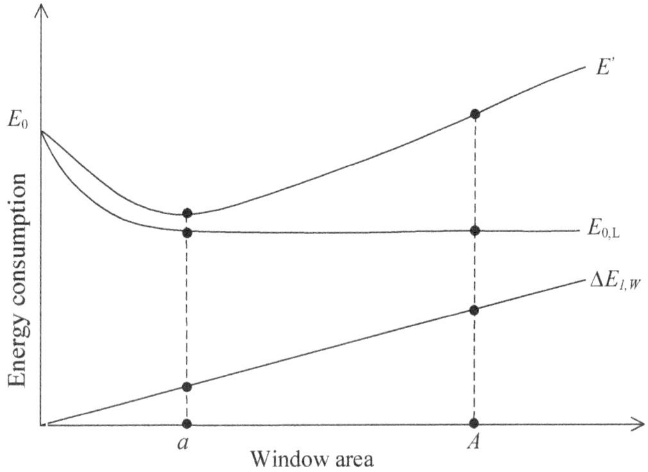

Figure 21. Influence of window area on energy consumption.

(1) Actual total energy consumption

The actual total energy consumption is calculated as $E'(A) = \Delta E_{1,W}(A) + E_{0,L}(A)$. After determining the window design scheme, the equivalent skylight characteristics and the actual total energy consumption at point A can be calculated, and then the actual total energy consumption $E'(A)$ is calculated by substituting the value of point A into the trade-off judgment formula.

(2) Lowest total energy consumption

The formula for calculating the lowest total energy consumption is $E'(a) = \Delta E_{1,W}(a) + E_{0,L}(a)$, find the position of point a with the lowest total energy consumption, and calculate the lowest total energy consumption $E'(a)$ by substituting the value of point a into the trade-off judgment formula.

$E_{0,L}$ is a curve changing with the daylight factor (C). To unify the independent variables of $\Delta E_{1,W}$ and $E_{0,L}$, $\Delta E_{1,W}$ which are recorded as a curve changing with the daylight factor (C), the slope of E' is

$$\frac{dE'}{dC} = \frac{d(\Delta E_{1,W})}{dC} + \frac{dE_{0,L}}{dC} \tag{18}$$

Since it is difficult to accurately fit the relation of $E_{0,L}$, the variation within a unit daylighting factor interval represents the average slope within the interval. To make the calculation results more accurate, 2% is taken as a unit daylighting factor:

$$\frac{dE'}{dC} \approx \frac{\Delta E'}{\Delta C} = 0.02 \times \frac{\Delta e_{1,W}{}'}{c'} + \Delta E_{0,L} \tag{19}$$

The derivative of the lowest point of the curve is 0. So let $\frac{dE'}{dC} = 0$, then the daylighting factor corresponding to the point a of the lowest total energy consumption can be calculated according to Equation (20).

$$0.02 \times \frac{\Delta e_{1,W}{}'}{c'} + \Delta E_{0,L} = 0 \tag{20}$$

(3) Judge the rationality of window area

In this paper, the right side of the lowest total energy consumption is less than or equal to 105% of the lowest total energy consumption as a reasonable window area zone, and the rationality judgment of window area is shown in Figure 22.

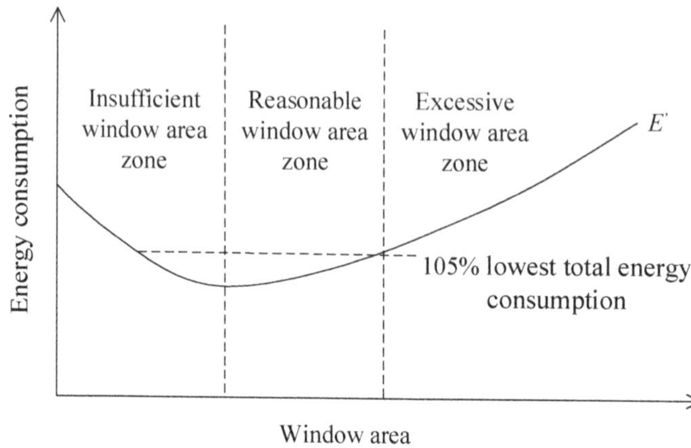

Figure 22. Schematic diagram of rational judgment of window area.

3.9. *The Progressive Optimization Calculation Process for Trade-Off Judgement*

Since the optimal window area and form cannot be directly given in the architectural design, the specific lighting design scheme can be weighed based on the above trade-off judgment of daylighting and energy consumption. If the window design scheme is unreasonable, then adjust the window design scheme and re-judge until it is reasonable. In this paper, this method is called the progressive optimization method for weighing daylighting and energy consumption. Its calculation process is shown in Figure 23.

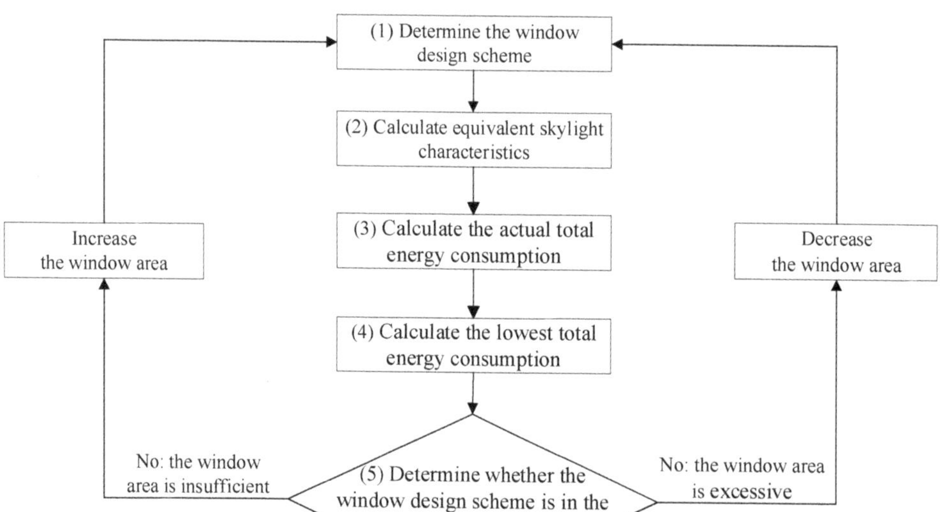

Figure 23. Flow diagram of trade-off judgment.

(1) Determine the window design scheme.

Determine the window area in each orientation and the VT, SHGC, and K of the exterior window glass, and substitute VT into Equations (9) and (10) to calculate the daylighting factor (c) formed by the skylight and side window per unit area in the indoor reference plane. Then, the SHGC and K are substituted into Equations (11)–(15) to calculate the $\Delta e_{1,W}$ in each orientation.

(2) Calculate equivalent skylight characteristics.

Substitute $\Delta e_{1,W}$ and c of each orientation into Equations (16) and (17), respectively, to calculate the $\Delta e_{1,W}{}'$ and the daylighting factor c' formed by the per-unit equivalent skylight area.

(3) Calculate the actual total energy consumption

$E'(A) = \Delta E_{1,W}(A) + E_{0,L}(A)$ is the formula for the actual total energy consumption. And the formula for $\Delta E_{1,W}(A)$ is $\Delta E_{1,W}(A) = \frac{A'}{A_0}\Delta e_{1,W}{}'$. According to the daylighting factor $C(A)$ of the actual total energy consumption, $E_{0,L}(A)$ can be obtained by referring to Figure 20, where the daylighting factor $C(A) = \frac{A'}{A_0}c'$.

(4) Calculate the lowest total energy consumption.

The lowest total energy consumption is calculated as $E'(a) = \Delta E_{1,W}(a) + E_{0,L}(a)$. Substitute $\Delta e_{1,W}{}'$ and c' into Equation (20) to calculate $\Delta E_{0,L}$. Accordingly $\Delta E_{0,L}$, the daylighting factor $C(a)$ corresponding to point a with the lowest total energy consumption can be obtained from Table 3. The variation in air conditioning energy consumption $\Delta E_{1,W}(a) = \frac{\Delta e_{1,W}{}'}{c'} \times C(a)$. And according to the daylighting factor $C(a)$ of the lowest total energy consumption, $E_{0,L}(a)$ can be obtained by referring to Figure 20.

(5) Determine whether the window design scheme is in a reasonable zone.

The rationality judgment of the window design scheme is divided into the following three situations:

When $C(A) < C(a)$, it is in the insufficient window area zone.

The actual total energy consumption point A is to the left of the lowest total energy consumption point a, and the actual total energy consumption is in the declining stage of the total energy consumption curve. At this time, if the window area is increasing, the actual total energy consumption will decrease, so it is in the insufficient window area zone.

When $C(A) \geq C(a)$ and $E'(A) \leq 1.05E'(a)$, it is in the reasonable window area zone.

The actual total energy consumption point A is on the right of the lowest total energy consumption point a, the actual total energy consumption is in the rising stage of the total energy consumption curve, and the actual total energy consumption is less than 105% of the lowest total energy consumption, so it is in the reasonable window area zone.

When $C(A) \geq C(a)$ and $E'(A) > 1.05E'(a)$, it is in the excessive window area zone.

The actual total energy consumption point A is on the right of the lowest total energy consumption point a, and the actual total energy consumption is in the rising stage of the total energy consumption curve. At this time, if the window area is increasing, the actual total energy consumption will increase, and the actual total energy consumption is more than 105% of the lowest total energy consumption, so it is in the excessive window area zone.

3.10. Example of Trade-Off Judgement

This section takes a gymnasium in Beijing as an example to illustrate how the progressive optimization method for weighing daylighting and energy consumption can be applied in practical situations.

The main lighting forms, size, side window to wall ratio, skylight ratio, VT, SHGC, and K of the gymnasium are shown in Table 4.

Table 4. Parameter for the gymnasium.

Name	Parameter
The main lighting forms	East and west side windows and skylights
Size	75 m × 55 m × 25 m
Side window-to-wall ratio	East and west side: 70%
Skylight ratio	5%
Glass material	VT = 0.639
	SHGC = 0.81
	$K = 1.50\ W/(m^2 \cdot K)$

(1) Determine the window design scheme.

By substituting the VT of the glass into Equations (9) and (10), the daylighting factor (c) formed by the skylight and side windows per unit area indoors can be calculated.

Skylight:

$$c_t = 1.92667 \times 0.639 + 0.09556 = 1.33\%;$$

East and west side windows:

$$c_e = c_w = 0.365 \times 0.639 + 0.14417 = 0.37\%.$$

By substituting SHGC and K of the glass into Equations (11)–(13), the variation of air conditioning energy consumption per unit area $\Delta e_{1,W}$ for orientation can be calculated.

Skylight:

$$\Delta e_{1,W}(t) = 1.4264 \times 0.81 + 0.17624 \times 1.5 + 0.11571 = 1.54\ \text{kwh/m}^2;$$

East side window:

$$\Delta e_{1,W}(e) = 0.21212 \times 0.81 + 0.04588 \times 1.5 + 0.01045 = 0.25\ \text{kwh/m}^2;$$

West side window:

$$\Delta e_{1,W}(w) = 0.31572 \times 0.81 + 0.05037 \times 1.5 + 0.03011 = 0.36\ \text{kwh/m}^2.$$

(2) Calculate equivalent skylight characteristics.

By substituting the daylighting factor (c) and $\Delta e_{1,W}$ into Equations (16) and (17), the total energy consumption variation $\Delta e_{1,W}{}'$ and the daylighting factor c' caused by the equivalent skylight per unit area can be calculated.

$$\Delta e_{1,W}{}' = \frac{0.05 \times 75 \times 55 \times 1.54 + 0.7 \times 55 \times 25 \times (0.25 + 0.36)}{0.05 \times 75 \times 55 + 2 \times 0.7 \times 55 \times 25} = 0.42\ \text{kwh/m}^2;$$

$$c' = \frac{0.05 \times 75 \times 55 \times 1.33 + 2 \times 0.7 \times 55 \times 25 \times 0.37}{0.05 \times 75 \times 55 + 2 \times 0.7 \times 55 \times 25} = 0.36\%;$$

$$\frac{A'}{A_0} = \frac{0.05 \times 75 \times 55 + 2 \times 0.7 \times 55 \times 25}{0.01 \times 75 \times 55} = 52.$$

(3) Calculate the actual total energy consumption

The daylighting factor corresponding to the actual total energy consumption at point A: $C(A) = \frac{A'}{A_0}c' = 52 \times 0.46 = 23.92\%;$

$$\Delta E_{1,W}(A) = \frac{A'}{A_0}\Delta e_{1,W}{}' = 52 \times 0.42 = 21.84\ \text{kwh/m}^2;$$

According to the daylighting factor C(A) of the actual total energy consumption, $E_{0,L}(A)$ can be obtained by referring to Figure 20: $E_{0,L}(A) = 211.75$ kwh/m^2;

The actual total energy consumption at point A:

$$E'(A) = \Delta E_{1,W}(A) + E_{0,L}(A) = 21.84 + 211.75 = 233.59 \text{ kwh/m}^2.$$

(4) Calculate the lowest total energy consumption.

Calculate $\Delta E_{0,L}$ by substituting $\Delta e_{1,W}'$ and c' into $0.02 \times \frac{\Delta e_{1,W}'}{c'} + \Delta E_{0,L} = 0$: $\Delta E_{0,L} = -1.83$ kwh/m^2;

According to $\Delta E_{0,L}$, the daylighting factor C(a) corresponding to point a with the lowest total energy consumption can be obtained from Table 3: $C(a) = 10\%$;

$$\Delta E_{1,W}(a) = \frac{\Delta e_{1,W}'}{c'} \times C(a) = \frac{0.42}{0.0046} \times 0.1 = 9.13 \text{ kwh/m}^2;$$

According to the daylighting factor C(a) of the lowest total energy consumption, $E_{0,L}(a)$ can be obtained by referring to Figure 20: $E_{0,L}(a) = 216.44$ kwh/m^2;

The lowest total energy consumption at point a:

$$E'(a) = \Delta E_{1,W}(a) + E_{0,L}(a) = 9.13 + 216.44 = 225.57 \text{ kwh/m}^2.$$

(5) Determine whether the window design scheme is in a reasonable zone.

From the above results, the trade-off judgment result between daylighting and energy consumption in the gymnasium can be determined. As shown in Figure 24, C(A) > C(a) and $E'(A) \leq 1.05E'(a)$. The actual total energy consumption is in the rising stage of the total energy consumption curve, and the actual total energy consumption is about 105% of the lowest total energy consumption. Therefore, the window design scheme of the gymnasium is in the reasonable window area zone.

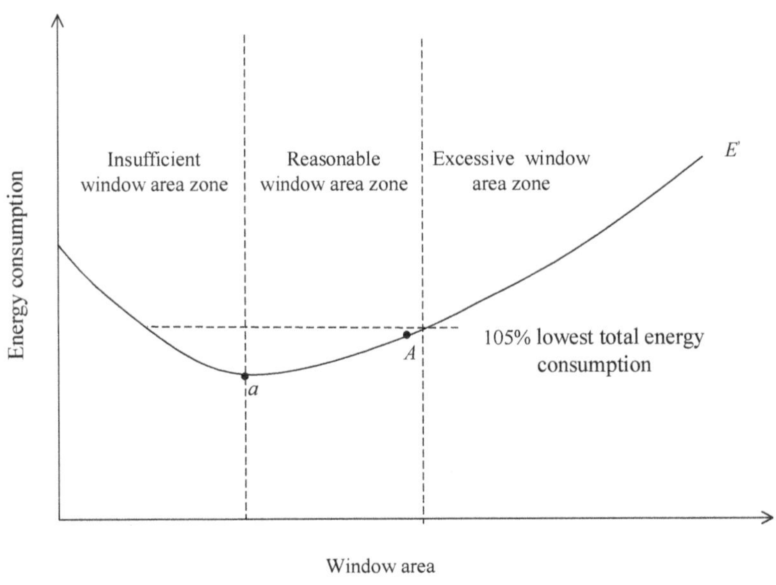

Figure 24. Trade-off judgment result chart.

4. Conclusions

This paper simulates the building energy consumption for the high and large space of the main stadium of the University of Science and Technology Beijing Gymnasium and makes a trade-off judgment on its daylighting and energy consumption. The main conclusions are as follows:

(1) The relationship between the daylight factor and VT formed by skylights and side window openings per unit area is obtained by simulation.

(2) The University of Science and Technology Beijing Gymnasium is a high and large space. If changing the SHGC, K, and VT under lighting control, the total energy consumption and air conditioning energy consumption in summer will first decrease and then increase with the rise of the skylight ratio. However, the variation in air conditioning energy consumption in winter is opposite to them. Air conditioning energy consumption in the summer is the dominant factor affecting the variation in total energy consumption.

(3) Under no lighting control in the room, by changing the areas of the side window and skylight, the total energy consumption of the gymnasium increases with the rise of the WWR, and the total energy consumption is in a linear relationship with the WWR, in which changing the area of the east and south side windows has little impact on the total energy consumption. If the daylighting effect is strengthened, the east and south side windows should be added.

(4) The SHGC and K of the exterior window have a significant impact on the total energy consumption. Through energy consumption simulation with different types of exterior windows, it is concluded that only the SHGC of the south-facing window is negatively correlated with the $\Delta e_{1,W}$, and its regression coefficient is -0.08436, while the other oriented windows are positively correlated. Compared with other orientations, the SHGC and K of the skylight have the most significant influence on the $\Delta e_{1,W}$.

(5) $E_{0,L}$ decreases rapidly as the daylight factor increases, between approximately 2–20%. After that, its rate of decrease gradually slows down, eventually reaching a constant value.

(6) The total energy consumption first decreases and then increases with the rise of the window area, and there is a lowest point, so the right side of the lowest total energy consumption is less than or equal to 105% of the lowest total energy consumption as a reasonable window area zone. When $C(A) < C(a)$, it is in the insufficient window area zone; when $C(A) \geq C(a)$ and $E'(A) \leq 1.05E'(a)$, it is in the reasonable window area zone; when $C(A) \geq C(a)$ and $E'(A) > 1.05E'(a)$, it is in the excessive window area zone. And a progressive optimization calculation process for weighing daylighting and energy consumption in university gymnasiums in Beijing is proposed. Use this calculation process to determine whether the specific window design scheme of university gymnasiums in Beijing is reasonable. If it is unreasonable, adjust the window design scheme and reevaluate until it is reasonable.

5. Prospect

The shortcomings of this paper and the suggestions for further research:

(1) Restricted by on-site requirements and objective measurement conditions, the lack of measurement data for verification is one of the shortcomings of this paper. In further research, a comparative analysis can be carried out based on the measurement data of other similar buildings. Although this method cannot completely replace the on-site measurement data, it can still provide a certain degree of support for the reliability of the simulation results.

(2) Although this paper explores the effect of different window design positions (WWR, SHGC, K, and VT) of the transparent envelope on the total energy consumption, the final foothold is the influence of the window area on the total energy consumption. In the next step of research, specific optimizations will be made for other thermal performance parameters.

(3) This paper only considers gymnasiums with enclosed windows, which use mechanical ventilation systems to control indoor airflow. In the following research, the use of windows for natural ventilation will be considered, and the ventilation aspects will be explained in detail.

(4) For modeling convenience, only one window is set up in this paper according to the WWR. In the following study, the impact of the number, shape, and distribution of windows on energy consumption will be considered.

Author Contributions: Conceptualization, Y.W.; methodology, M.J.; software, M.J.; formal analysis, Y.W. and M.J.; investigation, Y.W., T.Z. and M.J.; data curation, Y.W. and T.Z.; writing—original draft preparation, Y.W. and M.J.; writing—review and editing, Y.W. and M.J.; supervision, Y.W. and T.Z.; project administration, Y.W. and T.Z.; funding acquisition, Y.W. All authors have read and agreed to the published version of the manuscript.

Funding: This research was funded by the Beijing Natural Science Foundation (8202034) and the Central Guidance for Local Scientific and Technological Development Funding Project (236Z5202G).

Data Availability Statement: Data are contained within the article.

Conflicts of Interest: The authors declare no conflicts of interest.

Abbreviations

WWR	Window-to-Wall Ratio
SHGC	Solar Heat Gain Coefficient
K	Heat Transfer Coefficient
VT	Visible Light Transmittance
$\Delta E_{1,W}$	Variation in air conditioning energy consumption caused by the change in heat transfer performance of the envelope structure (indoor and outdoor temperature difference heat transfer and solar radiation heat gain) after opening the window
$\Delta e_{1,W}$	Variation of air conditioning energy consumption per unit area
$\Delta e_{1,W}{}'$	Total energy consumption variation caused by the equivalent window per unit area
c	Daylight factor formed by window per unit area
c'	Daylight factor formed by the unit equivalent skylight
$C(A)$	Daylighting factor of the actual total energy consumption
$C(a)$	Daylighting factor of the lowest total energy consumption
$E'(A)$	Actual total energy consumption
$E'(a)$	Lowest total energy consumption

References

1. Liu, S. Research on the Sustainable Design Strategy of Cold Area Gymnasium Considering the Efficient Utilization after Competition. Master's Thesis, Shenyang Jianzhu University, Shenyang, China, 2020. [CrossRef]
2. Shi, L.; Li, Y.; Tao, L.; Zhang, Y.; Jiang, X.; Yang, Z.; Qi, X.; Qiu, J. Sporters' visual comfort assessment in gymnasium based on subjective evaluation & objective physiological response. *Build. Environ.* **2022**, *225*, 109678. [CrossRef]
3. Zheng, Y.; Xie, H.; Wang, L.; Liang, W.; Zou, K. Energy consumption of one gymnasium building in Beijing. *Build. Energy Effic.* **2012**, *40*, 54–58. [CrossRef]
4. Bian, Y.; Luo, J.; Hu, J.; Liu, L.; Pang, Y. Visual discomfort assessment in an open-plan space with skylights: A case study with POE survey and retrofit design. *Energy Build.* **2021**, *248*, 111215. [CrossRef]
5. Ochoa, C.E.; Aries, M.B.C.; Van Loenen, E.J.; Hensen, J.L.M. Considerations on design optimization criteria for windows providing low energy consumption and high visual comfort. *Appl. Energy* **2012**, *95*, 238–245. [CrossRef]
6. Bunjongjit, S.; Ngaopitakkul, A. Feasibility study and impact of daylight on illumination control for energy-saving lighting systems. *Sustainability* **2018**, *10*, 4075. [CrossRef]
7. Li, D.H.W.; Tsang, E.K.W. An analysis of daylighting performance for office buildings in Hong Kong. *Build. Environ.* **2008**, *43*, 1446–1458. [CrossRef]
8. Peng, P. Study on the Energy Consumption Characteristics of Daylighting in Commercial Buildings. Master's Thesis, Chongqing University, Chongqing, China, 2006.
9. Ruck, N.C. International energy agency's solar heating and cooling task 31, 'Daylighting Buildings in the 21st Century'. *Energy Build.* **2006**, *38*, 718–720. [CrossRef]
10. Alghoul, S.K.; Rijabo, H.G.; Mashena, M.E. Energy consumption in buildings: A correlation for the influence of window to wall ratio and window orientation in Tripoli, Libya. *J. Build. Eng.* **2017**, *11*, 82–86. [CrossRef]
11. Yang, Q.; Liu, M.; Shu, C.; Mmereki, D.; Uzzal Hossain, M.; Zhan, X. Impact analysis of window-wall ratio on heating and cooling energy consumption of residential buildings in hot summer and cold winter zone in China. *J. Eng.* **2015**, *2015*, 538254. [CrossRef]
12. Dong, H.; Xu, S. Energy consumption of residential building influenced by the window-wall ratio and sunshade in Beijing. *Build. Energy Effic.* **2010**, *235*, 66–69. [CrossRef]

13. Futrell, B.J.; Ozelkan, E.C.; Brentrup, D. Bi-Objective optimization of building enclosure design for thermal and lighting performance. *Build. Environ.* **2015**, *92*, 591–602. [CrossRef]
14. Konis, K.; Gamas, A.; Kensek, K. Passive performance and building form: An optimization framework for early-stage design support. *Sol. Energy.* **2016**, *125*, 161–179. [CrossRef]
15. Motamedi, S.; Liedl, P. Integrative algorithm to optimize skylights considering fully impacts of daylight on energy. *Energy Build.* **2017**, *138*, 655–665. [CrossRef]
16. Fang, Y.; Cho, S. Design Optimization of building geometry and fenestration for daylighting and energy performance. *Sol. Energy.* **2019**, *191*, 7–18. [CrossRef]
17. Lakhdari, K.; Sriti, L.; Painter, B. Parametric optimization of daylight, thermal and energy performance of middle school classrooms, case of hot and dry regions. *Build. Environ.* **2021**, *204*, 108173. [CrossRef]
18. Yan, L. Coupling analysis of natural lighting and air conditioning energy consumption of small and medium-sized railway stations. *HV&AC* **2017**, *47*, 90–93.
19. Wu, D.; Liu, C. Study on the energy use pattern and energy saving potential of railway station in cold climate zone. *Urban Dev. Stud.* **2015**, *22*, 36–40.
20. Xing, H.; Bi, X.; Liu, C. Analysis on building space and energy consumption characteristics of gymnasiums in cold climate zone. *Urban Dev. Stud.* **2015**, *22*, 161–166.
21. *GB55015-2021*; General Code for Energy Efficiency and Renewable Energy Application in Buildings. China Architecture & Building Press: Beijing, China, 2021.
22. *GB5033-2013*; Standard for Daylighting Design of Buildings. China Architecture & Building Press: Beijing, China, 2013.
23. Liu, Y. Weigh of Natural Lighting and Air-Conditioning's Energy Consumption in Large Space of Railway Station's Public Area. Master's Thesis, Southwest Jiaotong University, Chengdu, China, 2012.

buildings

MDPI

Review

Integrated Systems of Light Pipes in Buildings: A State-of-the-Art Review

Yanpeng Wu *, Meitong Jin, Mingxi Liu and Shaoxiong Li

School of Civil and Resource Engineering, University of Science and Technology Beijing, Beijing 100083, China; m202310072@xs.ustb.edu.cn (M.J.); s20200075@xs.ustb.edu.cn (M.L.); m202110070@xs.ustb.edu.cn (S.L.)
* Correspondence: wuyanpeng@ustb.edu.cn

Abstract: Artificial lighting comprises nearly one-third of the total electrical load of buildings, resulting in significant carbon emissions. Reducing the carbon emissions caused by artificial lighting is one of the ways to achieve low-carbon buildings. To meet the demand for high-efficiency, energy-saving, and comfortable lighting, light pipes are increasingly used in buildings. This paper reviews the research and development of light pipes and integrated technology. Sky conditions as a dynamic factor always affect the performance of light pipes. The combination of light pipes and an artificial lighting system can effectively solve this problem. A light pipe can be integrated with a ventilation stack to achieve the ventilation and cooling or heating of a building. A lighting-heating coupled light guide can improve the energy efficiency and sustainability in buildings, such as where antimony tin oxide nanofluid is introduced to absorb additional heat and then provide domestic hot water. The application of a photocatalyst to light pipes can realize air purification and self-cleaning. The use of light pipes does not consume electricity and can reduce the time spent using artificial lighting, thus allowing for power savings. From a whole life cycle perspective, the use of light pipes can be a balance of cost and benefit. In conclusion, such information could be useful for engineers, researchers, and designers to assess the suitability of applying integrated light pipes in different building types and examine the potential of energy and cost savings.

Keywords: light pipes; daylighting; lighting; integrated systems; ventilation; air purification

1. Introduction

Buildings currently account for a third of global energy consumption and a quarter of CO_2 emissions [1]. Cooling, heating, and lighting are the main factors affecting building energy consumption. Artificial lighting accounts for about 33% of building electricity consumption [2]. Improper lighting can cause visual discomfort, reduce work efficiency, and even affect health [3–5]. In contrast, daylighting has good color rendering. It is one of the most ideal sources of biological circadian rhythm regulation, and its use can have a positive impact on indoor personnel and reduce building energy consumption [6–11].

Regarding the mechanism of collecting daylight, there are two types of light pipes, named active type and passive type. The active systems collect daylight by sun-tracking concentrators, such as the Himawari system in Japan, the HSL system in the USA, the Arthelio system in Europe, and the Heliobus system in Switzerland [12]. This type can collect daylight efficiently, but it is rarely used in construction because of the high cost. In contrast, passive light pipes are simpler and the production cost is lower. It is more and more widely used. This article is an overview of passive light pipes.

The light pipe is one of the lighting devices that introduce daylight into a space. It is generally composed of three parts: light collector, hollow tube, and diffuser (see Figure 1). The collector is produced from a transparent polycarbonate material, designed to collect lights from the sky. The light is reflected multiple times in the hollow tube with highly reflective material on the inner wall. A diffuser is generally produced from opal or prismatic material placed on the ceiling of a room. The light pipe lighting system gathers natural outdoor light through the light harvesting device (light collector) and channels it into the system. After strengthening and efficiently transmitting the light, the diffuser at the bottom of the system will evenly channel the light to any indoor places where light is needed. The propagation of the optical path within the light pipe is achieved by total reflection. When light enters from the side of the light pipe, it will undergo multiple total reflections on the walls inside the light pipe, which allows the light to be directed along the length of the light pipe. During propagation, the light is constantly reflected on the walls of the light pipe until it reaches the other end of the light pipe or is output to the place which needs to be illuminated. This all-reflective propagation allows the light pipe to transmit light efficiently, so that the light can be evenly distributed on the surface of the light pipe, thus achieving the illumination effect. Compared with traditional side windows and skylights, light pipes have better heat insulation performance and flexibility [13]. The influence of the light pipe on indoor temperature fields is not obvious, and the temperature rise at a similar illumination level is far less than that of artificial lighting [14]. It can avoid glare to ensure sufficient light comfort [15,16]. The light pipe is gradually being recognized for its levels of energy-saving and comfort.

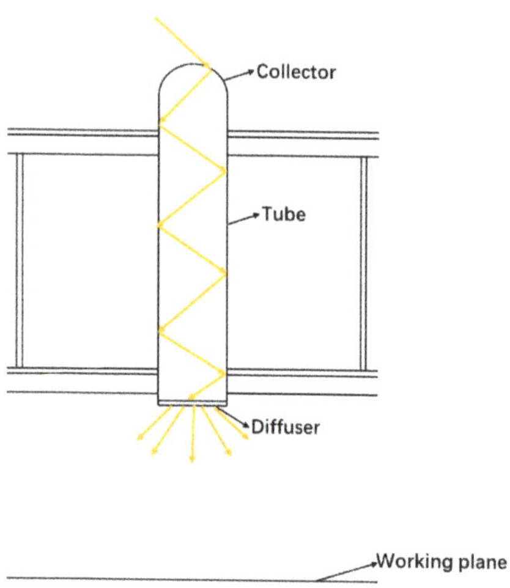

Figure 1. Components of a light pipe.

Based on the evaluation theory of tubular light guides, relevant studies provide a detailed description of a tubular light guide model with a flat glass cover under clear sky conditions with direct solar radiation [17]. Various physical aspects have been considered, including the elevation angle of incident light rays, the number of incident levels, the length and diameter of the light pipe, and the optical properties of the light pipe components. This theoretical framework allows for an in-depth exploration of how light penetrates the surface from a physics perspective and which incidence angles may be more suitable. Simultaneously, the study visually presents the propagation of light rays within the tubular light guide, including vertical plane projection, horizontal plane projection, and an axonometric scheme.

The chemical structures of light pipes encompass the arrangement and composition of materials utilized in the fabrication of these optical devices. The chemical structures play a pivotal role in dictating the optical properties and overall performance of light pipes. Organic polymers, such as polymethyl methacrylate (PMMA), are frequently employed in the construction of light pipes, offering a balance of transparency, flexibility, and ease of fabrication. The chemical structure of these polymers influences their refractive indices and, consequently, the guiding and transmission of light within the pipe. In the case of glass light pipes, the chemical composition of borosilicate glass or other specialized glass types is crucial. The unique arrangement of silicon, oxygen, and other elements in the glass matrix influences its optical transparency, thermal stability, and resistance to environmental factors. The continuous advancements in nanotechnology have introduced nanostructured materials into the realm of light pipe design. Nanoengineered surfaces, coatings, or additives contribute to tailoring the optical properties of light pipes, enabling precise control over light propagation and enhancing their performance [13,18].

The stability of light pipes refers to their ability to maintain consistent and reliable performance over time. Their stability is crucial in ensuring that they can withstand various environmental conditions and usage scenarios without significant degradation in their optical properties. Several factors affect the stability of light pipes: (1) Material durability: materials with high durability and resistance to environmental factors, such as temperature changes or humidity, contribute to the long-term stability of the light pipe [19]. (2) Mechanical integrity: a robust design and construction ensure that the light pipe can withstand physical stresses and maintain its performance under normal usage conditions. (3) Chemical resistance: chemical resistance of the materials used ensures that the light pipe remains stable even when subjected to potentially corrosive substances.

The preparation methods of light pipes involve specific techniques and processes employed in the fabrication of these optical devices designed for efficient light transmission. Several key methods are utilized: (1) Molding and extrusion: one prevalent technique involves molding or extruding materials, such as polymers, into the desired shape of the light pipe. This method is suitable for producing light pipes with straightforward geometries and is often employed for cost-effective mass production. (2) Glass forming: for light pipes requiring higher optical precision, glass forming techniques are utilized. This involves shaping glass through processes like blowing, drawing, or precision machining to achieve the desired optical properties [20]. (3) Coating and cladding: thin layers of materials with specific refractive indices may be applied to modify light transmission properties or improve efficiency [18]. (4) Fiber optic manufacturing: in the case of fiber optic light pipes, a complex process known as drawing is employed. This involves stretching a preformed rod of glass or other materials to create long, thin fibers. These fibers can transmit light over long distances with minimal loss.

The authors of this article carried out a state-of-the-art review on light pipes in buildings. The light pipe is an effective system to provide daylighting for remote and windowless spaces within buildings. Solar altitude, cloud volume, external illuminance, and aerosols in the atmosphere all affect daylighting performance. These factors can be uniformly referred to as sky conditions. Sky conditions are constantly changing over time. In addition to these dynamic factors, the components of the light pipe and the orientation of the light pipe also have an impact on daylighting performance. To enhance daylighting performance, the traditional light pipe has been modified. With the above factors, it is difficult to accurately predict the daylighting performance. Prediction methods have been continuously improved. Existing prediction methods tend to consider the transmission efficiency of the three main components of the light pipe separately, and then combine them to calculate the total efficiency. Based on meeting the requirements of lighting, light pipes can be combined with ventilation systems, artificial lighting, and air purification. Since buildings generate significant energy consumption, the change in electricity consumption and the economic benefits of using light pipes for lighting are also being considered.

In Section 2, light pipes combined with artificial lighting are presented. Light pipes combined with the ventilation system are described in Section 3. Light pipes combined with ATO (Antimony Tin Oxide) nanofluids and photocatalysis are described in Section 4. The energy consumption and the economic efficiency of light pipes are described in Section 5. Prediction methods for integrated systems and influence of orientation on the integrated system are discussed in Section 6. The major findings of the article are summarized in Section 7. Finally, Section 8 discusses further improvements of this study.

2. Light Pipes Combined with Artificial Lighting

Generally, solar altitude, cloud volume, and external illuminance are considered to be weather factors that affect the lighting performance of the light pipe. In a day, the luminous flux introduced by the light pipe increases with the increase of solar altitude and decreases with the increase of cloud cover [21]. In addition, the light distribution on the working plane is more uniform when the solar altitude is low [22]. The efficiency of the light pipe under overcast conditions is slightly higher than that under clear sky conditions [23]. With the change of seasons, the luminous flux output by the light pipe in summer is better than that in winter, but there will be obvious light and dark changes on the working plane in summer [24]. However, among solar altitude, cloud cover, and external illuminance, external illuminance has the greatest impact on internal illuminance. The influence of solar altitude and cloud cover on internal illuminance may be affected by changes in external illuminance [25]. The research by Vasilakopoulou [26] further pointed out that the average internal illuminance and the maximum internal illuminance have an exponential relationship with the external illuminance, respectively. In addition to the above factors that affect the daylighting performance of the light pipe, Kocifaj [27] has shown that aerosols in the atmosphere are also one of the factors. Different types of aerosols will have different effects on illuminance patterns and efficiency.

Light pipes cannot meet the indoor illuminance requirements in some situations, and artificial lighting equipment is needed for auxiliary lighting. A study on light pipe systems stated that these devices, when paired with electric lighting controls, could achieve 20% energy savings [28]. Thus, the combination of passive light pipes and artificial lighting equipment systems has come into being. In this integrated system, in most cases, luminaires are installed over individual workstations or defined visual task areas and equipped with, variously, integrated network controls, occupancy sensors, personal dimming, or daylight dimming. The two systems are not physically connected. This system can maximize the use of available daylight. However, the color rendering of the two light sources is significantly different, which may bring an uncomfortable visual experience. Unstable daylight makes the artificial lighting system open and close frequently or adds more complicated control systems, which may reduce the service life of the system or increase the cost. Görgülü and Ekren [29] illuminated a windowless room via a light pipe and dimmable electronic ballasts. During the operation of the system, the required 350 lux illumination level on the work plane was measured and retained by the controller throughout the day.

3. Light Pipes Combined with the Ventilation System

Integration of the passive stack ventilation system and light pipes would make both technologies more attractive. The basic idea is to guide daylight into the building and air out of the building using an integrated structure (see Figure 2). The integration of these technologies proved to be feasible [30]. The light pipe-natural ventilation system experimental apparatus is shown in Figure 3. To enhance the natural stack ventilation and prevent reverse flow, a winding terminal is installed on the top of the light-vent pipe [31–34]. Common forms of wind terminals are shown in Figure 4. To enhance the driving force of natural ventilation, Elmualim et al. [35] used dichroic materials to construct light pipes, which increased the flow of natural ventilation by 14%. L. Shao and Riffat [36] combined light-vent pipe with heat pipes and used the principle of thermosyphon, which can not only enhance ventilation but also be used for building heating. This system can be realized without relying on a mechanical driving force. Taengchum et al. [37] designed a light pipe integrated with a solar-heated ventilation stack that can be used for night ventilation and designed a method that can be used to design a system to achieve the required ventilation rate, transmit a given luminous flux, and evaluate the cost-effectiveness of the configuration. This system needs to use a hot water pump as the circulating power.

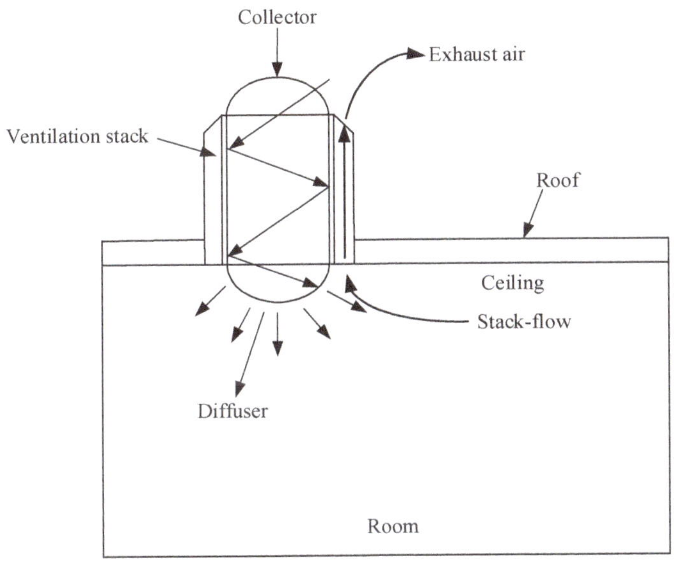

Figure 2. Combined the light pipe and passive stack ventilation system.

Figure 3. Light pipe-natural ventilation system experimental apparatus.

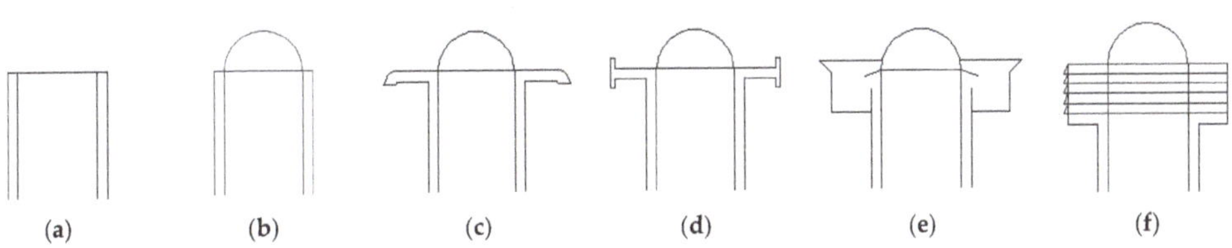

Figure 4. Different ventilation terminals: (**a**) open channel, flat top; (**b**) open channel, hemispherical dome; (**c**) umbrella type; (**d**) H-pot type 1; (**e**) H-pot type 2; (**f**) shutter type.

4. Light Pipes Combined with ATO Nanofluids and Photocatalysis

The physical properties of the three components (light collector, hollow tube, and diffuser) of the light pipe affect the optical properties. The research on the three components is summarized in Table 1. Mohelnikova et al. [17] and Robertson et al. [38] proved that the daylighting performance of the hemispherical dome is better than that of the flat glass cover. Collectors of the light pipes are installed outdoors. The surface of the collector is inevitably exposed to the sun and debris, which affects the lighting effect [39] and reduces the service life of the collector. Applying a super-hydrophobic layer to the collector can realize self-cleaning, which plays an important role in improving the lighting performance [40]. The major constraints to the development of superhydrophobic coatings are expensive superhydrophobic materials, nano-feature durability, coating stability, precipitation/condensation problems, impact problems, and emulsifier/oil wetting problems [41,42].

The length, diameter, and inner wall material of the hollow tubes have an impact on the light reflectivity of light pipes. The ratio of the diameter to the length of the reflecting tube is defined as the aspect ratio of the light pipe. Only the length and diameter of light pipes are changed to obtain different specifications of light pipe. Different specifications of light pipes with the same aspect ratio have the same Transmission Tube Efficiency (TTE) under a CIE (International Commission on Illumination) standard overcast sky [43]. Transmission Tube Efficiency (TTE) is defined as the ratio of the output light flux to the input light flux of the light pipe. The efficiency is a decreasing function of aspect ratio independent of whether overcast sky or clear sky conditions are taken into consideration [44]. Wu et al.'s [45] study showed that higher reflectance of the inner surface will lead to a higher efficiency of the straight pipe. For a light pipe with an elbow, under the same bending angle and light reflection ratio, the TTE of elbows with different pipe diameters is basically the same. The reflective material on the inner wall of the light pipe determines the color rendering properties of transmitted light [46,47]. With an increase of the number of reflections and the change of the angle of incident ray, the color characteristics of the optical tube will change. Furthermore, Nilsson et al. [47] used spectrophotometer measurement and ray-tracing simulation methods to quantitatively verify that highly reflective films with spectral variations of a few percent do not markedly affect the color of the transmitted light. Different geometrical shapes, thicknesses, and fabrication materials lead to different optical properties of the diffuser. We found that light pipes with "snow type" diffusers had better performance compared with the "diamond type" ones [48]. Swift et al. [49] experimentally studied 1 mm thick diffuser (the diffusing material used is Plexiglas), finding that it performed better when the uniformity of illumination was improved and the loss of luminous flux was small. Robertson et al. [38] compared three different diffuser types. From the perspective of luminous flux, flat Fresnel performed the best, followed by flat frosted, and curved frosted performed the worst. However, there is a modest disbenefit to the Fresnel when there is diffuse light only. Two-component glazing can solve this problem [50]. The inner parts of circular glazing work like Lambertian diffusers while the edges of the glazing are built of clear glass. This diffuser can reduce light loss and enhance the use of direct light. Ikuzwe and Sebitosi [51] designed a light collimator made of frosted aluminum, which improved the illuminance of the working surface under sunny conditions compared with commercial diffusers. When rough re-used aluminum cooking foil is used as the interior lining of the collimator, uniform spatial light distribution can also be obtained. Kocifaj and Petržala [52] proposed a new method of designing the diffuser. Under the premise of reducing light loss, the diffuser is designed according to the illuminance distribution required. It has been proven that such a slab can mimic the illuminance distribution required, while the particle sizes and refractive indices have been determined from optimization routines.

The incident angle of light affects the performance of the light pipe, and the number of reflections in the tube for light with different incident angles is different. The number of reflections of low-angle incident light is more than that of high-angle incident light. At the same time, the reflection coefficient of the reflective material on the inner wall of the tube will also affect the reflection of light. The larger the reflection coefficient, the smaller the light loss. Therefore, reducing the number of reflections of light in the tube is one of the ways to enhance the performance of light pipes. Optimization methods are described in Table 2. Sharma et al. [53] compared the performance of modified light pipes which have a slight difference in their designs in the upper 20 cm length of the pipe compared to a conventional light pipe. Wang et al. [54] added a non-phase concentrator inside the light pipe to improve the lighting performance of the light pipe system. Kim and Kim [13] developed a new type of light pipe. The south-facing optical device is placed in the dome. The inside of the tube is coated with thin prismatic material. The diffuser is made of acrylic. The efficiency of this new light pipe is 99%, and it is suitable for sunny and cloudy days. Robertson et al. [38] experimentally showed that a deflector can provide up to a 22% increase in illuminance when there is significant direct light present and reduce the illuminances by approximately 5% under lower, diffuse light conditions. Malet-Damour et al. [55] showed that the orientation of the deflector device placed in the dome needs to be adjusted according to sky conditions, otherwise the light transmittance will be affected. Similar experiments were also conducted. We simulated the effects of different structures of domes and deflectors on the performance of the light pipe, and the results showed that the structure of the lighting hood, the size of the reflector, and sky conditions all affect the lighting performance [56]. Edmonds et al. [57] enhanced the performance under clear skies in winter by adding LCP (Laser-Cut Light-Deflecting Panels) to light pipe domes. However, this system fails to redirect light down the pipe for greater azimuth angles. Venturi et al. [58] got a similar conclusion. Garcia Hansen et al. [59] studied the influence of a pyramid LCP, which provides panel area exposed to the incident light, though it has a reduced area of light collection at any one time. The system can increase the performance of light pipes for low elevation angles.

Table 1. Papers devoted to studying the influence of the three main components of light pipes on daylighting.

Authors	Collector		Tube				Diffuser		
	Geometrical Shape	Light Transmittance	Length	Diameter	Inner Wall Material	Bend	Geometrical Shape	Thicknesses	Fabrication Material
Swift and Smith [46]			√	√	√				
Swift et al. [49]			√	√				√	
Mohelnikova et al. [17]	√	√	√	√	√				
Wu et al. [48]							√		
Kocifaj [50]									√
Robertson et al. [38]	√		√	√	√	√			
Darula et al. [23]			√	√	√				
Nilsson et al. [47]					√				
Ikuzwe and Sebitosi [51]					√				
Gao et al. [43]			√	√					
Kocifaj and Petržala [52]									√
Wu et al. [45]			√	√	√	√			

Table 2. Methods of reducing the number of reflections of light.

Type	Illustration	Authors
Differences in the upper 20 cm length of the pipe with a conventional light pipe	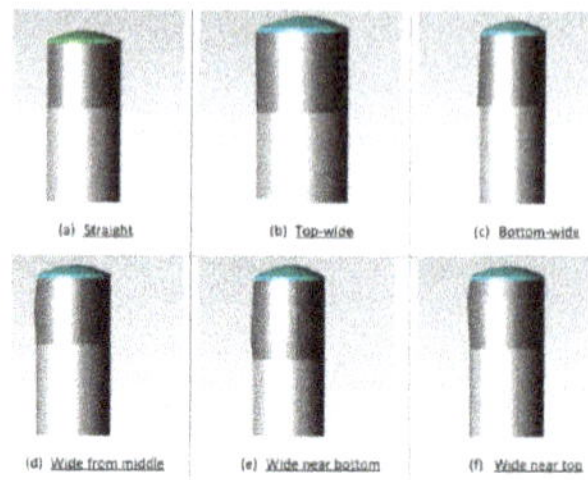	Sharma et al. [53]
Non-phase concentrator	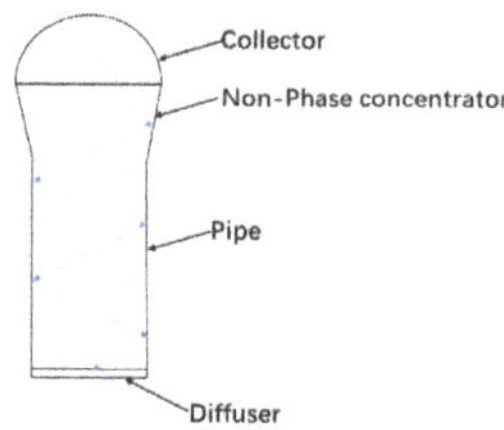	Wang et al. [54]
A south-facing optical device in the dome and a thin prismatic material inside of the tube		Kim and Kim [13]
Deflectors used in domes	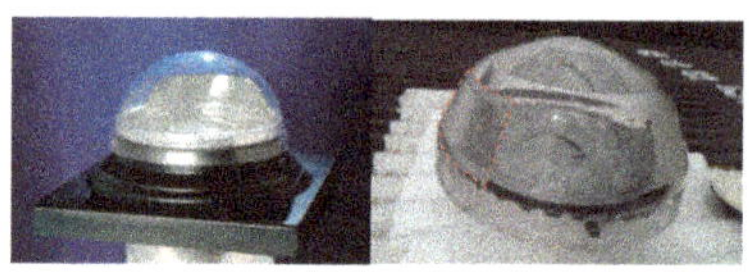	Robertson et al. [38]; Malet-Damour et al., 2019 [55]
Domes and reflectors of different structures	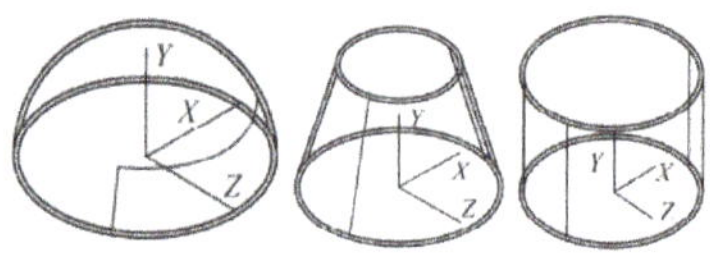	Wu and Wang [56]
Different types of LCP	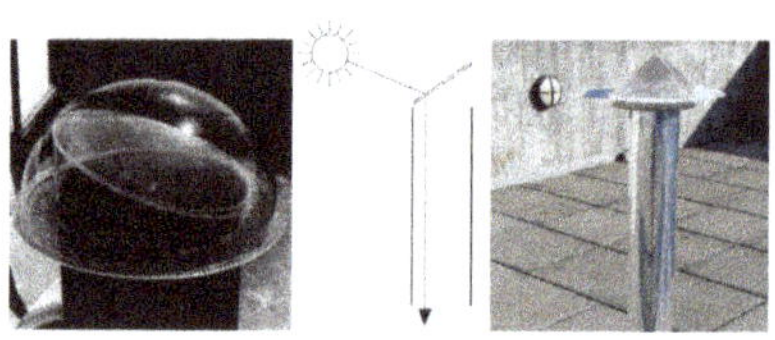	Edmonds et al. [57]; Venturi et al. [58]; Garcia Hansen et al. [59]

It was found that the fluid mixed with metal or metal oxide nanoparticles, named nanofluids, would have unique optical and thermal properties [60,61]. A new lighting-heating coupled TDDs (LH-TTDs) system was developed based on ATO nanofluids which can absorb thermal energy from solar IR radiation to heat domestic hot water, with almost no effects on the visible solar radiation, thus having both lighting and heating functions [62]. The designed LH-TDD is composed of a receiving dome, lighting pipe, and ATO-nanofluid-contained diffuser. Results showed that the ATO nanofluids of 100 ppm can absorb 50% solar radiation coming along with more efficient visible lighting. In a case study in Beijing, the total energy-saving performance improved by 10%.

We developed a system combining a light pipe with photocatalysis which could not only introduce sunlight into the room but also serve as an air purifier [63,64]. A certain amount of photocatalyst sprayed on the outside of the diffuser of the light pipe can have a purifying effect on the air (see Figure 5). However, this still has some technical problems, such as low efficiency and film curing of the catalyst. A new efficient catalyst will be pursued.

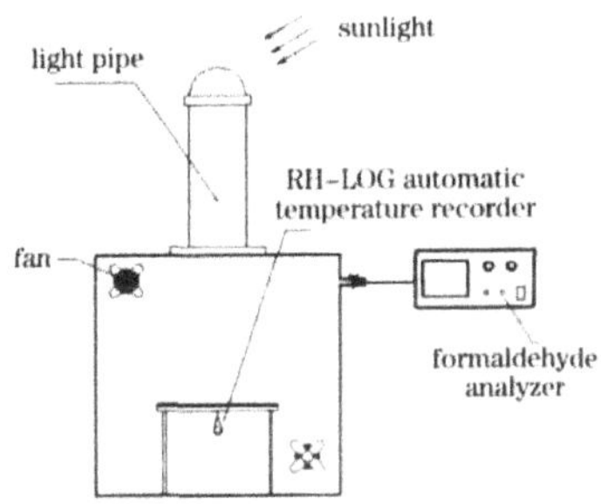

Figure 5. Experimental apparatus of the light pipe combined with photocatalysis.

5. Energy Consumption and Economic Efficiency of Light Pipes

The impact of light pipes on indoor energy consumption is mainly in three aspects: the influence of radiation and heat conduction of light pipes on indoor energy consumption, and the influence of light pipes on the energy consumption of artificial lighting. Compared to windows, the light pipe transfers less heat [65]. In the case of excessive heat loss, the heating energy required in winter may be greater than the saved lighting energy consumption. Large-diameter light pipes cause higher heat loss than small-diameter light pipes [66]. Our research showed that under the same brightness conditions, the increased indoor heat of light pipes is less than that of artificial lighting equipment in Beijing [48]. However, the increase in heat cannot be ignored in tropical regions [67,68]. Pirasaci [69] showed that the overall heat transfer coefficient of the light pipe can be decreased by using an acrylic separator plate in the light pipe. However, solar radiation was not considered. For the installation of light pipes on insulated roofs of low-energy buildings, thermal bridge effects may occur. Šikula et al. [70] indicated that additional glasses units installed in light pipes have a positive effect on solving thermal bridging and condensation problems. However, this can also result in lower overall optical transmittance.

The light distribution of the working plane is composed of two parts. A large part of the light energy is directly irradiated to the working plane through the diffuser, and the rest of the light energy is refracted to the working plane through the ceiling, walls, furniture, floor, etc. The increase in the average reflection coefficient will lead to a significant increase in the average indoor illuminance, which will reduce the time spent on artificial lighting [55]. According to Section 2, perfecting the hybrid lighting system can also effectively increase the energy-saving rate.

The investment of the light pipe system includes initial investment and operation investment. The initial investment includes the cost of the light pipe system and the artificial lighting equipment. The initial investment of the light pipe system is much higher than that of the pure power system. To reduce the cost of the product, the polymer acrylonitrile butadiene styrene (ABS), coated with aluminum by physical vapor deposition (ionization), was evaluated for some tests [71]. Although the market cost was reduced by about 50%, the reflectivity was affected. Operational investment is mainly electricity and maintenance costs. Mayhoub and Carter [72] demonstrated that the light pipe system is generally not economical using conventionally accepted measures of both cost and benefit. They used whole life cycle costing (WLCC) to analyze the costs and benefits of using the two main classes of daylight guidance to light offices as an alternative to conventional electric lighting. A more favorable balance of cost and benefit is obtained. Feng et al. [73]

took an underground garage with a combined lighting scheme of a light pipe system and LED lighting system as the research object. This project designs and installs 20 sets of spherical light pipe systems and 15 sets of flat plate light pipe systems. The simulation results showed that the scheme met the standard illuminance requirements. This project can save 16,306.5 kWh of electricity throughout the year and save about 14,000 yuan in electricity bills.

6. Discussion

6.1. Prediction Methods for Integrated Systems

For the light pipes combined with artificial lighting, a study aimed at optimizing lighting and energy saving, and an algorithm for the integrated light pipe and artificial lighting system, was proposed [74]. It should be noted that an integrated system in which light pipes and artificial sources are not physically related, but the artificial lighting is controlled by daylight sensing systems. In addition, there are few theoretical prediction methods for integrated systems. In the existing prediction methods, the performance of light pipes can be predicted by changing the spatial parameters and performance indicators (aspect ratio, inner wall reflectivity, the light transmittance of collectors and diffusers, common sky models, etc.). This type of prediction is based on the assumption of the three components of light pipes and the transmission process. To obtain more accurate, consistent, and actual performance prediction results, different geometric shapes and optical characteristics of light pipe components are also considered. More simulation and verification are needed. Various forms of light pipe components and complete climate observation data also need to be provided and referenced. For a light pipe with an elbow, the existing method of calculating light transmission efficiency is proved to be wrong [75].

Zhang et al. [76,77] used the Daylight Penetration Factor (DPF) to predict the performance. DPF is defined as the ratio of a given point's internal illuminance to the total external illuminance. Carter [78] used commercial lighting analysis software. These methods of calculating 'static' daylight can be used for the assessment of the performance regarding the same room without solar light pipes. Jenkins et al. [79,80] in the UK developed a model that uses the cosine law of illuminance to describe the distribution of light from the solar light pipes taking into account pipe elbow pieces or bends, named luxplot. Kocifaj et al. [81] proposed a physical model to study interior daylight illuminance distribution based on ray-tracing, named HOLIGILM. The results of this study suggest that the daylight factor should be replaced by the Useful Daylight Illuminance and that more research was needed to establish appropriate criteria for acceptable luminance ratios in the case of well-daylit buildings. The above studies have taken the three main parts of the light pipe as a whole and studied their performance under specific sky conditions. These methods may not give consistent results when the form of the daylighting cover changes or when the sky conditions are constantly changing. Verso et al. [82] presented an approach to characterize the daylighting performance of solar light pipes in terms of light transmission method: the global light transmission efficiency is determined as the product of the efficiencies of the three individual components (collector, pipe, and diffuser). Based on the daylight coefficient method, Wang et al. [83,84] analyzed the transmission characteristics of light in a cylindrical light pipe and established a mathematical model of the light transmission characteristics of the light pipe. This method can calculate the efficiency of the light pipe at any time and under any sky conditions. This mathematical model filled in the blanks of daylighting analysis software. Chen et al. [85] set a new method to evaluate the optical performance according to the output luminous flux. The method uses position (α, d) and angular distribution of luminous flux (Φ, θ, φ) information to describe the optical properties of the light pipe's dome. An artificial neural network used for the predicted performance was introduced [86].

The advantages and disadvantages of some approaches proposed before 2004 have been summarized [87]. The methods proposed after 2004 are summarized in Table 3.

Table 3. Prediction methods of light pipes.

Equations	Nomenclatures	Characteristic	Advantage	Disadvantage	References
–	–	Artificial neural network is used to calculate the illumination of the light pipe.	It can avoid the complicated process of light propagation without assumptions.	Need to collect a large number of light pipe learning samples to achieve the purpose of wide application.	[86]
$E_i = 0.406\prod_{\varphi_1}^{\varphi_n} e^{-0.0052\varphi} \frac{E_{ex}e^{-0.111A}\pi r^2 \cos^4\theta}{V^2}$	E_i: internal illuminance (lux); E_{ex}: external illuminance (lux); φ: angle of pipe elbow (°); r: pipe radius (m); V: vertical distance of pipe diffuser; to point of measurement/prediction (m); A: pipe aspect ratio;	The distribution of light from the light pipe is described by the cosine law of illuminance (a quartic cosine model).	Fewer input data requirements. The room area and the number of light pipes is not restricted. Consider bending at any angle. Model building is based on a larger range of tests.	There is a lack of descriptions of parameters such as solar dimensions and sky conditions. The types of light pipes that can be used are limited. The influence of factors such as light collector and diffuser is not considered.	[80]
Collector performance: $\eta_{b,collector(\gamma_s)} = \frac{\Phi_{b,with\ collector(\gamma_s)}}{\Phi_{b,without\ collector(\gamma_s)}}$ $\eta_{d,collector(\gamma_s)} = \frac{\Phi_{d,with\ collector(\gamma_s)}}{\Phi_{d,without\ collector(\gamma_s)}}$ $\eta_{collector(\gamma_s)} = \frac{E_{b,out}\cdot\eta_{b,collector(\gamma_s)} + E_{d,out}\cdot\eta_{d,collector(\gamma_s)}}{E_{g,out}}$ Similarly, pipe performance: $\eta_{pipe(\gamma_s)} = \frac{E_{b,out}\cdot\eta_{b,pipe(\gamma_s)} + E_{d,out}\cdot\eta_{d,pipe(\gamma_s)}}{E_{g,out}}$ Diffuser performance: $\eta_{diffuser} = \tau_{l,diffuser}$	$\Phi_{b,with\ collector}$ and $\Phi_{b,without\ collector}$, $\Phi_{d,with\ collector}$ and $\Phi_{d,without\ collector}$ the luminous flux values (lm) which are measured (or calculated) in correspondence of the output section of a pipe associated to the collector to analyze, with reference respectively to direct sunlight (beam component) and diffuse skylight (diffuse component), with and without the collector; $E_{b,out}$, $E_{d,out}$, $E_{g,out}$ are respectively the direct, diffuse and global external horizontal illuminance due to an obstructed sky (lux); $\tau_{l,diffuser}$ luminous transmittance value of the diffuser	An approach to characterizing photometric performances of the light pipe in terms of light transmission efficiency was presented.	This method can determine the overall efficiency based on external conditions.	Lack of component diversity and climate database.	[82]

Table 3. *Cont.*

Equations	Nomenclatures	Characteristic	Advantage	Disadvantage	References
$\tau_{TDD} = \frac{E(x)_{in-total} \cdot \pi \cdot R^2}{E_{exterior-total} \cdot \pi \cdot R^2} = \frac{\sum_1^n \eta DC_{a\gamma}(x) L_{a\gamma} \Delta S_{a\gamma} + DC_{sun}(\gamma_{sun}) E_{sun}(\gamma_{sun})}{E_{diffuse}(\gamma_{sun}) + E_{sun}(\gamma_{sun})}$	τ_{TDD}: the efficiency of the light pipe; $E(x)_{in-total}$: the output lumen of the light pipe at a specified situation; $E_{exterior-total}$: the illuminance entering the light pipe; R: the radius of the pipe;n: frequency of reflection; $DC_{a\gamma}(x)$: The daylight coefficient; $DC_{sun}(\gamma_{sun})$: the corresponding daylight coefficient of the sun; $L_{a\gamma}$: the luminance of the element (cd/m^2); $\Delta S_{a\gamma}$: is the angular size of the sky element (sr); $E_{diffuse}$: is the horizontal diffuse illuminance value; E_{sun}: the measured horizontal direct illuminance; γ_{sun}: the elevation angle of the sun; $\eta = 1$	A new mathematical model on the basis of the daylight coefficient method for light pipes' light transfer characteristics was presented.	Combining different sky brightness models, can calculate the transmission efficiency of light pipes under various sky conditions.	Need complete climate observation data.	[83,84]
$E_{LED} = \frac{\Phi_{LED} \cdot N \cdot UF \cdot MF}{A}$	E_{LED}: the necessary LED lumens; Φ_{LED}: the flux emitted by all the LED lamps included in one light pipe; N: the number of light pipes in the room; UF: is the utilization factor of the luminaire; MF: is the maintenance factor of the luminaire; A: the area of the working/reference plane	A methodology for optimizing the design of light pipes integrated with artificial lighting was presented.	This method gives a rough estimate for the energy savings that result from the use of light pipes and artificial lighting controlled to supplement the daylight systems, employing only the necessary wattage.	Currently suitable for small-sized rooms with only one light pipe.	[74]
$L = d \times \begin{cases} \sin(\alpha - \varphi)\varphi < \pi \\ \sin(\alpha - \varphi + \pi)\varphi > \pi \end{cases}$ $\Phi_{out} = \Phi_{in} \times R^n$ $R^n = f(L, \theta)$	L: the distance from the center to the incident point of light on the incident plane; φ: the horizontal angle of spherical coordinate in the testing direction in luminous intensity of a point source into a 3D surface; Φ_{out}: output flux; Φ_{in}: input flux; R: reflectivity; n: frequency of reflection 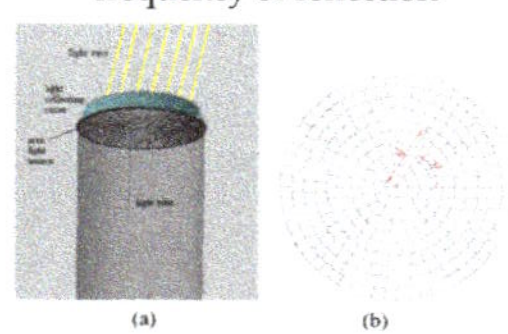	The collector is divided into many small light-emitting pieces. Each piece can be regarded as a point source, described by an intensity distribution curve.	Collector and light pipe are calculated separately. The original data are measurable by commercially available instruments, so it can be used in actual engineering practice.	The influence of sky conditions and elbows are not considered.	[85]

6.2. The Influence of the Orientation of Light Pipes on the Integrated System

The appropriate integrated system should be selected according to the orientation of light pipes. Architectural forms affect the orientation of the light pipe. The orientation of the light pipe includes vertical direction, horizontal direction, and inclined direction. Different installation forms can be connected through elbows to meet different needs. Courret et al. [88] compared the daylighting performance of a test room equipped with a horizontal light pipe and the same test room equipped with double facade glass. Measurements have established that the daylight factor on the work plane 5 m from the window is more than doubled. Chirarattananon et al. [89] proposed that the use of a light pipe in the plenum above the ceiling in multi-storied buildings for transmission of sunlight is one such viable configuration. Further, Heng et al. [90] studied the influence of the cross-sectional shape and the number of openings of the horizontal light pipe on deep plan buildings. Some buildings with sloping roofs are more suitable for installing light pipes with inclined bends. Darula et al. [76] analyzed the illuminance distributions under the diffuser and the light distribution of the indoor working surface of bent light pipes. The maximum light efficiency of bent light pipes can be expected in the situation when sunbeams entering the tube will be quasi-parallel with the upper tube axis. Kocifaj et al. [91] proved that light pipes can transmit lighter if bent rather than being straight under a temperate climate. Baroncini et al. [92] designed a double light pipe that can lead daylight into a multi-story room. The main difficulty of using light pipes in multi-story buildings is to ensure that each layer extracts and distributes the same amount of light. Garcia-Hansen and Edmonds [93] installed transparent plastic cones with a base angle of 37.5° on the transparent part of the light pipe to extract and distribute equal amounts of light at each level. Kennedy and O'Rourke [94] achieved the purpose of multi-story building lighting by opening side apertures of the light pipe.

7. Conclusions

This paper reviewed the research and development of integrated systems for light pipes. This research classified and introduced the development of integrated systems for light pipes from the aspects of artificial lighting, the ventilation system, ATO nanofluids and photocatalysis, energy consumption, and economic efficiency.

The major survey findings are as follows:

(1) The combination of light pipes and artificial lighting can meet illuminance levels in poor daylight conditions. The combination of light pipes with other technologies is feasible. Integration of a passive stack ventilation system and light pipes allows daylight to enter the building while exhausting indoor air. The combination of light pipes and photocatalytic systems can realize lighting and air purification at the same time. The use of superhydrophobic material on the outer surface of A collector can improve the service life of a light pipe. From a whole life cycle perspective, the use of light pipes can be a balance of cost and benefit.

(2) There are few theoretical prediction methods for integrated systems. In the existing prediction methods, the performance of the light pipe can be predicted by changing the space parameters of the room and the performance index (aspect ratio, reflectivity of pipes, transmittance of collector and diffuser, common sky model, etc.) of the light pipes.

8. Future Perspectives

According to the state-of-the-art achievements, the following research directions are recommended:

(1) The establishment of overall efficiency and heat transfer model of integrated system is important for the practical application of light-vent pipe systems and light-solar water heater systems. The light-vent pipe may increase a building's occupied area, and requirements such as fire prevention should be considered.

(2) The integrated systems of light pipes need to be studied via the technical and economic analysis of the light pipes combined with ventilation systems, and light pipes combined with ATO nanofluids and photocatalysis.

Author Contributions: Conceptualization, Y.W.; formal analysis, M.J.; funding acquisition, Y.W.; investigation, M.J. and M.L.; project administration, Y.W.; supervision, Y.W.; validation, M.J. and M.L.; writing—original draft, M.L.; writing—review and editing, M.J. and S.L.; visualization, Y.W. All authors have read and agreed to the published version of the manuscript.

Funding: This research was funded by the Central Guidance for Local Scientific and Technological Development Funding Project (236Z5202G) and the Beijing Natural Science Foundation (8202034).

Data Availability Statement: Data are contained within the article.

Conflicts of Interest: The authors declare no conflicts of interest.

Nomenclature

ATO	Antimony Tin Oxide
TTE	Transmission Tube Efficiency
CIE	International Commission on Illumination
LCP	Laser-Cut Light-Deflecting Panels
ABS	Acrylonitrile Butadiene Styrene
WLCC	Whole Life Cycle Costing

References

1. González-Torres, M.; Pérez-Lombard, L.; Coronel, J.F.; Maestre, I.R.; Yan, D. A Review on Buildings Energy Information: Trends, End-Uses, Fuels and Drivers. *Energy Rep.* **2022**, *8*, 626–637. [CrossRef]
2. Mujan, I.; Anđelković, A.S.; Munćan, V.; Kljajić, M.; Ružić, D. Influence of Indoor Environmental Quality on Human Health and Productivity—A Review. *J. Clean. Prod.* **2019**, *217*, 646–657. [CrossRef]
3. Kwong, Q.J. Light Level, Visual Comfort and Lighting Energy Savings Potential in a Green-Certified High-Rise Building. *Build. Eng.* **2020**, *29*, 101198. [CrossRef]
4. Lamb, S.; Kwok, K.C.S. A Longitudinal Investigation of Work Environment Stressors on the Performance and Wellbeing of Office Workers. *Appl. Ergon.* **2016**, *52*, 104–111. [CrossRef] [PubMed]
5. Bellia, L.; Bisegna, F.; Spada, G. Lighting in Indoor Environments: Visual and Non-Visual Effects of Light Sources with Different Spectral Power Distributions. *Build. Environ.* **2011**, *46*, 1984–1992. [CrossRef]
6. Carter, D.J. Tubular Guidance Systems for Daylight: UK Case Studies. *Build. Res. Inf.* **2008**, *36*, 520–535. [CrossRef]
7. Bellia, L.; Pedace, A.; Barbato, G. Daylighting Offices: A First Step toward an Analysis of Photobiological Effects for Design Practice Purposes. *Build. Environ.* **2014**, *74*, 54–64. [CrossRef]
8. Hraska, J. Chronobiological Aspects of Green Buildings Daylighting. *Renew. Energy* **2015**, *73*, 109–114. [CrossRef]
9. Yu, X.; Su, Y. Daylight Availability Assessment and Its Potential Energy Saving Estimation—A Literature Review. *Renew. Sust. Energ. Rev.* **2015**, *52*, 494–503. [CrossRef]
10. Bellia, L.; Acosta, I.; Campano, M.Á.; Fragliasso, F. Impact of Daylight Saving Time on Lighting Energy Consumption and on the Biological Clock for Occupants in Office Buildings. *Sol. Energy* **2020**, *211*, 1347–1364. [CrossRef]
11. Wirz-Justice, A.; Skene, D.J.; Münch, M. The Relevance of Daylight for Humans. *Biochem. Pharmacol.* **2021**, *191*, 114304. [CrossRef]
12. Yuan, Z.; Zhang, X.; Zhan, Q. Application research overview on active tubular daylight guidance system based on light transmission technology. *China Illum. Eng. J.* **2009**, *20*, 13–19+23. [CrossRef]
13. Kim, J.T.; Kim, G. Overview and New Developments in Optical Daylighting Systems for Building a Healthy Indoor Environment. *Build. Environ.* **2010**, *45*, 256–269. [CrossRef]
14. Alrubaih, M.S.; Zain, M.F.M.; Alghoul, M.A.; Ibrahim, N.L.N.; Shameri, M.A.; Elayeb, O. Research and Development on Aspects of Daylighting Fundamentals. *Renew. Sust. Energ. Rev.* **2013**, *21*, 494–505. [CrossRef]
15. Canziani, R.; Peron, F.; Rossi, G. Daylight and Energy Performances of a New Type of Light Pipe. *Energy Build.* **2004**, *36*, 1163–1176. [CrossRef]
16. Baglivo, C.; Bonomolo, M.; Beccali, M.; Congedo, P.M. Sizing Analysis of Interior Lighting Using Tubular Daylighting Devices. *Energy Procedia* **2017**, *126*, 179–186. [CrossRef]
17. Mohelnikova, J. Tubular Light Guide Evaluation. *Build. Environ.* **2009**, *44*, 2193–2200. [CrossRef]
18. Sreelakshmi, K. Daylight Performance of Collector-Diffuser Combinations in Light Pipe Systems at Different Geographical Locations. *Sol. Energy* **2024**, *267*, 112254. [CrossRef]

19. Katsumasa, I.; Hiroyuki, T.; Mitsunobu, M.; Yi-Wei, S.; Xiao-Song, Z.; Yuji, M. Transmission properties of dielectric-coated hollow optical fibers based on stainless tube. *Opt. Fibers Sens. Med. Diagn. Treat.* **2018**, *10488*, 1048804. [CrossRef]
20. Dogan, Y.; Morrison, M.; Hu, C.; Atkins, R.A.; Solmaz, M.E.; Madsen, C.K. Fabrication of Advanced Glass Light Pipes for Solar Concentrators. *Optifab* **2017**, *10448*, 104482G. [CrossRef]
21. Wang, W.; Li, S.; Ma, C. Experimental study on lighting performance of light pipe in winter. *Acta Energiae Solaris Sin.* **2008**, *29*, 1489–1493. [CrossRef]
22. Tsang, E.K.W.; Kocifaj, M.; Li, D.H.W.; Kundracik, F.; Mohelníková, J. Straight Light Pipes' Daylighting: A Case Study for Different Climatic Zones. *Sol. Energy* **2018**, *170*, 56–63. [CrossRef]
23. Darula, S.; Kocifaj, M.; Mohelníková, J. Hollow Light Guide Efficiency and Illuminance Distribution on the Light-Tube Base under Overcast and Clear Sky Conditions. *Optik* **2013**, *124*, 3165–3169. [CrossRef]
24. Li, S.; Wang, W.; Ma, C. Experimental study and analysis on lighting performance of light pipe. *Acta Energiae Solaris Sin.* **2009**, *30*, 586–590. [CrossRef]
25. Yun, G.Y.; Shin, H.Y.; Kim, J.T. Monitoring and Evaluation of a Light-Pipe System Used in Korea. *Indoor Built Environ.* **2010**, *19*, 129–136. [CrossRef]
26. Vasilakopoulou, K.; Kolokotsa, D.; Santamouris, M.; Kousis, I.; Asproulias, H.; Giannarakis, I. Analysis of the Experimental Performance of Light Pipes. *Energy Build.* **2017**, *151*, 242–249. [CrossRef]
27. Kocifaj, M.; Kómar, L.; Kohút, I. Modeling the Aerosol Effects on the Light Field below a Tubular-Pipe: A Case of Clear Sky Conditions. *Sol. Energy* **2014**, *107*, 122–134. [CrossRef]
28. Sharp, F.; Lindsey, D.; Dols, J.; Coker, J. The Use and Environmental Impact of Daylighting. *J. Clean. Prod.* **2014**, *85*, 462–471. [CrossRef]
29. Görgülü, S.; Ekren, N. Energy Saving in Lighting System with Fuzzy Logic Controller Which Uses Light-Pipe and Dimmable Ballast. *Energy Build.* **2013**, *61*, 172–176. [CrossRef]
30. Oakley, G.; Riffat, S.B.; Shao, L. Daylight Performance of Lightpipes. *Sol. Energy* **2000**, *69*, 89–98. [CrossRef]
31. Lu, S.; Lu, Y.; Sheng, J.; Wang, W.; Ma, C. CFD research on effects of natural ventilation in light pipe system. *J. Eng. Thermophys-Rus.* **2007**, *28*, 5–8. [CrossRef]
32. Varga, S.; Oliveira, A.C. Ventilation Terminals for Use with Light Pipes in Buildings: A CFD Study. *Appl. Therm. Eng.* **2000**, *20*, 1743–1752. [CrossRef]
33. Oliveira, A.C.; Silva, A.R.; Afonso, C.F.; Varga, S. Experimental and Numerical Analysis of Natural Ventilation with Combined Light/Vent Pipes. *Appl. Therm. Eng.* **2001**, *21*, 1925–1936. [CrossRef]
34. Sirén, K.; Helenius, T.; Shao, L.; Smith, S.; Ford, B.; Diaz, C.; Oliveira, A.; Varga, S.; Borth, J.; Zaccheddu, E. Chapter 76—Combining Light Pipe and Stack Ventilation—Some Development Aspects. In *World Renewable Energy Congress VI*; Sayigh, A.A.M., Ed.; Pergamon: Oxford, UK, 2000; pp. 395–400. [CrossRef]
35. Elmualim, A.A.; Smith, S.; Riffat, S.B.; Shao, L. Evaluation of Dichroic Material for Enhancing Light Pipe/Natural Ventilation and Daylighting in an Integrated System. *Appl. Energy* **1999**, *62*, 253–266. [CrossRef]
36. Shao, L.; Riffat, S.B. Daylighting Using Light Pipes and Its Integration with Solar Heating and Natural Ventilation. *Light. Res. Technol.* **2000**, *32*, 133–139. [CrossRef]
37. Taengchum, T.; Chirarattananon, S.; Exell, R.H.B.; Kubaha, K.; Chaiwiwatworakul, P. A Study on a Ventilation Stack Integrated with a Light Pipe. *Appl. Therm. Eng.* **2013**, *50*, 546–554. [CrossRef]
38. Robertson, A.P.; Hedges, R.C.; Rideout, N.M. Optimisation and Design of Ducted Daylight Systems. *Light. Res. Technol.* **2010**, *42*, 161–181. [CrossRef]
39. Wu, Y.P.; Wang, X.D.; Chen, Z.G.; Zhang, C.Y. Experimental Study on the Influence of Daylighting Performance of Solar Light Pipes by Dusts and Condensation. In *Sustainable Development of Urban Environment and Building Material*; Advanced Materials Research; Trans Tech Publications Ltd.: Bäch, Switzerland, 2012; Volume 374, pp. 1096–1099. [CrossRef]
40. Wu, Y.; Lei, X.; Lu, Y.; Chen, H. Research progress of superhydrophobic anti-reflection films applied on transparent surfaces of solar devices. *CIESC J.* **2021**, *72*, 21–29. [CrossRef]
41. Hooda, A.; Goyat, M.S.; Pandey, J.K.; Kumar, A.; Gupta, R. A Review on Fundamentals, Constraints and Fabrication Techniques of Superhydrophobic Coatings. *Prog. Org. Coat.* **2020**, *142*, 105557. [CrossRef]
42. Mehmood, U.; Al-Sulaiman, F.A.; Yilbas, B.S.; Salhi, B.; Ahmed, S.H.A.; Hossain, M.K. Superhydrophobic Surfaces with Antireflection Properties for Solar Applications: A Critical Review. *Sol. Energy Mat. Sol.* **2016**, *157*, 604–623. [CrossRef]
43. Gao, M.; Xu, G.; Cao, G.; Li, D.; Yu, J.; Liu, Q. Simulation analysis of efficiency and illumination distribution of light guide under CIE Overcast Sky condition in Dalian area, Liaoning. *Acta Energiae Solaris Sin.* **2017**, *38*, 2303–2308.
44. Darula, S.; Kocifaj, M.; Kittler, R.; Kundracik, F. Illumination of Interior Spaces by Bended Hollow Light Guides: Application of the Theoretical Light Propagation Method. *Sol. Energy* **2010**, *84*, 2112–2119. [CrossRef]
45. Wu, X.; Wang, C.; Ouyang, J. Analysis of transmission tube efficiency of light pipe based on TracePro. *China Illum. Eng. J.* **2020**, *31*, 145–150.
46. Swift, P.D.; Smith, G.B. Cylindrical Mirror Light Pipes. *Sol. Energy Mat. Sol.* **1995**, *36*, 159–168. [CrossRef]
47. Nilsson, A.M.; Jonsson, J.C.; Roos, A. Spectrophotometric Measurements and Ray Tracing Simulations of Mirror Light Pipes to Evaluate the Color of the Transmitted Light. *Sol. Energy Mat. Sol.* **2014**, *124*, 172–179. [CrossRef]

48. Wu, Y.; Yue, Z. Experimental Investigation on Light-Thermal Effects of Solar Light Pipes Used in USTB Gymnasium under Sunny Conditions in Beijing. In Proceedings of the 2009 International Conference on Energy and Environment Technology, Guilin, China, 16–18 October 2009; Volume 1, pp. 147–150. [CrossRef]
49. Swift, P.D.; Smith, G.B.; Franklin, J. Hotspots in Cylindrical Mirror Light Pipes: Description and Removal. *Light. Res. Technol.* **2006**, *38*, 19–28. [CrossRef]
50. Kocifaj, M. Efficient Tubular Light Guide with Two-Component Glazing with Lambertian Diffuser and Clear Glass. *Appl. Energy* **2009**, *86*, 1031–1036. [CrossRef]
51. Ikuzwe, A.; Sebitosi, A.B. A Novel Design of a Daylighting System for a Classroom in Rural South Africa. *Sol. Energy* **2015**, *114*, 349–355. [CrossRef]
52. Kocifaj, M.; Petržala, J. Designing of Light-Pipe Diffuser through Its Computed Optical Properties: A Novel Solution Technique and Some Consequences. *Sol. Energy* **2019**, *190*, 386–395. [CrossRef]
53. Sharma, L.; Ali, S.F.; Rakshit, D. Performance Evaluation of a Top Lighting Light-Pipe in Buildings and Estimating Energy Saving Potential. *Energy Build.* **2018**, *179*, 57–72. [CrossRef]
54. Wang, W.; Li, S.; Ma, C. Experimental study and analysis on lighting performance light pipe with a conic concentrator. *J. Beijing Univ. Technol.* **2009**, *35*, 1675–1679. [CrossRef]
55. Malet-Damour, B.; Bigot, D.; Guichard, S.; Boyer, H. Photometrical Analysis of Mirrored Light Pipe: From State-of-the-Art on Experimental Results (1990–2019) to the Proposition of New Experimental Observations in High Solar Potential Climates. *Solar Energy* **2019**, *193*, 637–653. [CrossRef]
56. Wu, Y.; Wang, Z. Simulation and experimental analysis of light pipes with reflector under direct sunlight. *J. China Coal Soc.* **2019**, *44*, 1941–1948. [CrossRef]
57. Edmonds, I.R.; Moore, G.I.; Smith, G.B.; Swift, P.D. Daylighting Enhancement with Light Pipes Coupled to Laser-Cut Light-Deflecting Panels. *Light. Res. Technol.* **1995**, *27*, 27–35. [CrossRef]
58. Venturi, L.; Wilson, M.; Jacobs, A.; Solomon, J. Light Piping Performance Enhancement Using a Deflecting Sheet. *Light. Res. Technol.* **2006**, *38*, 167–179. [CrossRef]
59. Garcia Hansen, V.; Edmonds, I.; Bell, J. Improving daylighting performance of mirrored light pipes Passive vs. active collection systems. In Proceedings of the PLEA2009—26th Conference on Passive and Low Energy Architecture, Quebec City, QC, Canada, 22–24 June 2009.
60. Taylor, R.; Coulombe, S.; Otanicar, T.; Phelan, P.; Gunawan, A.; Lv, W.; Rosengarten, G.; Prasher, R.; Tyagi, H. Small Particles, Big Impacts: A Review of the Diverse Applications of Nanofluids. *J. Appl. Phys.* **2013**, *113*, 011301. [CrossRef]
61. Qureshi, M.Z.A.; Ashraf, M. Computational Analysis of Nanofluids: A Review. *Eur. Phys. J. Plus.* **2018**, *133*, 71. [CrossRef]
62. Liu, X.; Shen, C.; Wang, J. Investigation on the Lighting/Heating Performance of Tubular Daylighting Devices (TDDs) Based on Nanofluids. *Energy Build.* **2022**, *263*, 112028. [CrossRef]
63. Wu, Y.; Ma, C. Status of studies and applications of photocatalytic technology in building environment and facility field. *Heat. Vent. Air Cond.* **2006**, *36*, 29–36. [CrossRef]
64. Wu, Y.; Wang, X.; Ma, C. Solar light pipe combined with photocatalysis to decompose formaldehyde under sunny conditions in summer in Beijing. *Acta Opt. Sin.* **2008**, *28*, 2408–2415. [CrossRef]
65. Oakley, G.; Smith, S.J.; Shao, L.; Riffat, S.B. TripleSave-the investigation and monitoring of a combined natural daylighting and stack ventilation system. In Proceedings of the World Renewable Energy Congress VII, Dublin, Germany, 29 June–5 July 2002.
66. Šikula, O.; Mohelníková, J.; Plášek, J. Thermal CFD analysis of tubular light guides. *Energies* **2013**, *6*, 6304–6321. [CrossRef]
67. Williams, D.A.; Dorville, J.M. Investigating the thermal and lighting performance of light pipes for sunny and cloudy conditions in insular tropical climate. *J. Elec. Eng.* **2014**, *2*, 221–227. [CrossRef]
68. Darula, S.; Kittler, R.; Kocifaj, M. Luminous Effectiveness of Tubular Light-Guides in Tropics. *Appl. Energy* **2010**, *87*, 3460–3466. [CrossRef]
69. Pirasaci, T. Investigation of laminar natural convection heat transfer within tubular daylighting devices for winter conditions. *J. Build. Eng.* **2015**, *4*, 52–59. [CrossRef]
70. Šikula, O.; Mohelníková, J.; Plášek, J. Thermal analysis of light pipes for insulated flat roofs. *Energy Build.* **2014**, *85*, 436–444. [CrossRef]
71. Spacek, A.D.; Neto, J.M.; Biléssimo, L.D., Jr.; Santana, M.V.F.D.; Malfatti, C.D.F. Proposal of the tubular daylight system using Acrylonitrile Butadiene Styrene (ABS) metalized with Aluminum for reflective tube structure. *Energies* **2018**, *11*, 199. [CrossRef]
72. Mayhoub, M.S.; Carter, D.J. The costs and benefits of using daylight guidance to light office buildings. *Build. Environ.* **2011**, *46*, 698–710. [CrossRef]
73. Feng, X.; Wu, X.; Xu, Y.; Zhang, L.; Qi, S. Analysis of the Composition of Optical Guide System and Multi-light Source Combined Lighting Scheme. *Jiangxi Build. Mater.* **2022**, *3*, 211–213.
74. Vasilakopoulou, K.; Synnefa, A.; Kolokotsa, D.; Karlessi, T.; Santamouris, M. Performance Prediction and Design Optimisation of an Integrated Light Pipe and Artificial Lighting System. *Intl. J. Sust. Energy* **2016**, *35*, 675–685. [CrossRef]
75. Wang, C.; Gao, Q.; Gao, W.; Ouyang, J. Discussion about Calculation Method of Light Transmission Efficiencies of Elbows in Cylindrical Light Pipes. *Sol. Energy* **2022**, *238*, 39–43. [CrossRef]
76. Zhang, X.; Muneer, T. Mathematical Model for the Performance of Light Pipes. *Light. Res. Technol.* **2000**, *32*, 141–146. [CrossRef]

77. Zhang, X.; Muneer, T.; Kubie, J. A Design Guide for Performance Assessment of Solar Light-Pipes. *Light. Res. Technol.* **2002**, *34*, 149–168. [CrossRef]
78. Carter, D.J. The Measured and Predicted Performance of Passive Solar Light Pipe Systems. *Light. Res. Technol.* **2002**, *34*, 39–51. [CrossRef]
79. Jenkins, D.; Muneer, T. Modelling Light-Pipe Performances—A Natural Daylighting Solution. *Build. Environ.* **2003**, *38*, 965–972. [CrossRef]
80. Jenkins, D.; Muneer, T.; Kubie, J. A Design Tool for Predicting the Performances of Light Pipes. *Energy Build.* **2005**, *37*, 485–492. [CrossRef]
81. Kocifaj, M.; Darula, S.; Kittler, R. HOLIGILM: Hollow Light Guide Interior Illumination Method—An Analytic Calculation Approach for Cylindrical Light-Tubes. *Sol. Energy* **2008**, *82*, 247–259. [CrossRef]
82. Verso, V.R.M.L.; Pellegrino, A.; Serra, V. Light Transmission Efficiency of Daylight Guidance Systems: An Assessment Approach Based on Simulations and Measurements in a Sun/Sky Simulator. *Sol. Energy* **2011**, *85*, 2789–2801. [CrossRef]
83. Wang, S.; Li, L.; Zhang, B. Daylight coefficient computational method-based study on calculation method of tubular daylight device efficiency. *Build. Sci.* **2013**, *29*, 12–15. [CrossRef]
84. Wang, S.; Zhao, J.; Wang, L. Research on Energy Saving Analysis of Tubular Daylight Devices. *Energy Procedia* **2015**, *78*, 1781–1786. [CrossRef]
85. Chen, B.; Wei, Y.; Li, X.; Cao, R.; Jin, P. Numerical Modeling of Tubular Daylighting Devices. *Optik* **2017**, *145*, 95–98. [CrossRef]
86. Wang, A. Artificial neural network applicable for light guide lighting calculation. *China Illum. Eng. J.* **2000**, *36*, 21–24. [CrossRef]
87. Jenkins, D.; Muneer, T. Light-Pipe Prediction Methods. *Appl. Energy* **2004**, *79*, 77–86. [CrossRef]
88. Courret, G.; Scartezzini, J.-L.; Francioli, D.; Meyer, J.-J. Design and Assessment of an Anidolic Light-Duct. *Energy Build.* **1998**, *28*, 79–99. [CrossRef]
89. Chirarattananon, S.; Chedsiri, S.; Renshen, L. Daylighting through Light Pipes in the Tropics. *Sol. Energy* **2000**, *69*, 331–341. [CrossRef]
90. Heng, C.Y.S.; Lim, Y.-W.; Ossen, D.R. Horizontal Light Pipe Transporter for Deep Plan High-Rise Office Daylighting in Tropical Climate. *Build. Environ.* **2020**, *171*, 106645. [CrossRef]
91. Kocifaj, M.; Kundracik, F.; Darula, S.; Kittler, R. Availability of Luminous Flux below a Bended Light-Pipe: Design Modelling under Optimal Daylight Conditions. *Sol. Energy* **2012**, *86*, 2753–2761. [CrossRef]
92. Baroncini, C.; Boccia, O.; Chella, F.; Zazzini, P. Experimental Analysis on a 1:2 Scale Model of the Double Light Pipe, an Innovative Technological Device for Daylight Transmission. *Sol. Energy* **2010**, *84*, 296–307. [CrossRef]
93. Garcia-Hansen, V.; Edmonds, I. Methods for the Illumination of Multilevel Buildings with Vertical Light Pipes. *Sol. Energy* **2015**, *117*, 74–88. [CrossRef]
94. Kennedy, D.M.; O'Rourke, F. Experimental Analysis of a Scaled, Multi-Aperture, Light-Pipe, Daylighting System. *Sol. Energy* **2015**, *122*, 181–190. [CrossRef]

buildings

MDPI

Article

The Impact of Indoor Air Humidity on the Infiltration of Ambient Particles

Jiayi Qiu [1], Haixi Zhang [2], Jialu Liu [3] and Yanhua Liu [1,*]

1 School of Human Settlements and Civil Engineering, Xi'an Jiaotong University, Xi'an 710049, China; bore911@stu.xjtu.edu.cn
2 China Northwest Architectural Design and Research Institute Co., Ltd., Xi'an 710018, China
3 Jiangnan Shipyard (Group) Co., Ltd., Shanghai 201913, China
* Correspondence: yhliu@mail.xjtu.edu.cn

Abstract: Ambient particles contribute to occupant exposure as they infiltrate indoor environments through building envelope cracks, impacting indoor air quality. This study investigates the impact of indoor air humidity on the infiltration of ambient particles, highlighting humidity's crucial role in influencing particle dynamics in indoor environments. Employing a controlled chamber system, we conducted experiments to quantify the infiltration of size-resolved particles under varying relative humidity (RH) conditions. Both the total number and the mass concentration of particles increased with RH in the experimental chamber. The smallest particles (0.3–0.4 μm) experienced reduced infiltration at higher RH levels due to hygroscopic growth, while intermediate-sized particles showed increased infiltration, resulted from coagulation effects. Large particles (>1.0 μm) demonstrated reduced infiltration factors, caused by lower penetration and higher deposition rates, with minimal impact from RH changes. Our findings reveal that RH influences particle hygroscopic growth, deposition rate, and coagulation process, thereby affecting indoor particle size distribution and concentration.

Keywords: indoor air quality; indoor air humidity; relative humidity (RH); infiltration; ambient particles

1. Introduction

Ambient fine particulate matter ($PM_{2.5}$) poses a well-documented risk to human health [1–3]. Considering that individuals now spend over 90% of their time indoors, a trend amplified by the COVID-19 pandemic [4,5], the infiltration of particles into indoor spaces through building envelopes is a critical concern [6–8]. Understanding and quantifying this infiltration process is essential to assessing indoor particulate loads and devising effective control measures.

The infiltration factor (F_{inf}), which denotes the proportion of ambient particles that penetrate indoor environments and remain suspended [9], serves as a critical estimator in this domain. Research indicates that F_{inf} values vary across different buildings and particle sizes [10,11], influenced by factors such as meteorological conditions (including air pressure, temperature, and humidity), building characteristics (e.g., envelope tightness, volume), and aerosol dynamics behavior (e.g., penetration, deposition, coagulation) [12–14].

Field studies have extensively explored these variables. For instance, Wallace and Williams [15] monitored 37 residents across four seasons to study personal, indoor, and outdoor exposure to $PM_{2.5}$. They calculated F_{inf} values based on indoor/outdoor sulfur ratios, concluding that the infiltration factor was the lowest in summer when air conditioners were in use. MacNeill et al. [16] collected continuous data for two years to estimate daily infiltration factors, finding that the daily F_{inf} of $PM_{2.5}$ was higher in summer than in winter. They also linked daily infiltration to window-opening behaviors, air conditioning, meteorological variables, and home age. Huang et al. [17] measured indoor and outdoor $PM_{2.5}$

To measure the impact of indoor air humidity on particle infiltration, we analyzed the particle dynamic behavior using a mass balance model. Within an indoor environment, ventilation, penetration, and deposition primarily drive particle decay. Additionally, considering the role of indoor humidity, we must also account for hygroscopicity and coagulation mechanisms, as these can significantly influence particle number concentration fluctuations.

The size distribution of particles from incense burning and other common indoor aerosol particles can be fitted with a lognormal distribution [26]. The distribution function for the number density can be expressed as:

$$f(d_p) = \frac{1}{d_p \ln\sigma_g \sqrt{2\pi}} \exp\left[-\frac{(\ln d_p - \ln d_g)^2}{2(\ln\sigma_g)^2}\right], \tag{3}$$

where d_p is the particle diameter, σ_g is the geometric standard deviation based on particle diameter, and d_g is the geometric mean particle diameter.

In a humid indoor environment, water-soluble compounds in incense-burning-derived particles will absorb water vapor, potentially forming a liquid shell [27]. The ratio between the humidified and the dry particle diameters is known as the hygroscopic diameter growth factor (*Gf*). *Gf* is measured and parameterized as a function of *RH* (in %) [28], expressed as:

$$Gf = \frac{d_w}{d_d} = \left(1 + A \cdot \frac{RH}{1 - RH}\right)^{\frac{1}{3}}, \tag{4}$$

where d_w is the humidified particle diameter, d_d is the dry particle diameter, and A is a fitted parameter.

We introduce the hygroscopic number change ratio (β) to quantify changes in particle number concentrations due to hygroscopicity. This ratio compares particle concentrations at a specific relative humidity to those in a dry state, illustrating the impact of water vapor absorption and subsequent hygroscopic growth on particle number concentrations. For each particle size category, this ratio is denoted as β_i. In this study, β_i is proposed and was formulated based on the following assumptions: (1) the water vapor–particle equilibrium in the hygroscopic process is achieved in seconds [29]; (2) the total number and geometric standard deviation (GSD) of particles remain constant during hygroscopicity [30]; (3) particles from incense burning are "nearly hydrophobic", with the hygroscopic diameter growth factor (*Gf*) being independent of particle size [31]. Thus, the formula for β_i can be derived using Equations (3) and (4):

$$\beta_i = \frac{N_{iw}}{N_{id}} = \frac{f_w(d_i)}{f(d_i)} = \frac{\frac{1}{Gf} f\left(\frac{d_i}{Gf}\right)}{f(d_i)} = \exp\frac{(\ln d_i - \ln d_g - \ln\sqrt{Gf})\ln Gf}{\ln^2\sigma_g}, \tag{5}$$

where N_{iw} is the number concentration of humidified particles in the *i*th size category, N_{id} is the number concentration of dry particles in the *i*th size category, f_w is the distribution function of humidified particles, and d_i is the average diameter of the *i*th size category.

In our experiments, particles from incense burning penetrated the indoor environment through the cracks, along with the airflow, between Chambers A and B, without any human activity or other sources in the chambers. Considering the relative humidity of indoor air, the mass balance equation for particles indoors (i.e., Chamber B) is:

$$\frac{dN_i(t)}{dt} = aP_i\beta_i N_{i,\text{out}} - aN_i(t) - k_i N_i(t) - \gamma_i N_i(t)^2, \tag{6}$$

where $N_i(t)$ is the number concentration (cm^{-3}) of particles in an indoor environment at time t (min), the subscript i denotes the particle size range, a is the air exchange rate (min^{-1}), P_i is the particle penetration factor, $N_{i,\text{out}}$ is the particle number concentration

Zeiss Group, Oberkochen, Germany) was placed in the center of Chamber B to collect the deposited particles.

Before each test, Chambers A and B were cleaned by an air cleaner with a high-efficiency particulate aerosol (HEPA) filter for 60 min. The particle source in this study was the combustion of incense, ignited in a smoke generator, with the smoke pumped into Chamber A. The concentration of particles in Chamber A was controlled by four paralleled pumps (JQ-12V, Jinqi Technology Co., Ltd., Shenzhen, China) and maintained at a steady level of 400 ± 40 $\mu g/m^3$. The incense-burning-derived particles from Chamber A infiltrated Chamber B through the window cracks, with the airflow. The infiltration process lasted 480 min in each test, with no other particle sources in either chamber.

The air exchange rate between Chambers A and B was determined by the tracer gas concentration decay method [25]. Carbon dioxide (CO_2) was injected as the tracer gas into Chamber B, and its concentration decay was monitored continuously by a CO_2 sensor (TES-1370, TES Electrical Electronic Corp., Taiwan, China) with an accuracy of ±3% or ±50 ppm (whichever was greater), ranging from 0 to 6000 ppm. One sensor was placed in the center of Chamber B, and another was positioned outside the experimental system to record the background concentration. The tracer gas concentration decay method is based on the following mass conservation equation:

$$C_t - C_{bg} = \left(C_0 - C_{bg}\right) \exp(-at), \tag{1}$$

where C_t is the tracer gas concentration at time t, C_{bg} is the background tracer gas concentration, C_0 is the tracer gas concentration at the initial moment, and a is the air exchange rate. The tightness of Chamber B was verified when the window cracks were sealed. With the cracks unsealed, the air exchange rate between Chambers A and B was calculated using Equation (1), resulting to be 0.144 h^{-1}.

To assess the impact of indoor air humidity on particle infiltration, seven target RH levels were maintained in Chamber B across different tests: 30%, 40%, 50%, 60%, 70%, 80%, and 85%. The air relative humidity in Chamber A was kept at 35 ± 5% in each test. Humidity was regulated by a wet-membrane humidifier (XH-805, OPV Electric Co., Ltd., Ji'nan, China), and RH probes (WSZY-1, Tianjian Huayi Technology Development Co., Ltd., Beijing, China) with an accuracy of ±2% were positioned in the center of the chambers at a height of 1.2 m. RH data were recorded every two minutes. Each humidity level test was repeated three times to ensure reliability.

The air temperature in the chamber was maintained at approximately 25 °C by the insulated walls and an external air-conditioning system. The accuracy of the temperature probe (AT4320, Applent Instruments Ltd., Changzhou, China) was ±0.3 °C in the range from 0 to 50 °C.

The pressure difference between Chambers A and B was monitored by a low differential pressure transducer (Model 264, Setra, Boxborough, MA, USA) with an accuracy of ±0.4 Pa over the range from 0 to 100 Pa. It was positioned in a corner of Chamber A, with its openings fixed in the center of each chamber. The air pressure in Chamber A was slightly higher, resulting in pressure differences between Chambers A and B of 1.86 ± 0.40 Pa.

2.3. Analytical Model

The infiltration factor (F_{inf}) represents the proportion of outdoor particles that penetrate indoors and remain suspended. In our experiments, particles from incense burning penetrated through the cracks, along with the airflow, between Chamber A and Chamber B, without any human activity or other sources present in the experimental chambers. Thus, the general expression of the infiltration factor can be written as:

$$F_{inf} = \frac{C_{in}}{C_{out}}, \tag{2}$$

where C_{in} and C_{out} represent the indoor and outdoor particle concentrations, respectively.

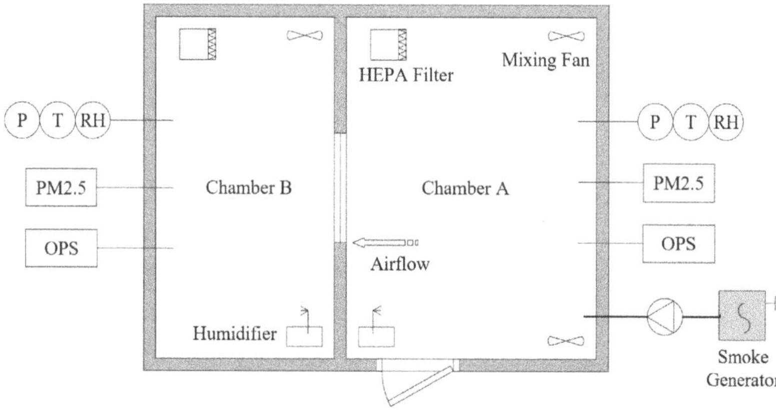

Figure 1. Experimental chamber system. PM2.5—$PM_{2.5}$ monitoring system; OPS—optical particle sizer; RH—relative humidity probe; T—temperature probe; P—differential pressure transducer.

The partition wall between Chambers A and B featured a polyurethane foam core, enclosed by gypsum boards with a polyurethane coating, providing insulation against heat and moisture. A stainless-steel window was installed in the partition wall. The specialized window, rectangular in shape (0.9 m in width and 0.6 m in height), had a windowsill at a height of 0.9 m. Its crack sizes were accurately measured and adjusted. The cracks between the window and its frame were straight, consisting of two horizontal cracks and two vertical cracks. Figure 2 provides a schematic representation of the crack's architecture. The length of the crack along the airflow, denoted by the symbol z, was set at 90 mm. The height of the crack was small, only 1 mm, represented symbolically by d, and the width of the crack was denoted by w. It was assumed that the inner face of the crack was smooth, as a study [14] showed that the penetration factor is not sensitive to crack roughness.

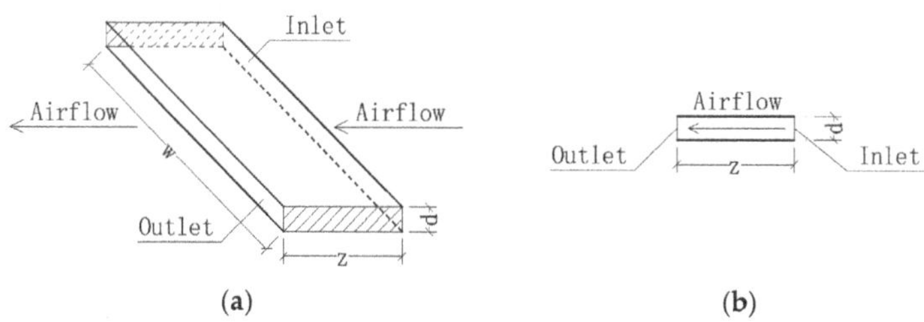

Figure 2. Straight crack between window and frame: (**a**) three-dimensional view; (**b**) sectional view.

2.2. Instrumentations and Test Conditions

The mass concentrations of particles ($PM_{2.5}$) in the chambers were continuously monitored by a $PM_{2.5}$ monitoring system (QD-X1-A, Green Built Environment Technology Co., Ltd., Beijing, China), which had six measuring probes with an accuracy of $\pm 10\%$ in the range of 10 to 500 $\mu g/m^3$. The number concentration and size distribution of particles were measured with two optical particle sizers (OPSs, Model 3330, TSI Incorporated, Shoreview, MN, USA), offering a size resolution of 5% at 0.5 μm. Before each test, zero checks and flow rate verifications were performed on both OPS units to ensure measurement accuracy. The OPS operated to count particles in 9 size categories: 0.3–0.35 μm, 0.35–0.4 μm, 0.4–0.5 μm, 0.5–0.7 μm, 0.7–1.0 μm, 1.0–1.3 μm, 1.3–1.6 μm, 1.6–2.0 μm, and 2.0–2.5 μm in diameter. The sampling ports were positioned in the center of chambers at a height of 1.2 m. A sample holder for the field-emission scanning electron microscope (FE-SEM, Gemini SEM 500, mass concentrations in 41 urban residences and examined the connections between F_{inf} and 78 environmental factors, showing that F_{inf} was influenced by building characteristics, traffic, wall and floor coverings, and human behavior.

The literature reveals significant variability in F_{inf}, pointing to the need for further investigation into the physical mechanisms dictating the behavior of indoor particles originating from outdoor sources. This variability is also evident in studies on the effects of relative humidity (RH) on particle behavior, including hygroscopic growth, coagulation, and deposition rate, suggesting an interaction between indoor RH and particulate dynamics. Hänel [18,19] computed the dependency of gravitational settling and Brownian diffusion on relative humidity for particles of different sizes. The results confirmed that RH significantly affects the deposition of aerosol particles. Dockery and Spengler [20] were the first to develop a steady-state mass balance equation for indoor and outdoor concentrations, assuming that the air exchange rate (a), penetration efficiency (P), and deposition rate (k) remained constant over the study period. They estimated F_{inf} values for respirable particles (PM_{10}) to be 0.70, using indoor and outdoor concentrations of PM_{10}. However, their assumptions may not correspond to real conditions, where the values of a, P, and k can vary. Li et al. [21] showed both theoretically and experimentally that RH, rather than temperature, plays the dominant role in determining the hygroscopic growth of aerosol particles. Jamriska and Morawska [22] continuously measured the evolution of particle volume and number concentration in an experimental chamber, concluding that approximately 64–84% of the total particle number loss was caused by coagulation mechanisms in their experiments, while surface deposition dominated the mass reduction of particles. Faulkner et al. [23] built a reduced-scale chamber with a ventilation system to measure particle concentration, showing that increasing air change rates reduced the particulate penetration factor for small particles but had little effect on large particles. Wang et al. [24] investigated the effect of RH on the deposition and coagulation of particles in an environmental test chamber, revealing that lower and higher RH conditions tended to enhance the deposition rate of nanoparticles; the coagulation coefficient of the nanoparticles increased with increased humidity due to strong inter-particle adherence.

Building on these findings, our study focuses on the impact of indoor air humidity on the infiltration of size-resolved particles. Utilizing an experimental chamber system, we conducted a series of experiments under varying RH conditions to examine the dynamic behavior of polydisperse aerosols. Through a mass balance model accounting for penetration, hygroscopicity, coagulation, deposition, and ventilation, we aimed to provide a nuanced quantitative assessment of F_{inf} in relation to air humidity. This assessment is anticipated to contribute to the broader discussion on indoor air quality, offering insights that could inform future modeling approaches, indoor particulate load calculations, and building energy consumption estimations.

2. Experimental Setup and Methods

Experiments were performed in room-sized chambers to determine the infiltration of fine particles under different humidity conditions. The mass concentration of total particles (i.e., $PM_{2.5}$) and the number concentration of size-resolved particles were monitored simultaneously and continuously. Based on the collected data, a mass balance model was used to analyze the infiltration process.

2.1. Experimental Chambers

An experimental chamber system was designed to study the impact of indoor air humidity on the infiltration of size-resolved particles (Figure 1). It was a room measuring 3.9 m in length, 2.9 m in width, and 2.6 m in height, divided by a wall into two sections to simulate an outdoor environment, named Chamber A, and an indoor environment, named Chamber B. The mass concentration of particles, particle size distribution, air relative humidity, air temperature, and pressure difference in the experimental chambers were controlled and measured.

(cm^{-3}) outdoors (i.e., Chamber A), k_i is the particle deposition rate (min^{-1}), and γ_i is the coagulation rate ($cm^3\ min^{-1}$).

The first term on the right-hand side of Equation (6) reflects the process of outdoor particles penetrating with the airflow and absorbing water vapor from indoor air. The penetration factor can be approximated considering the mechanisms of gravitational settling and Brownian diffusion [15]. The second and third terms account for the particle loss due to ventilation and deposition onto surfaces in Chamber B, respectively. The last term represents the number change for particles in the ith size category caused by coagulation, in which pairs of particles collide and stick together.

The steady-state condition of indoor particle concentration is achieved when the dynamic balance between particle penetration, deposition, and coagulation is established:

$$N_i = \frac{-(a+k_i) + \sqrt{(a+k_i)^2 + 4aP_i\beta_i\gamma_i N_{i,out}}}{2\gamma_i}. \tag{7}$$

From this state, the size-resolved infiltration factor ($F_{inf,i}$) can be deduced, providing a means to analyze the impact of indoor air humidity on infiltration efficiency:

$$F_{inf,i} = \frac{-(a+k_i) + \sqrt{(a+k_i)^2 + 4aP_i\beta_i\gamma_i N_{i,out}}}{2\gamma_i N_{i,out}} \tag{8}$$

The experimental data, including particle concentrations in the chambers over time, were analyzed using the above approaches to understand the dynamics of particle infiltration and behavior in humid indoor environments.

3. Results and Discussion

3.1. Particle Characteristics Measured in the Experimental Chamber

In this study, we continuously measured the size-resolved number concentrations of the test particles in the experimental chambers, Chamber A and Chamber B. Chamber A simulated an outdoor environment, maintaining a relative humidity of 35 ± 5% and a steady particle concentration of 400 ± 40 μg/m^3 throughout the various tests. Meantime, seven sets of tests were conducted under different RH conditions in Chamber B, simulating the indoor environment. The detailed RH conditions and particle characteristics of each test in Chambers A and B are provided in Table 1.

Table 1. Test conditions and characteristics of test aerosols in Chambers A and B.

Test	Chamber	RH (%)	N_{total} (cm^{-3})	M_{total} (μg m^{-3})	F_{inf}	GMD (μm)	GSD	R^2
1	A	35 ± 5	5.543×10^3	400.3	0.342	0.24	1.66	0.968
	B	30 ± 4	2.571×10^3	136.9		0.22	1.65	0.972
2	A	35 ± 5	5.494×10^3	399.7	0.347	0.24	1.66	0.956
	B	40 ± 3	2.582×10^3	138.7		0.22	1.65	0.927
3	A	35 ± 5	5.522×10^3	399.7	0.353	0.24	1.66	0.943
	B	50 ± 2	2.583×10^3	141.1		0.23	1.65	0.966
4	A	35 ± 5	5.577×10^3	400.3	0.362	0.24	1.66	0.954
	B	60 ± 2	2.587×10^3	144.9		0.24	1.64	0.951
5	A	35 ± 5	5.527×10^3	399.5	0.378	0.24	1.66	0.961
	B	70 ± 3	2.608×10^3	151.0		0.26	1.64	0.955
6	A	35 ± 5	5.532×10^3	399.8	0.404	0.24	1.66	0.944
	B	80 ± 2	2.653×10^3	161.5		0.28	1.64	0.938
7	A	35 ± 5	5.560×10^3	400.5	0.429	0.24	1.66	0.979
	B	85 ± 3	2.706×10^3	171.8		0.29	1.63	0.967

The total number concentration (N_{total}) and the total mass concentration (M_{total}) in both chambers are average values, calculated over the last 20 min of the infiltration process.

Overall, the concentrations of particles in Chamber B were positively correlated with the increase in RH, which may be attributed to increased particle hygroscopicity and growth due to water uptake. Additionally, the infiltration factor (F_{inf}), which represents the extent of particle penetration from Chamber A to Chamber B, also escalated with increasing RH.

The particle size distributions in both chambers followed a unimodal and lognormal pattern. The geometric mean diameter (GMD) and the geometric standard deviation (GSD) of the particles were derived using the lognormal distribution function (Equation (3)). The R-square values were relatively high (>0.9), indicating a good fit of the function to the experimental data. An analysis of the GMD values revealed an upward trend with increasing RH, from 0.22 μm at 30% RH to 0.29 μm at 85% RH. In comparison, the GMD of the source particles in Chamber A remained consistent at an average value of 0.24 μm. Interestingly, when RH in Chamber B was lower than 60%, the GMD of the particles in Chamber B was smaller than in Chamber A, while in high RH conditions (RH > 60%), the GMD in Chamber B increased as RH rose. This observation indicates that the penetration process dominated the particle size change in low RH conditions, and the hygroscopic mechanism took over when the particles absorbed water vapor and grew larger under high RH conditions. The GSD values displayed a minor declining trend with increasing RH, from 1.65 at 30% RH to 1.63 at 85% RH. Based on the self-preserving size distribution theory of Friedlander [32], we treated the GSD as constant in our analytical approach.

The time-varying experimental data for particle concentrations at the lowest and highest RH levels were selected for further analysis. In Figure 3, each point represents an average value of the size-resolved number concentrations of particles over a twenty-minute period. To analyze the influence of air humidity on infiltration, we fitted the data using the solution of the humidity model proposed in Equation (6). The data for the driest condition (RH = 30 $\pm$ 4%) are shown in Figure 3a. The number of particles in each size range increased over time, while the growth rate dropped to near zero after 480 min of infiltration. As the particle size range increased, the number concentration of the particles generally decreased. At the end of the measurement period, the concentration scales decreased with the increasing particle diameter. Finally, the concentration of PM_1 was about 2490 cm^{-3}, while for the 1.0–2.5 μm size range, it was only 10 cm^{-3}. This suggests that particles larger than 1 μm were less prevalent in the chamber. For comparison, Figure 3b presents the evolution of particle concentrations at 85% RH. The general trend of increasing particle concentration over time was similar in the two cases. However, notable differences in particle number concentrations between the two RH conditions were observed. For particles within the size ranges of 0.3–0.35 μm and 0.35–0.4 μm, their number concentration and growth rate were higher at RH = 85% than at RH = 30% in the first half of the infiltration experiment. However, in the final stage of the experiment, when the particle concentrations were high, the particle numbers were lower at RH = 85% than at RH = 30%, and their concentrations even showed negative growth in the last hour. Meanwhile, for particles in the 0.4–0.5 μm and larger-size ranges, concentrations were higher at RH = 85% than at RH = 30%. It is evident that under high humidity conditions, the increased availability of water vapor led to the hygroscopic growth of particles across all size ranges. Furthermore, it can be deduced that higher humidity conditions and/or higher particle concentrations may facilitate the coagulation of small particles.

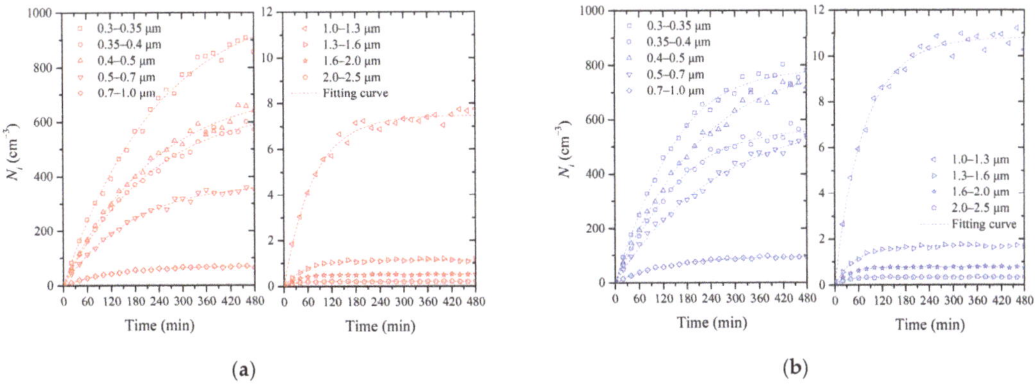

Figure 3. The evolution in the number concentration of size-resolved particles in Chamber B: (**a**) RH = 30 ± 4%; (**b**) RH = 85 ± 3%.

3.2. Infiltration Factor

The experimental data presented in Figure 3 indicate that particle infiltration varied significantly with both indoor RH conditions and particle size. We calculated the infiltration factor for particles in different size ranges separately using Equation (2) and present the results in Figure 4.

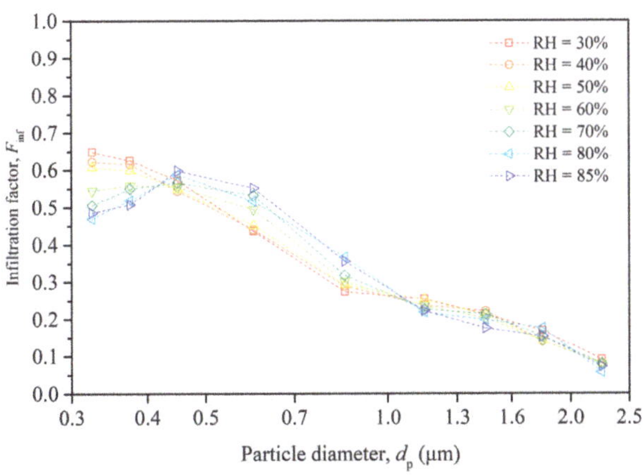

Figure 4. Size-resolved infiltration factor of particles under different humidity conditions.

The infiltration factor generally decreased as the particle size increased, except for particles in the smallest-size range (0.3–0.35 μm). This is because larger particles have lower penetration factors and higher deposition rates than smaller particles, which reduces their infiltration efficiency.

The relationship between F_{inf} and RH is complex, exhibiting size-dependent variations. For smaller particles (0.3–0.4 μm), increased RH correlates with a diminished F_{inf}, due to enhanced hygroscopic growth which elevates the effective particle size. In contrast, the F_{inf} of intermediate-sized particles (0.4–1.0 μm) increases with RH, suggesting coagulation processes that increase particle size and collision efficiency. Larger particles (>1.0 μm) show minimal sensitivity to RH shifts, indicating relative stability in their infiltration behavior. Moreover, at very high humidity levels, particles may become stickier and adhere more to the building envelope, which can increase their penetration factor. The overall effect of humidity depends on the balance of these factors.

The figure highlights the importance of considering both particle size and indoor RH when assessing infiltration factors, with significant implications for managing indoor air quality and formulating mitigation strategies. Additionally, it brings to light the need for a nuanced understanding of the chemical composition of ambient aerosols. The hygroscopic properties of aerosol particles, dictated by their chemical makeup, significantly govern their growth dynamics in response to variations in RH. This growth, in turn, influences particle deposition and coagulation processes, ultimately affecting particle infiltration efficiency.

To assess the detailed effect of indoor air humidity on particle infiltration, we analyzed the dynamics of particles using the mass balance model in Equation (6). We fitted the solution of Equation (6) to the time-varying and size-resolved particle concentrations under seven different humidity conditions and obtained the deposition rate (k_i), hygroscopic number change ratio (β_i), and coagulation rate (γ_i) of the particles. These coefficients are quantitatively analyzed in the following sections.

3.3. Deposition Rate of the Particles

Based on the changes in measured particle concentrations, we obtained the size-resolved deposition rates of the particles in Chamber B, which are plotted in a logarithmic coordinate system in Figure 5. The points in the figure, represented by different colors, correspond to seven datasets under different humidity levels, and the dashed line illustrates the regression of these seven datasets of particle deposition rate. It is apparent that the deposition rate increased with the increasing particle size for particles ranging from 0.3 to 2.5 μm, regardless of the relative humidity level. For smaller particles, specifically within the 0.3–0.4 μm range, the deposition rate experienced a slight increase with a rise in relative humidity. However, the correlation between deposition rate and relative humidity was neither significant nor uniform across different particle sizes.

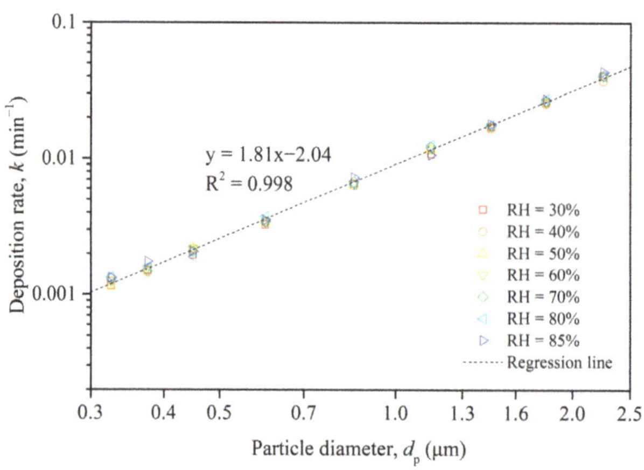

Figure 5. Size-resolved deposition rate of particles in Chamber B and regression of 7 different humidity datasets.

As a result, our investigation concentrated on the relationship between deposition rate and particle size, combining the data from all humidity levels and performing a regression analysis to establish a mathematical connection. As demonstrated in Figure 5, the liner regression equation implied a strong power law relationship between the deposition rate and the particle diameter across the seven humidity levels. The deposition rate increased with an increasing particle diameter, and the regression fitted the data very well, as evidenced by the high R-squared value.

3.4. Hygroscopicity of the Particles

The hygroscopic number change ratio (β) quantifies the potential variation in particle number concentration resulting from the hygroscopic growth of particles, a process driven by water vapor absorption. Figure 6 illustrates the β values corresponding to various particle sizes under different humidity conditions, with individual colors representing the humidity levels. The value of β showed a notable increase with the rising RH across all particle sizes. This trend reflects the fundamental concept of hygroscopic growth, whereby particles absorb more water vapor at higher humidity levels. Interestingly, higher values of β were observed for larger particles compared to smaller ones at equivalent relative humidity. This phenomenon might be attributed to the transformation of smaller particles into larger ones as they absorbed water, subsequently shifting to the larger-size categories, thereby escalating the corresponding number concentrations and β values. Moreover, the rate of increase in the β values appeared to grow with the rising relative humidity, suggesting that the effects of hygroscopic growth became more pronounced at higher humidity levels. This underscores the profound influence of relative humidity on particle number concentrations and the overall dynamics of aerosol populations.

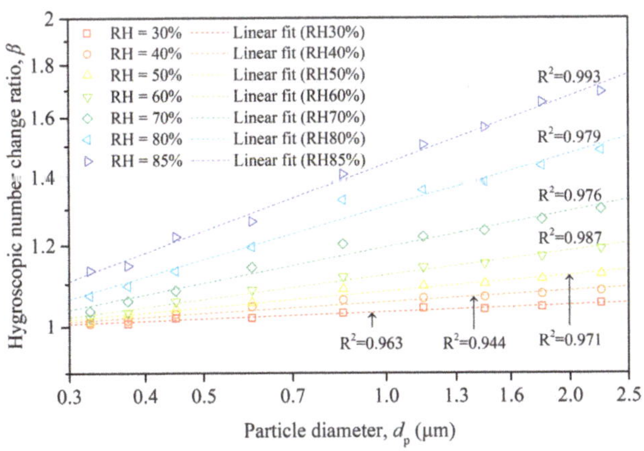

Figure 6. Size-resolved hygroscopic number change ratio of particles and fitting lines under different humidity conditions.

In addition to the data points, Figure 6 also portrays the linear regression for each dataset within a logarithmic coordinate system. The regression analysis revealed a linear association between the natural logarithm of β and that of the particle diameter (d_p) under constant humidity. This power law relationship between β and d_p aligns with the model introduced in Equation (5), suggesting potential inherent mechanisms that operated proportionally.

3.5. Coagulation of the Particles

In the experimental chamber, most particles were smaller than 1 μm, with coagulation events primarily occurring among these particles. Figure 7 illustrates the varying trends of the coagulation rate for different particle size ranges and humidity conditions. The coagulation rate for the ith size category (γ_i) measures the extent to which particles collide and stick together. A positive γ_i value indicates that the number of particles in that size category decreases due to coagulation, while a negative γ_i value signifies an increase in the number of particles in that size category due to coagulation.

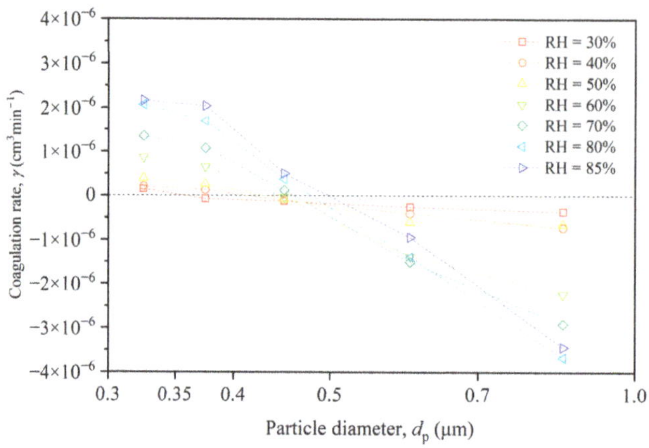

Figure 7. Size-resolved coagulation rate of particles under different humidity conditions.

For the smallest particles (0.3–0.4 μm), the coagulation rate increased with the rising humidity, reaching a maximum at RH = 85%. This suggests that these particles were more prone to coagulate as humidity increased, possibly because water molecules on the particle surfaces enhanced the attractive forces between them. For intermediate particles (0.4–0.5 μm), the coagulation rate was negative under low humidity conditions (RH < 50%), indicating that these particles were less likely to coagulate than smaller particles. However, as humidity increased, the coagulation rate became positive, suggesting these particles became more prone to coagulation with rising humidity. For the largest particles (0.5–1.0 μm), the coagulation rate was negative across all humidity conditions, indicating that the particles in smaller-size ranges coagulated and then were counted in the larger-size ranges.

We derived the values of the coagulation rate from Equation (6), using the time-varying and size-resolved particle concentrations. This approach accounted for enhanced collision rates between particles caused by factors such as turbulence or charge, as well as the polydispersity of the particle size distribution.

4. Conclusions

Our experimental setup, employing a controlled chamber system, allowed for the precise quantification of particle infiltration under varying RH conditions. Through experimentation and analysis, we established that variations in indoor RH levels distinctly influenced the behavior of infiltrating particles, affecting their hygroscopic growth, deposition rates, and coagulation processes. Notably, our findings underscore the critical role of RH in modulating the size distribution and concentration of indoor particles, thereby altering indoor air quality and potentially impacting human health. The main conclusions are as follows:

(1) In Chamber B, simulating indoor conditions at different RH levels, both the total number and the mass concentrations of particles increased with RH, indicating a clear correlation between RH and the particle hygroscopic growth. This growth led to an increase in the infiltration factor (F_{inf}) as RH rose. The particle size distributions suggest that hygroscopic growth and coagulation processes dominated under different RH conditions, affecting particle size and concentration dynamics.

(2) Particle infiltration was significantly influenced by indoor relative humidity (RH) and particle size, showing distinct behaviors across different size ranges. Larger particles exhibited decreased infiltration factors due to lower penetration and higher deposition rates. In the size range of 0.3–0.4 μm, the particles experience reduced infiltration at higher RH levels due to hygroscopic growth, while intermediate-sized particles

showed increased infiltration, resulted from coagulation effects. Large particles (>1.0 μm) were less affected by RH changes.

(3) There was a clear power law relationship between particle size and deposition rate across various humidity levels, showing an increase in the deposition rate with the particle size. While humidity impacted the deposition rate for smaller particles (0.3–0.4 μm), this effect was not uniformly observed across all particle sizes. A comprehensive regression analysis underscored a strong correlation between deposition rate and particle diameter, validated by a high R-squared value.

(4) The hygroscopic number change ratio (β) increased with relative humidity (RH) across all particle sizes, indicating enhanced water vapor absorption by particles at higher RH levels. Larger particles exhibited higher β values than smaller ones at the same RH, suggesting a size transformation due to hygroscopic growth. The linear regression within the logarithmic coordinate system showed a power law relationship between β and particle diameter, supporting the model that describes proportional mechanisms of hygroscopic growth.

(5) Particle coagulation varied with size and humidity, showing an increased possibility of coagulation among smaller particles (<1 μm) as humidity rose. This was attributed to enhanced attractive forces from water molecules on particle surfaces. The coagulation rate transitioned from negative to positive for intermediate-sized particles with increasing humidity. Larger particles exhibited consistently negative coagulation rates across all humidity levels, suggesting a redistribution towards larger sizes due to the coagulation of smaller particles.

This investigation contributes to a deeper understanding of the environmental factors affecting indoor air quality, emphasizing the indispensability of RH in the dynamics of particle infiltration. However, because of its reliance on a controlled chamber system, it may not fully capture the complexity of actual indoor environments. Focusing solely on RH could overlook the comprehensive effects of other environmental variables and indoor particle sources.

Future research should broaden the examination scope to include the synergistic effects of multiple environmental variables, such as temperature and indoor sources. Additionally, directly analyzing the chemical composition of particles will clarify their contribution to hygroscopic growth, deposition, and coagulation.

Author Contributions: Conceptualization, J.Q. and H.Z.; methodology, J.Q.; software, J.Q. and J.L.; validation, J.Q. and H.Z.; formal analysis, J.Q.; resources, Y.L.; data curation, J.Q.; writing—original draft preparation, J.Q.; writing—review and editing, Y.L.; supervision, J.L.; funding acquisition, Y.L. All authors have read and agreed to the published version of the manuscript.

Funding: This research was funded by the National Key Research and Development Program of China under Grant No. 2018YFB0604101.

Data Availability Statement: The data used to support the findings of this study are included in the article.

Conflicts of Interest: Author Haixi Zhang was employed by the company China Northwest Architectural Design and Research Institute Co., Ltd. Author Jialu Liu was employed by the company Jiangnan Shipyard (Group) Co., Ltd. The remaining authors declare that the research was conducted in the absence of any commercial or financial relationships that could be construed as a potential conflict of interest.

References

1. Pope, C.A.; Ezzati, M.; Dockery, D.W. Fine-Particulate Air Pollution and Life Expectancy in the United States. *N. Engl. J. Med.* **2009**, *360*, 376–386. [CrossRef]
2. Bell, M.L.; Zanobetti, A.; Dominici, F. Evidence on Vulnerability and Susceptibility to Health Risks Associated with Short-Term Exposure to Particulate Matter: A Systematic Review and Meta-Analysis. *Am. J. Epidemiol.* **2013**, *178*, 865–876. [CrossRef] [PubMed]

3. Lu, F.; Xu, D.; Cheng, Y.; Dong, S.; Guo, C.; Jiang, X.; Zheng, X. Systematic Review and Meta-Analysis of the Adverse Health Effects of Ambient $PM_{2.5}$ and PM_{10} Pollution in the Chinese Population. *Environ. Res.* **2015**, *136*, 196–204. [CrossRef]
4. Adgate, J.L.; Ramachandran, G.; Pratt, G.C.; Waller, L.A.; Sexton, K. Spatial and Temporal Variability in Outdoor, Indoor, and Personal $PM_{2.5}$ Exposure. *Atmos. Environ.* **2002**, *36*, 3255–3265. [CrossRef]
5. Awada, M.; Becerik-Gerber, B.; Hoque, S.; O'Neill, Z.; Pedrielli, G.; Wen, J.; Wu, T. Ten Questions Concerning Occupant Health in Buildings During Normal Operations and Extreme Events Including the COVID-19 Pandemic. *Build. Environ.* **2021**, *188*, 107480. [CrossRef]
6. Wallace, L. Indoor Particles: A Review. *J. Air Waste Manag. Assoc.* **1996**, *46*, 98–126. [CrossRef]
7. Chen, C.; Zhao, B. Review of Relationship Between Indoor and Outdoor Particles: I/O Ratio, Infiltration Factor and Penetration Factor. *Atmos. Environ.* **2011**, *45*, 275–288. [CrossRef]
8. Meier, R.; Schindler, C.; Eeftens, M.; Aguilera, I.; Ducret-Stich, R.E.; Ineichen, A.; Davey, M.; Phuleria, H.C.; Probst-Hensch, N.; Tsai, M.-Y.; et al. Modeling Indoor Air Pollution of Outdoor Origin in Homes of Sapaldia Subjects in Switzerland. *Environ. Int.* **2015**, *82*, 85–91. [CrossRef]
9. Wilson, W.E.; Mage, D.T.; Grant, L.D. Estimating Separately Personal Exposure to Ambient and Nonambient Particulate Matter for Epidemiology and Risk Assessment: Why and How. *J. Air Waste Manag. Assoc.* **2000**, *50*, 1167–1183. [CrossRef]
10. Bi, J.; Wallace, L.; Sarnat, J.; Liu, Y. Characterizing Outdoor Infiltration and Indoor Contribution of PM2.5 with Citizen-Based Low-Cost Monitoring Data. *Environ. Pollut.* **2021**, *276*, 116763. [CrossRef]
11. Barkjohn, K.K.; Bergin, M.H.; Norris, C.; Schauer, J.J.; Zhang, Y.; Black, M.; Hu, M.; Zhang, J. Using Low-Cost Sensors to Quantify the Effects of Air Filtration on Indoor and Personal Exposure Relevant $PM_{2.5}$ Concentrations in Beijing, China. *Aerosol Air Qual. Res.*. [CrossRef]
12. Diapouli, E.; Chaloulakou, A.; Koutrakis, P. Estimating the Concentration of Indoor Particles of Outdoor Origin: A Review. *J. Air Waste Manag. Assoc.* **2013**, *63*, 1113–1129. [CrossRef] [PubMed]
13. Lai, A.C.K. An Experimental Study of Indoor and Outdoor Concentrations of Fine Particles Through Nonisothermal Cracks. *Aerosol Sci. Technol.* **2013**, *47*, 1009–1016. [CrossRef]
14. Yang, T.Y.; Riskowski, G.L.; Chang, A.C.-Z. Effects of Air Relative Humidity and Ventilation Rate on Particle Concentrations Within a Reduced-Scale Room. *Indoor Built Environ.* **2019**, *28*, 335–344. [CrossRef]
15. Wallace, L.; Williams, R. Use of Personal-Indoor-Outdoor Sulfur Concentrations to Estimate the Infiltration Factor and Outdoor Exposure Factor for Individual Homes and Persons. *Environ. Sci. Technol.* **2005**, *39*, 1707–1714. [CrossRef] [PubMed]
16. MacNeill, M.; Wallace, L.; Kearney, J.; Allen, R.W.; Van Ryswyk, K.; Judek, S.; Xu, X.; Wheeler, A. Factors Influencing Variability in the Infiltration of $PM_{2.5}$ Mass and Its Components. *Atmos. Environ.* **2012**, *61*, 518–532. [CrossRef]
17. Huang, L.; Hopke, P.K.; Zhao, W.; Li, M. Determinants on Ambient $PM_{2.5}$ Infiltration in Nonheating Season for Urban Residences in Beijing: Building Characteristics, Interior Surface Coverings and Human Behavior. *Atmos. Pollut. Res.* **2015**, *6*, 1046–1054. [CrossRef]
18. Hänel, G. The Properties of Atmospheric Aerosol Particles as Functions of the Relative Humidity at Thermodynamic Equilibrium with the Surrounding Moist Air. *Adv. Geophys.* **1976**, *19*, 73–188. [CrossRef]
19. Hänel, G. Humidity Effects on Gravitational Settling and Brownian Diffusion of Atmospheric Aerosol Particles. *Pure Appl. Geophys.* **1977**, *115*, 775–797. [CrossRef]
20. Dockery, D.W.; Spengler, J.D. Indoor-Outdoor Relationships of Respirable Sulfates and Particles. *Atmos. Environ.* **1981**, *15*, 335–343. [CrossRef]
21. Li, W.; Montassier, N.; Hopke, P.K. A System to Measure the Hygroscopicity of Aerosol Particles. *Aerosol Sci. Technol.* **1992**, *17*, 25–35. [CrossRef]
22. Jamriska, M.; Morawska, L. Quantitative Assessment of the Effect of Surface Deposition and Coagulation on the Dynamics of Submicrometer Particles Indoors. *Aerosol Sci. Technol.* **2003**, *37*, 425–436. [CrossRef]
23. Faulkner, W.B.; Memarzadeh, F.; Riskowski, G.; Kalbasi, A.; Chang, A.C.Z. Effects of Air Exchange Rate, Particle Size and Injection Place on Particle Concentrations Within a Reduced-Scale Room. *Build. Environ.* **2015**, *92*, 246–255. [CrossRef]
24. Wang, Y.; Chen, L.; Chen, R.; Tian, G.; Li, D.; Chen, C.; Ge, X.; Ge, G. Effect of Relative Humidity on the Deposition and Coagulation of Aerosolized SiO_2 Nanoparticles. *Atmos. Res.* **2017**, *194*, 100–108. [CrossRef]
25. Cui, S.; Cohen, M.; Stabat, P.; Marchio, D. CO_2 Tracer Gas Concentration Decay Method for Measuring Air Change Rate. *Build. Environ.* **2015**, *84*, 162–169. [CrossRef]
26. Kumar, P.; Robins, A.; Vardoulakis, S.; Britter, R. A Review of the Characteristics of Nanoparticles in the Urban Atmosphere and the Prospects for Developing Regulatory Controls. *Atmos. Environ.* **2010**, *44*, 5035–5052. [CrossRef]
27. Li, W.; Hopke, P.K. Initial Size Distributions and Hygroscopicity of Indoor Combustion Aerosol Particles. *Aerosol Sci. Technol.* **1993**, *19*, 305–316. [CrossRef]
28. Rissler, J.; Vestin, A.; Swietlicki, E.; Fisch, G.; Zhou, J.; Artaxo, P. Size Distribution and Hygroscopic Properties of Aerosol Particles from Dry-Season Biomass Burning in Amazonia. *Atmos. Chem. Phys.* **2006**, *6*, 471–491. [CrossRef]
29. Chan, M.N.; Chan, C.K. Mass Transfer Effects in Hygroscopic Measurements of Aerosol Particles. *Atmos. Chem. Phys.* **2005**, *5*, 2703–2712. [CrossRef]

30. Swietlicki, E.; Hansson, H.-C.; Hämeri, K.; Svenningsson, B.; Massling, A.; McFiggans, G.; McMurry, P.H.; Petäjä, T.; Tunved, P.; Gysel, M.; et al. Hygroscopic Properties of Submicrometer Atmospheric Aerosol Particles Measured with H-TDMA Instruments in Various Environments—A Review. *Tellus B* **2008**, *60*, 432–469. [CrossRef]
31. Vu, T.V.; Ondracek, J.; Zdímal, V. Physical Properties and Lung Deposition of Particles Emitted from Five Major Indoor Sources. *Air Qual. Atmos Health* **2017**, *10*, 1–14. [CrossRef]
32. Friedlander, S.K.; Wang, C.S. The Self-Preserving Size Distribution for Coagulation by Brownian Motion. *J. Colloid Interface Sci.* **1966**, *22*, 126–132. [CrossRef]

buildings

MDPI

Article

Simulation Study on Student Residential Energy Use Behaviors: A Case Study of University Dormitories in Sichuan, China

Yingzi Zhang, Bo Zhang and Jingwen Hou *

School of Architecture, Southwest Jiaotong University, Chengdu 611756, China; zhangyingzi@swjtu.edu.cn (Y.Z.); bozhang@my.swjtu.edu.cn (B.Z.)
* Correspondence: yumengli@my.swjtu.edu.cn

Abstract: Studying the energy use behavior of occupants is crucial for accurately predicting building energy consumption. However, few studies have considered the impact of occupant behaviors on energy consumption in university dormitories. The objective of this study is to establish an agent-based model of energy consumption for university dormitories based on energy use behavior. The dormitories of a typical university in Sichuan, China, were subdivided into three clusters using a two-step cluster analysis. Subsequently, the energy use behaviors of occupants in each type of dormitory were characterized to establish a stochastic energy use behavior model. On the basis of the above, NETLOGO was used to construct an agent-based model for dormitories' energy consumption to dynamically simulate energy use behavior. The accuracy of the model was verified by comparing the simulated values with the measured data. Finally, a building-energy-friendly retrofit scheme was proposed, and it was found that the optimized dormitory reduced energy consumption by 16.07%. Therefore, the results can provide information support for energy-saving decisions during the early design and retrofit phases of buildings. With the popularity of centralized supply, the research methodology may provide an extensive reference for energy management policies and sustainable strategies in the building sector.

Keywords: university dormitory; building energy consumption; occupant behavior; agent-based model; energy conservation retrofit design

1. Introduction

More than 30% of primary energy is consumed in the building sector [1]. As an important element of society, universities, with large building volumes and high staff densities, contribute to about 8% of total social energy consumption [2]. However, current studies mainly focus on residential and office buildings [3,4], and research on the energy consumption of universities is limited. Studies show that the per capita energy consumption of Chinese universities is three times higher than the national per capita energy consumption. Moreover, the energy consumption of dormitories accounts for 18 percent of the total energy consumption of school buildings. Therefore, there is a strong energy-saving potential for Chinese universities' dormitories [5]. Occupant energy use behaviors have a significant impact on building energy consumption. Research has revealed that energy consumption per capita varies up to 14 times in the same building due to the differences in energy use behavior [6]. University dormitories in China follow the rule of "multiple occupants per room". The interaction of students with the environment, appliances, and other students is much more complex. The study of Poirazis et al. [7] demonstrated that ignoring the difference in energy use patterns in the early stage of building energy conservation design and renovation may result in greater energy consumption. However, few studies have considered the potential impact of students' stochastic energy use behaviors on energy consumption in university dormitories.

Energy use behavior generally refers to the usage of energy-consuming appliances in a building, and it is related to the occupant's movement state and indoor environment [8].

Many researchers have recognized the importance of occupant behavior for building energy consumption [9–11]. The randomness, regularity, and spatio-temporality of the movement are described by studies such that the dynamic decision-making behavior of occupants can be represented via energy simulation. Hong et al. (2015) studied energy-related occupant behavior in buildings and proposed a framework for standardized representation of occupant behavior in buildings [12]. Fixed schedules are currently the most widely used method for locating occupant motion in building energy simulation, i.e., the position of the occupant at each moment is set according to a schedule [13]. However, this method cannot capture the randomization and individual differences between occupants. Statistical methods based on data such as the Markov-Chain technique, Monte Carlo method, Bernoulli process, and survival analysis are applied in simulations to describe the stochastic motion of the occupants [14–16]. However, these models are only suitable for movement descriptions of simple events and are not applicable to the applications involving multiple types of events in dormitory buildings [17–19].

When it comes to devices controlled by multiple types of events, air conditioners, and lighting are the main contributors to energy consumption [20,21]. The simulation is performed by the functional relationship between the environmental factors (i.e., indoor temperature [22,23] and illumination [24,25]) and energy use behaviors, mainly including the statistical regression model. The simulation model was constructed based on data statistics and probabilistic fitting. An operation model of the device was established to characterize the different behaviors, including the number of times the appliance switched on per day, and the probability distribution function for the duration [26–28]. Researchers developed a personalized stochastic behavioral model for air conditioners that combined measurement data, data analysis, and logistic regression [29,30]. However, stochastic models based on state descriptions are unable to examine individual energy use behavior [31]. Therefore, there is a need to accurately describe behavior-based differences in the energy use needs of individual university students.

Benefiting from the development of computer technology, agent-based modeling (ABM) has been developed to address the behavior of occupant's complex interaction and randomness, and it can simulate the dynamic interaction of occupants in a building system. Liao et al. (2010) conducted the first study to apply the ABM to simulate the behavior of individual occupants in a building [32]. Ding et al. (2019) established an agent-based energy consumption simulation model based on student behavior and explored the energy-saving effect of four different strategy scenarios [33]. Azar et al. (2012) developed a commercial building energy model by considering the interactions of occupant energy use behavior with the indoor environment and with each other [34]. Kashif et al. (2013) proposed a four-step simulation method to assess the impact of occupant behavior on energy consumption [35]. Jia et al. developed a novel ABM to explore occupant behaviors in commercial buildings [36] and investigated the impact of occupant behavior information on building performance simulation [37]. In summary, ABM has become one of the most powerful approaches for studying the impact of occupant behaviors on building performance simulation. Therefore, ABM was selected for this simulation research.

There are two main applications of occupant behavior studies in building performance simulation. The first application is to reduce the gap between the simulated energy consumption and actual values [38,39]. The second is to provide useful information for retrofit solutions of building energy conservation to replace the original design [40,41]. These findings support the idea that the occupant's behavioral pattern is an important factor in evaluating building design and retrofit design [20]. Therefore, the impact of occupant behavior on improving the energy performance of buildings should be fully considered in the decision-making process for the schematic design and retrofit design of buildings.

There are still some obstacles in the current study. First of all, although occupants may exhibit different energy use features with time, current energy estimation tools are predicated on behavioral constancy [34,42,43]. A study conducted by Clevenger and Haymaker [44] quantified the impact of different occupant behaviors on building energy use.

Hoes et al. (2009) evaluated the impact of user behavior on building performance [45]. The researchers assumed that all occupants have similar energy consumption patterns. They did not consider different energy use behaviors. In the actual situation, occupants will adjust their behaviors to interact with their surroundings and devices. Moreover, these interactions and behaviors are dynamically autonomous for different purposes. The changes in these behaviors ultimately affect the prediction of building energy consumption. Second, existing studies have focused on the simulation of residential and office or commercial buildings [36,37]. Chen et al. (2020) developed the simulation method of residential energy use behavior [13]. Azar et al. [34] and Jia et al. [36,37] investigated the relationship between occupant behavior and building simulation performance. However, fewer studies have been conducted on the energy use behavior of university dormitories. Unlike residential buildings, the occupants of university dormitories come from different regions and exhibit different equipment usage habits. Therefore, it is necessary to study the features of students' energy use behavior as the basic work for dormitory building simulations. Given these current problems, this paper focused on the following areas:

(1) to understand the underlying patterns of students' energy use behavior;
(2) to establish an agent-based model that can simulate the energy consumption in university dormitories based on the energy use behavior occurrence rules.

2. Research Method

The research technical route is shown as Figure 1. First of all, we investigated the energy consumption of university dormitories via questionnaires. The dormitories were subdivided into different clusters using a two-step cluster analysis based on the duration of operation and the energy consumption of electrical appliances. Subsequently, a method of describing energy use behaviors of different types of dormitories was proposed by recording students' energy use behaviors. The description was expressed as a probability function, including occupancy rates and the behaviors of turning on the electrical appliances in different seasons. After that, an agent-based simulation model was established to visually simulate students' energy use behavior in NETLOGO. The accuracy of the model was verified by comparing the agent's energy use behavior between the simulation pattern and the surveyed situation and comparing the average energy consumption between the simulation values and the surveyed results. Finally, we proposed a renovation optimization design for the student dormitory building. The reasonableness of the proposed renovation design was verified by comparing the energy consumption of the dormitory between the existing and proposed dormitory design.

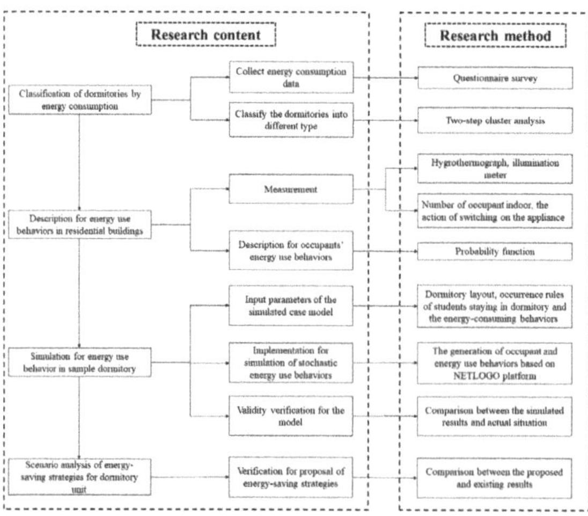

Figure 1. The research technical route.

3. Classification of Dormitory Status via Energy Use Behavior

3.1. Questionnaire Data

Chen et al. (2015) showed that there are significant differences in the characteristics of energy-using behaviors and energy consumption among different types of dormitories [46]. Therefore, it is necessary to classify the dormitories and study the students' behaviors in different types of dormitories separately. In this study, a random sampling questionnaire survey was conducted in 2022 at Southwest Jiaotong University in Chengdu (Chengdu, China) to collect data on energy consumption, energy use behaviors, and duration in different seasons. These data can be used for the subsequent cluster analysis of energy consumption in dormitories. The questionnaires were distributed anonymously without any communication between participants to eliminate expectation bias. Participants were randomly selected from Chengdu universities. The reliability of the questionnaire ranged from 0.7 to 0.9, which was tested by Cronbach's alpha coefficient. A total of 422 questionnaires were distributed, and 388 effective questionnaires were collected, with an effective rate of 92%. There were 125, 131, and 132 questionnaires for summer (in June, July, and August), winter (in December, January, and February), and transition seasons (from March to May and from September to November), respectively.

3.2. Cluster Analysis of Occupants' Energy Consumption-Related Behaviors

Figure 2 displays the percentage of the monthly average energy consumption of student dormitories. The percentages of monthly average energy consumption higher than 166.67 kWh are 11.36%, 21.93%, and 23.67% in the transition season, summer, and winter, respectively. This indicates that it is difficult for the indoor environment to meet the needs of daily activities in summer and winter. The building energy consumption will increase further with the use of devices. According to the questionnaire, the energy-consuming appliances in university dormitories mainly include air conditioner, lighting, and computers.

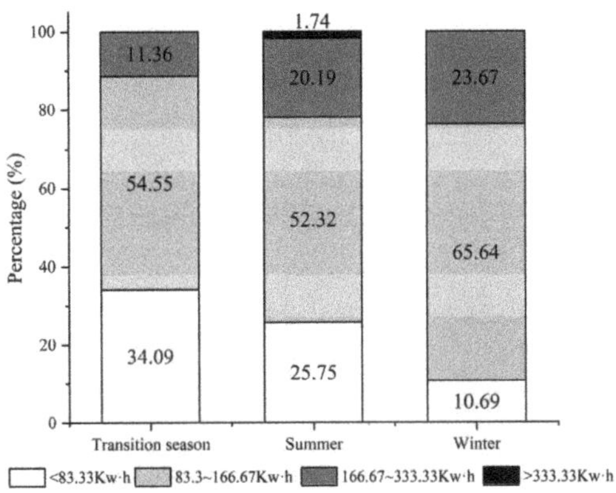

Figure 2. The percentage of monthly mean energy consumption in different seasons.

A two-step cluster can subdivide occupant energy-consumption-related behaviors into multiple clusters to characterize energy use [47,48]. This is a commonly used method in field studies to subdivide a set of observations into subsets where identical clusters have high similarity [49]. Therefore, a two-step cluster analysis was used to cluster the dormitories in this study.

The result reveals that when the dormitories are divided into three groups, the differences in the dormitories between different clusters are apparent, and the energy use features of dormitories within the same cluster are similar. The results of the cluster analysis are shown in Table 1. The three clusters are thrifty group (Cluster A), moderate group (Cluster B), and luxury group (Cluster C). The usage time of air conditioners for Cluster

A is less than 5 h a day in hot and humid summers, and air conditioners are rarely used for heating in winter. In terms of Cluster B, the air conditioner is operated in summer and winter for approximately 5–8 h per day, and the duration of light use is longer than that in Cluster A. While Cluster C relies heavily on air conditioning, the usage time of air conditioner may be up to 8 h and 12 h per day in winter and summer, respectively. The energy consumption of Cluster C is 38.67% and 88.48% higher than that of Cluster A in summer and winter, respectively. This finding is similar to the results of a study carried out in Shanghai [46], which demonstrated that the three clusters vary greatly in terms of the usage time of air conditioners. Different types of dormitories exhibit different electricity usage features in different seasons. Therefore, it is necessary to conduct a cluster analysis to establish the energy usage patterns for different clusters.

Table 1. Results of the cluster analysis of student dormitories.

Cluster	Season	Energy Consumption (kw·h/Month)	Cases/Proportion (%)
A	Summer Transition season Winter	28.93 16.78 22.40	106/27.32
B	Summer Transition season Winter	34.57 27.63 36.63	142/36.60
C	Summer Transition season Winter	40.03 32.33 42.22	140/36.08

4. Case Study to Simulate Energy Use Behaviors in Student Residences

Southwest Jiaotong University (in Chengdu city), located in western China, was selected as the study object. According to the Chinese Building Thermal Design Code [50], Chengdu belongs to the hot summer and cold winter zone. An average dry bulb temperature (values read on a dry-bulb temperature gauge exposed to air without direct sunlight) of 32.4 °C and 1.3 °C are recorded for the hottest month in summer and the coldest month in winter, respectively. Therefore, the region of Southwest Jiaotong University has a strong demand for building heating and refrigeration. In this section, six typical dormitories of Southwest Jiaotong University were selected to measure the indoor environment and record the state of operation for electrical appliances. According to the results of students' rules of movement and the occurrence pattern of energy use behavior, the energy consumption model of the dormitory was built using NETLOGO software (version 6.2.0). NETLOGO is a system engineering software for simulating complex systems that change with time. It can simulate the behavior of the agents in a chronological order and the rules of operation. The agents interact with each other and the environment, thus contributing to the progress of the whole system [51]. In this study, NETLOGO performed agents' energy use behavior and ultimately exported the operation status of different appliances with time.

4.1. Data Collection

A data collection method of using smart devices and a paper-based survey was used to acquire actual environment parameters and behavior records [36]. We chose two dormitories of each cluster for comprehensiveness, and gender was one of the considerations in this study. Finally, six typical dormitories were selected as the objects for the preliminary survey.

To analyze the variability of energy use behavior on weekends and weekdays, we conducted a total of six experiments in July, October, and December, respectively. Smart devices (e.g., hygrothermographs and illuminance meters) were placed in dormitories to record environmental parameters, i.e., the indoor temperature (°C) and illuminance (lux). For occupant behavior information, a paper-based survey sheet was distributed to the dormitory occupants. The survey sheet included 1 h consecutively from 0:00 to 24:00

in each row and the number of students in the dormitory and the open/close statuses for air conditioners, lights, and computers. The occupants were asked to complete the survey sheet manually. Additionally, they need to make an "x" in the corresponding box whenever he/she conducts a behavior at a certain time. Based on the regression analysis, the probability functions for the occurrence of different types of behaviors for summer, transition season, and winter on weekends and weekdays were constructed. The validation of the model was conducted by comparing the energy use behavior and energy consumption between the observed data and the simulation results.

4.2. Construction of the Physical Model

The "patch" and "turtle" were created in NETLOGO representing the dormitory and the interaction agent, respectively. The interactive agents were the students and appliances in the model. The student staying in the room would generate data according to the occurrence rules and interaction with the appliances by updating the operation status in the simulation.

According to the results of the surveys and the literature, three types of basic equipment including lights, air conditioners, and computers, which account for about 85–90% of the total dormitory energy consumption, were considered in this study [52,53]. It is worth noting that air conditioners and computers, connected to the power supply without running, will also generate a certain amount of power, i.e., standby power. Lights do not consume energy when they are not running. The standby power of an air conditioner is 2.8 W, which was used in the energy consumption simulation. The dimensions of the dormitory floor plan are 3.6 m × 7.2 m, and several visualized electrical components were created based on the sample dormitory. The nameplate power demands of the air conditioner, light, and computer is 2000 W, 40 W, and 50 W, respectively. The numbers of air conditioners, lights, and computers for a typical university dorm are 1, 2, and 4, respectively.

4.3. Input in Simulation

4.3.1. The Rules for Students Staying in a Dormitory

During weekdays, the probability of occupancy is affected by the curriculum. As shown in Figure 3, almost all students are in the dormitory between 0:00 and 8:00. However, there is a significant decrease in the occupancy rate due to the curriculum between 9:00 and 12:00. The students return to dormitories for breaks after lunch. The occupancy rates are the lowest from 17:00 to 19:00 as most students choose to go out for dinner. After that, the students gradually return to the dormitory. It is worth noting that students are likely to go out in the evening, especially in summer. The occupancy rate is only 50% in summer, even at 22:00. Students would spend a long period of time going out for socialization due to the cooler evening. Figure 4 shows the result of occupant rates on weekends. Students spend a significant amount of time in the dormitory on weekends studying or playing games, and the occupancy rate briefly decreases at lunch and dinner time. The probability of student activity in the room is shown in Table 2. The activity is related to the schedule. For example, the student is sleeping between 0: 00 and 7:00, the greatest probability of the student activity in this period of time is lying down. It is obvious that students are likely to sit in the daytime, and they seldom stand for communication.

Table 2. The probability of activity in the room.

Action	0:00–7:00	7:00–10:00	10:00–13:00	13:00–16:00	16:00–21:00	21:00–24:00
Lie	0.93	0.60	0.23	0.40	0.10	0.65
Sit	0.07	0.30	0.70	0.50	0.77	0.20
Stand	0.00	0.10	0.07	0.10	0.13	0.15

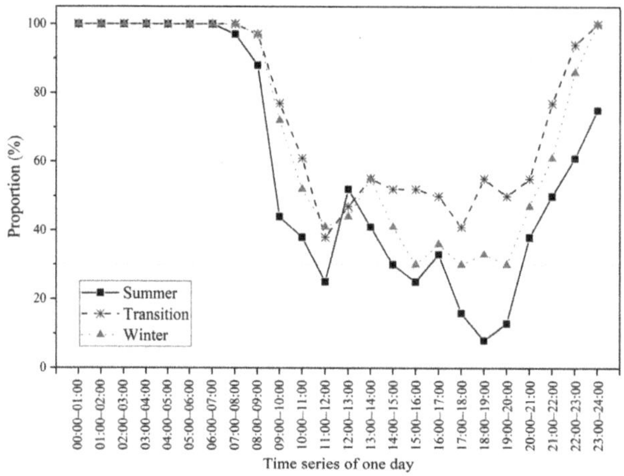

Figure 3. Proportion of students who stayed in dormitories on weekdays.

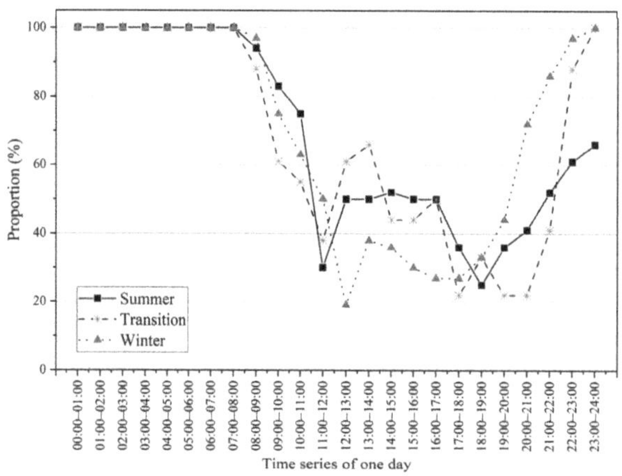

Figure 4. Proportion of students who stayed in dormitories on weekends.

There are four occupants in the simulated dormitory. The simulation process for occupants in the dormitory depends on Section 4.2. The hour-by-hour probabilities of students being in the room are shown in Figures 3 and 4. NETLOGO stochastically simulates the state of occupants based on the proportion of students who stayed in dormitories on weekdays or weekends during the summer, transition season, and winter. The duration of simulation is one year with 1 h time step size.

4.3.2. The Occurrence Rules of Energy-Consuming Behaviors

According to the survey, the air conditioners were rarely used for Cluster A (thrifty group), even in summer or winter. Therefore, the air conditioning usage patterns of Cluster B (moderate group) and Cluster C (luxury group) are summarized in this section. In terms of Cluster A, the usage patterns in NETLOGO software were "part-time" and "part-space".

The usage patterns of air conditioners are listed in Table 3. The control logic for turning on air conditioners was environment-related, while the logic for switching off air conditioners was event-related. The indoor temperature at which the air conditioner switched on was recorded, and the fitting was conducted to observe the relationship between the probability and indoor temperature. For the probability function, the R^2 of over 0.9 was achieved. The corresponding probability functions can be used to simulate different clusters of students switching on their air conditioners in different seasons. For example, students tended to switch off the appliances when the last one leaves the room, which was a key indicator leading to switching off the air conditioners. Similarly, the lighting behavior use was monitored in sample dormitories. In terms of the illumination appliance, the probability functions of switching on the lighting were fitted (Table 4). The control logic for switching off the lighting was event-related, which was the occurrence of the last one leaving the room or sleeping.

Table 3. The usage patterns of air conditioners.

Action	Type	Cluster	Probability Function
Switch on	Environment-related	Cluster C	$P = 0.1555x - 3.765$ (summer), $R^2 = 0.97$ $P = 0.009x^2 - 0.37x + 3.879$ (winter), $R^2 = 0.92$
		Cluster B	$P = 0.0185x^2 - 0.92x + 11.422$ (summer), $R^2 = 0.95$ $P = -0.0004x^2 - 0.005x + 0.242$ (winter), $R^2 = 0.92$
Switch off	Event-related	/	$P = \begin{cases} 1, & \textit{the last one leaving the room} \\ 0, & \textit{others} \end{cases}$
x: indoor temperature			

Table 4. The usage patterns of illumination appliances.

Action	Type	Probability Function
Switch on	Environment-related	$P = -0.22\ln(x) + 1.17$ (summer), $R^2 = 0.99$ $P = -0.02x + 0.95$ (transition season), $R^2 = 0.97$ $P = -0.02x + 0.68$ (winter), $R^2 = 0.94$
Switch off	Event-related	$P = \begin{cases} 1, & \textit{the last one leaving the room or sleeping} \\ 0, & \textit{others} \end{cases}$
x: indoor illumination		

According to the measurements, the usage patterns of computers do not show significant differences among different seasons, but differences are seen between weekdays and weekends (as shown in Table 5). To be specific, the probability of students operating the computer is much greater on weekends than on weekdays. Students are more likely to operate the computer in the afternoon or at night due to their living habits at the university. In terms of operation duration (as shown in Table 6), the average duration of computer use on weekdays is longer than that on weekends, as they need to do homework for a long time.

Table 5. The probability of computer switching on.

Action	Date	Probability (%)			
		0:00–6:00	6:00–12:00	12:00–18:00	18:00–24:00
Switch on	Weekdays	4.0	14.3	25.0	28.2
	Weekends	3.8	16.3	38.8	35.8

Table 6. The probability of computer operation duration.

Indicator	Date	Probability (%)				
		<2 h	2–5 h	5–8 h	8–12 h	12–18 h
Duration	Weekdays	43.2	25.0	17.3	9.7	3.8
	Weekends	47.5	27.0	15.8	7.9	1.8

It should be noted that one of the prerequisites for appliance control is the presence of students in the room, so the indoor rate is one of the key indicators affecting electrical control in a dormitory. For air conditioning and lighting, the prerequisite for controlling the appliance is at least one person in the dormitory. As for computers, in addition to the number of occupants in the dormitory, it is necessary to take time into consideration before controlling the appliance. In addition to the fact that the control logic of the energy-using device is affected by the temperature and light environment, the last student would turn off the electrical switch before leaving the dormitory. Therefore, the appliances are switched off regardless of changes in the indoor environment when there is no one in the dormitory.

It is necessary to consider other people's opinions when controlling public equipment, such as in a collection of college dormitories, which is expressed in the form of the majority rule. For the simulation, two students agree to regulate the appliances at least when the number of indoor students is three or four. One student agrees to regulate the appliances at least when the number of indoor students is two. The appliances are controlled by his/her own will when he/she is the only one in the dormitory [54]. The probability of controlling logic is shown in Table 7.

Table 7. Probability of an action occurring with different numbers of occupants in dormitory.

No.	Probability of Agreement	Probability of Disagreement
1	$P_{ag.} = p$	$P_{disag.} = 1 - p$
2	$P_{ag.} = C_2^1 \times p \times (1-p) + p^2$	$P_{disag.} = (1-p)^2$
3	$P_{ag.} = C_3^1 \times p^2 \times (1-p) + p^3$	$P_{disag.} = C_3^2 \times (1-p)^2 \times p + (1-p)^3$
4	$P_{ag.} = C_4^1 \times p^3 \times (1-p) + C_4^2 \times (1-p)^2 \times p^2 + p^4$	$P_{disag.} = C_4^1 \times (1-p)^3 \times p + (1-p)^4$

The simulation process is determined by a combination of the indoor rate described in Section 4.3.1, and the energy control behavior is described in Section 4.3.2. In this study, the duration is one year with a total of 8760 h, and the time step size is one hour. The indoor environmental variables (i.e., indoor temperature and indoor illumination) should be initialized. Autodesk Ecotect Analysis v5.2 software is used for the building thermal environment analysis (e.g., time-to-time temperature analysis of buildings) and building light environment analysis [55]. The software focuses on sustainable building design and analysis and is widely used in similar studies [56]. Therefore, Autodesk Ecotect Analysis software is selected to simulate the indoor environment for one year, and the values of indoor temperature and indoor illuminance of the dormitory are obtained. During the simulation, the default indoor temperature and indoor illuminance for each time step are the output values of the ECOTECT. The indoor environment values change to the standard values set in NETLOGO as soon as the appliances are switched on and, then, revert to the output values of the ECOTECT software (version 5.2) when the devices are switched off.

4.4. Simulation Results and Model Validation

The NETLOGO software allows visual simulations to demonstrate the state of the four students in the dormitory and their energy use behavior with human icons of different colors representing each of the four students in the dormitory. As shown in Figure 5, NETLOGO can perform the presence of students in the dormitory at different periods with a certain amount of randomization. There is a total of 2190 h per season in NETLOGO, of which 1564 h are used for weekdays and 626 h are used for weekends. Figures 6 and 7 show the hour-by-hour cumulative time of students in the dormitory on weekdays and weekends, respectively. The results show that the simulated situation is highly consistent with the actual situation (Figures 3 and 4). It is worth noting that the cumulative time between spring and autumn also exhibits a slight difference. The difference reflects the fact that NETLOGO software can well simulate the randomness and autonomy of the person's behavior.

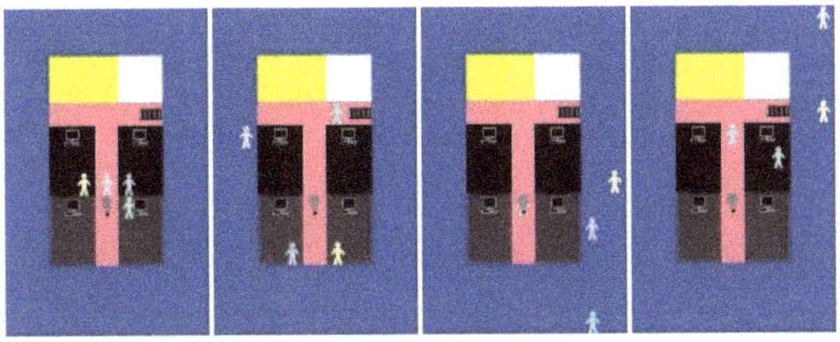

Figure 5. Visualization of behavior changes in dormitory.

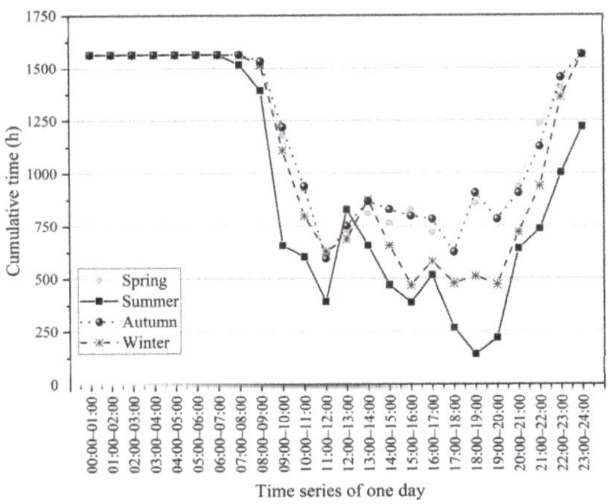

Figure 6. Cumulative time of students who stayed in dormitories on weekdays.

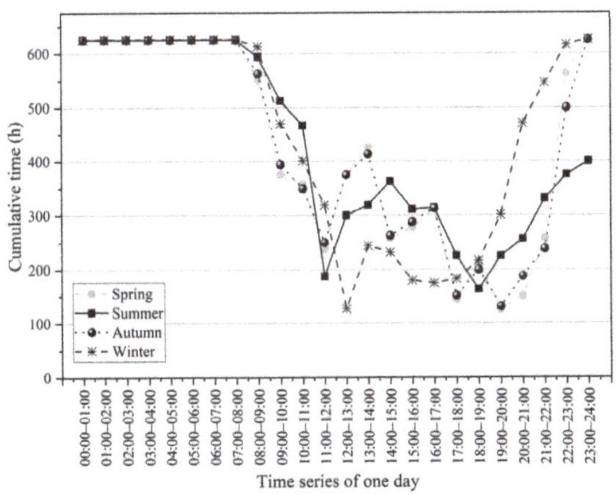

Figure 7. Cumulative time of students who stayed in dormitories on weekends.

Taking a student as an example, a sequence of behaviors in the dormitory covering the whole year is generated by NETLOGO. Figure 8 shows a sequence of behaviors within two days, where the abscissa denotes the hourly time points, and the ordinate lists the typical activities of the student, while the bars indicate the starting and ending time, which represent the duration of the activities. The simulation results show that all activities can be filled for 48 h. For the activities, the time of action and the duration are analogous to the survey results. Thus, the applicability of the method, describing and modeling the indoor behavior of students, has been further validated.

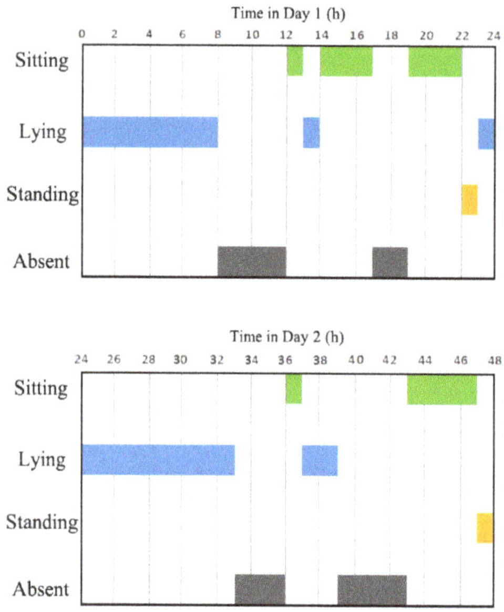

Figure 8. A sequence of activity simulated in two days.

The accuracy between the simulation situation and the actual results should be further checked. Table 8 shows the duration of air conditioners by simulation. The difference in cumulative operation time and daily operation time for air conditioners among seasons is quite significant, while there is less variability in computers and lights. In terms of Cluster A, the cumulative operation time of air conditioners over the year is 53 h, accounting for only 0.61%. This is similar to the actual situation. The air conditioning was rarely used during the measurements, while a small amount of air conditioning operation time occurred during the simulation, due to the fact that the measurements were taken on a typical weather day, while the air conditioning was still operated under extremely cold and hot conditions. The cumulative operation time of air conditioners for Cluster B is 705.5 h, with 316 h for refrigeration and 389.5 h for heating. However, the number of days for heating is greater than that for refrigeration, resulting in the daily operation time being greater in the summer than in the winter. The result of simulation situation is in line with the measured results. The cumulative operation time of air conditioners for Cluster C is 64% greater than that for Cluster B. According to the statistics from NETLOGO output, the cumulative operation time of lights and computers is 2974 h and 968.75 h, respectively. These data will form the basis of the energy consumption of the dormitory throughout the year.

Table 8. The duration of air conditioners by simulation.

Cluster	Cumulative Operation Time (h)/Daily Operation Time (h)	
	Refrigeration	Heating
Cluster A	27.0/3.86	26.0/1.63
Cluster B	316/5.10	389.5/2.89
Cluster C	638/5.85	519.5/4.64

Figure 9 shows the energy consumption of different clusters in three seasons, calculated by the cumulative operation time and the rated power of the corresponding appliances. The energy consumption in summer and winter is significantly greater than that in the transition season, especially for air conditioners. It indicates that the indoor environment in Chengdu is much poorer in summer and winter. The building size of the simulated dormitory is 7.2 m × 3.6 m, with a floor area of 25.92 m^2. The average annual energy consumption is 15.27, 40.67, and 66.70 kWh·m^{-2} for Cluster A, B, and C, respectively. The actual annual energy consumption, obtained by multiple measurements, is 16.54, 42.63, and 62.75 kWh·m^{-2} for Cluster A, B, and C, respectively. The difference between the simulation results and actual situation is very small, which proves the validity of model.

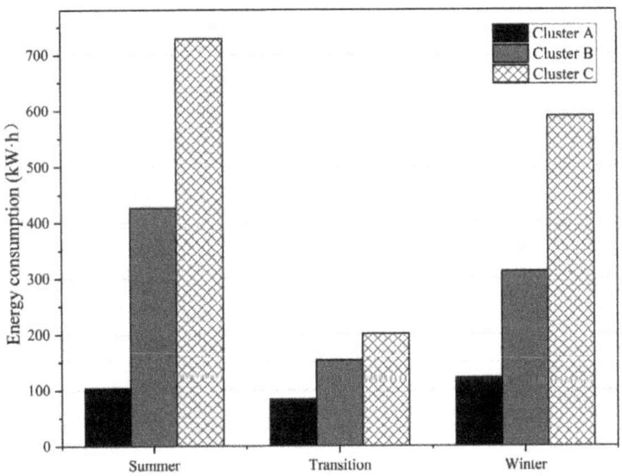

Figure 9. Energy consumption of dormitory by simulation.

5. Scenario Analysis of Energy-Saving Strategies for Dormitory Units

In order to demonstrate the impact of energy use behavior on energy consumption, a typical dormitory is chosen as the baseline for the comparative analysis to illustrate an energy-friendly design proposal alternative to the status quo. The existing dormitory is shown in Figure 10a. The simulation results and actual measurements demonstrate the dormitory is difficult to meet the needs of daily life and study, especially in summer and winter. Therefore, appliances are used to improve the indoor environment, which result in great energy consumption. The existing dormitories exhibit several defects: (1) poor indoor ventilation and high indoor temperature result in great energy consumption for refrigeration in summer. (2) Excessive indoor heat is lost due to oversized north-facing balconies, which leads to great energy consumption for heating. (3) The light intensity of the room is not sufficient to meet the needs of study. Effective lighting area of the dormitory is insufficient and the sunlight is not fully utilized.

A hypothetical new design with the same function is proposed as an alternative for this precinct. As shown in Figure 10b, the proposed south-facing dormitory connects the balconies of the neighboring dormitories to enlarge the wind-absorbing surface, which can promote the effect of lighting in winter. On the other hand, it can enhance indoor ventilation in summer, which is beneficial for removing excessive heat from the dormitory. In terms of the south-facing dormitory shown in Figure 10c, measures are taken to merge the balconies of the neighboring dormitories and appropriately reduce the area of the merged balconies to reduce indoor heat loss in winter, due to the small amount of light available for north-facing dormitories and the prevalence of cold northerly winds in Chengdu during winter.

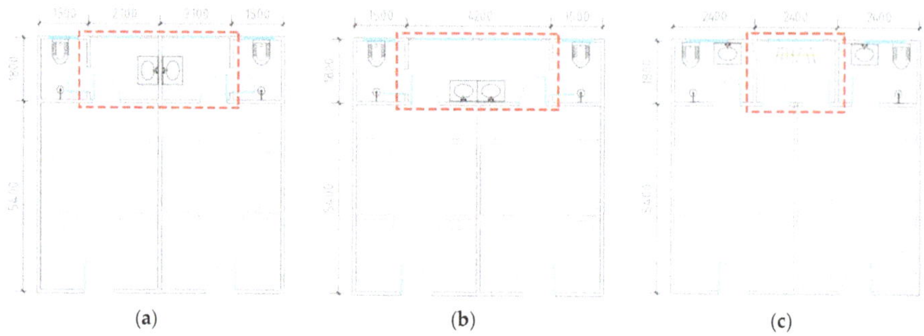

Figure 10. Dormitory layout plan of (**a**) the existing dormitory; (**b**) the proposed south-facing dormitory; (**c**) the proposed north-facing dormitory.

Figure 11 shows the comparison of daily average dormitory energy consumption in summer and winter for the existing and proposed dormitories. The results show that the average energy consumption of the proposed dormitory unit for Cluster C reduces 1.35 kW·h/day, realizing the cost saving by 246.38 yuan/year in summer. In winter, the energy consumption of proposed dormitory unit can save 1.61 kW·h/day and reduce costs by 293.83 yuan/year. The energy consumption for the proposed design is 16.61% and 23.83% less than that of existing dormitories in summer and winter, respectively. In terms of Cluster B, the energy consumption for the proposed design is 8.97% and 11.86% less than that of existing dormitories in summer and winter, respectively. In general, the energy consumption of the proposed design is 14.24% and 17.85% less than that of the existing dormitory in summer and winter, respectively. It demonstrates that the proposed energy-friendly design exhibits great energy-saving potential, especially for Cluster C. The cost reduction for the proposed dormitory unit is up to 310.13 yuan/year and 383.25 yuan/year in summer and winter, respectively. The energy-saving potential for winter is greater than summer. Therefore, optimizing the architectural design of university dormitories can effectively reduce energy consumption.

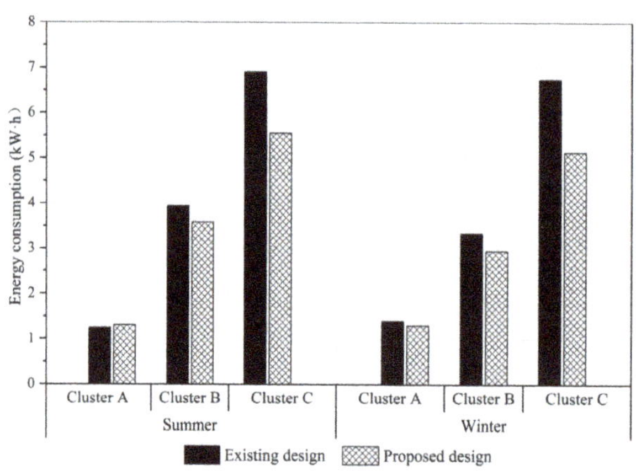

Figure 11. Daily average energy consumption of dormitory for the existing and proposed design.

6. Conclusions

Occupant energy use behavior plays a significant role in predicting building energy consumption. Current building energy simulations usually assume that occupants use electrical equipment (e.g., air conditioner, light) according to the deterministic schedule or environmental values. However, occupants' behavior is uncertain and varies with the surrounding environment and other occupants, especially for college students. Universities are characterized by high densities of occupants and students from different regions and cities, which result in complex energy use behaviors. Taking university dormitories in Sichuan, China as an example, this study established an agent-based model to explore the relationship between university occupant behaviors and dormitory energy consumption.

Firstly, the student residences are subdivided into three clusters by a two-step cluster analysis method to characterize the features of appliance use behavior. Secondly, a field survey of measuring the indoor environmental data (e.g., indoor temperature and illumination) and recording the energy use behavior is conducted on typical climate days. The control logic of energy use behavior is derived using statistical analysis and logistic regression. Then, a simulation model of energy consumption for college dormitories based on energy use behavior is established. The simulation is conducted on NETLOGO software, and the control logic is the occurrence of various energy use behaviors. The validation results show that the model can be used to simulate the university residential energy use behaviors with sufficient simulation accuracy. Finally, a hypothetical new design is proposed as an alternative for this dormitory unit. The simulation results show the energy consumption of the proposed design is 14.24% and 17.85% less than that of the existing dormitory in summer and winter, respectively. The retrofitting of university dormitories exhibits great potential for energy saving, and it is important for reducing energy consumption and promoting low-carbon sustainable development. In addition, building operators can develop more scientific and reasonable energy supply programs based on this study on student residential energy use behaviors. The method of this research is expected to provide a basis for decision-making and theoretical guidance for energy management policies.

This study has three limitations that must be noted for future studies. Firstly, the energy use behavior model proposed in this study is more representative for Chengdu. However, logistic regression coefficients and behavioral models may differ among different climate zones. Therefore, while considering behavioral models for building energy simulation, buildings in other cities should be carefully considered. The method and the logic of study can serve as a reference for readers. Secondly, the time step in this study is 1 h, which is usually used in energy consumption simulations. Although the simulation results indicate that the model had sufficient accuracy, a more accurate model is needed for other purposes, i.e., the step time of 15 min and even 1 min. Thirdly, the study does not separate the data into training and test sets for the development and validation of ABM. Assumptions on occupant energy use behavior was only based on survey data. The model may not allow generalizing to future data or data from different sources. Cross-validation or additional test sets should be used to improve the generalization performance of the model in future studies. In the next development, we will collect a wider sample size and conduct the simulation multiple times to build a more extensive and accurate occupant behavior model for energy conservation and carbon reduction.

Author Contributions: Conceptualization, Y.Z.; methodology, B.Z. and J.H.; software, B.Z. and J.H.; formal analysis, B.Z. and J.H.; investigation, Y.Z. and J.H; resources, Y.Z. and J.H.; data curation, B.Z. and J.H.; writing—original draft preparation, B.Z. and J.H.; writing—review and editing, Y.Z.; visualization, B.Z. and J.H.; supervision, Y.Z.; project administration, Y.Z.; funding acquisition, Y.Z. All authors have read and agreed to the published version of the manuscript.

Funding: This research was funded by Chengdu Science and Technology Bureau, grant number 2020-YF05-00595-SN and Key Technologies Research and Development Program, grant number 2022YFC3802702.

Data Availability Statement: Data are contained within the article.

Conflicts of Interest: The authors declare no conflicts of interest.

References

1. Yan, D.; Hong, T.; Dong, B.; Mahdavi, A.; Gaetani, I.; Feng, X. IEAEBC Annex 66: Definition and simulation of occupant behavior in buildings. *Energy Build.* **2017**, *156*, 258–270. [CrossRef]
2. Liu, Q.; Ren, J. Research on the building energy efficiency design strategy of Chinese universities based on green performance analysis. *Energy Build.* **2020**, *224*, 110242. [CrossRef]
3. Zhang, Y.; Bai, X.; Mills, F.; Pezzey, J. Rethinking the role of occupant behavior in building energy performance: A review. *Energy Build.* **2018**, *172*, 279–294. [CrossRef]
4. Salimi, S.; Liu, Z.; Hammad, A. Occupancy prediction model for open-plan offices using real-time location system and inhomogeneous Markov chain. *Build. Environ.* **2019**, *152*, 1–16. [CrossRef]
5. Zhu, S.; Tan, H.; Chen, S.; Zhi, Z. Situation and analysis of green campus culture construction in China. *Build. Energy Effic.* **2014**, *42*, 95–99.
6. Crawley, D.; Lawrie, L.; Winkelmann, F.; Buhl, W.; Huang, Y.; Pedersen, C.; Strand, R.; Liesen, R.; Fisher, D.; Witte, M.; et al. EnergyPlus: Creating a newgeneration building energy simulation program. *Energy Build.* **2001**, *33*, 319–331. [CrossRef]
7. Poirazis, H.; Blomsterberg, A.; Wall, M. Energy Simulations for Glazed Office Buildings in Sweden. *Energy Build.* **2008**, *40*, 1161–1170. [CrossRef]
8. Li, S.; Wang, Y. Summarization of present building energy consumption and corresponding strategies in China. *Environ. Sci. Manag.* **2008**, *33*, 6–9.
9. Qian, J.; Qi, L.; Zhang, C.; Wang, J.; Dou, Z.; Chen, A.; Yang, Y.; Ren, H.; Zhang, L. Excavation of building energy conservation in university based on energy use behavior analysis. *Energy Build.* **2023**, *280*, 112726.
10. Balvedi, B.; Ghisi, E.; Lamberts, R. A review of occupant behaviour in residential buildings. *Energy Build.* **2018**, *174*, 495–505. [CrossRef]
11. Pisello, A.; Asdrubali, F. Human-based energy retrofits in residential buildings: A cost-effective alternative to traditional physical strategies. *Appl. Energy* **2014**, *133*, 224–235. [CrossRef]
12. Hong, T.; D'Oca, S.; Turner, W.; Sarah, T. An ontology to represent energy-related occupant behavior in buildings. Part I: Introduction to the DNAs framework. *Build. Environ.* **2015**, *92*, 764–777. [CrossRef]
13. Chen, S.; Wu, J.; Pan, Y. Simulation and case study on residential stochastic energy use behaviors based on human dynamics. *Energy Build.* **2020**, *223*, 110182. [CrossRef]
14. Richardson, I.; Thomson, M.; David, I. A high-resolution domestic building occupancy model for energy demand simulations. *Energy Build.* **2008**, *40*, 1560–1566. [CrossRef]
15. Macdonald, I.; Strachan, P. Practical application of uncertainty analysis. *Energy Build.* **2001**, *33*, 219–227. [CrossRef]
16. Zhang, C.; Ma, L.; Han, X.; Zhao, T. Reconstituted data-driven air conditioning energy consumption prediction system employing occupant-orientated probability model as input and swarm intelligence optimization algorithms. *Energy* **2024**, *288*, 129799. [CrossRef]
17. Wang, C. Simulating Research on Occupant Energy-Related Behaviors in Building. Ph.D. Thesis, Tsinghua University, Beijing China, 2014.
18. Ebuy, H.T.; Bril El Haouzi, H.; Benelmir, R.; Pannequin, R. Occupant Behavior Impact on Building Sustainability Performance: A Literature Review. *Sustainability* **2023**, *15*, 2440. [CrossRef]
19. Jun, L.; Zhun, Y.; Fariborz, H.; Zhang, G. Development and improvement of occupant behavior models towards realistic building performance simulation: A review. *Sustain. Cities Soc.* **2019**, *50*, 101685.
20. Yan, D.; Brien, W.; Hong, T.; Feng, X.; Gunay, H.; Tahmsaebi, F.; Mahdavi, A. Occupant behavior modeling for building performance simulation: Current state and future challenges. *Energy Build.* **2015**, *107*, 264–278. [CrossRef]
21. Lou, S.; Lin, Z.; Zou, Y.; Xia, D.; Huang, Y.; Li, Z.; Gu, Z. Investigation on occupant presence and appliance operation schedules for university campus in south China sub-tropical area. *Build. Sim.* **2024**, *17*, 301–318. [CrossRef]
22. Liu, H.; Sun, H.; Mo, H.; Liu, J. Analysis and modeling of air conditioner usage behavior in residential buildings using monitoring data during hot and humid season. *Energy Build.* **2021**, *250*, 111297. [CrossRef]
23. Xia, D.; Lou, S.; Huang, Y.; Zhao, T. A study on occupant behaviour related to airconditioning usage in residential buildings. *Energy Build.* **2019**, *203*, 109446. [CrossRef]
24. Nicol, F.; Wilson, M.; Chiancarella, C. Using field measurements of desktop illuminance in European offices to investigate its dependence on outdoor conditions and its effect on occupant satisfaction, and the use of lights and blinds. *Energy Build.* **2006**, *38*, 802–813. [CrossRef]
25. Richardson, I.; Thomson, M.; Infield, D.; Delahunty, A. Domestic lighting: A high-resolution energy demand model. *Energy Build.* **2009**, *41*, 781–789. [CrossRef]
26. Jukka, V.; Lund, P. A model for generating household electricity load profiles. *Int. J. Energy Res.* **2006**, *30*, 273–290.
27. Jin, Y.; Yan, D.; Sun, H. Household Appliance Behavior Model Based on Statistical Method. *Build. Sci.* **2018**, *34*, 91–102.
28. Yamaguchi, Y.; Shimoda, Y. Validation of an Energy Demand Model of Residential Buildings. In Proceedings of the 2nd Asia Conference on International Building Performance Simulation Association, Nagoya, Japan, 28–29 November 2014.
29. Jian, Y. Modelling and simulating occupant behaviour on air conditioning in residential buildings. *Energy Build.* **2018**, *175*, 1–10.
30. Ren, X.; Yan, D.; Wang, C. Air-conditioning usage conditional probability model for residential buildings. *Build. Environ.* **2014**, *81*, 172–182. [CrossRef]

31. Zhang, Y.; Luo, X.; Sun, Y.; Liu, X. Impact of air-conditioning usage behavior of student groups on air-conditioning load simulation in university dormitories. *Energy Build.* **2023**, *303*, 113765. [CrossRef]
32. Liao, C.; Barooah, P. An integrated approach to occupancy modeling and estimation in commercial buildings. In Proceedings of the 2010 American Control Conference (ACC), Baltimore, MD, USA, 30 June–2 July 2010.
33. Ding, Z.; Hu, T.; Li, M.; Xu, X.; Zou, P. Agent-based model for simulating building energy management in student residences. *Energy Build.* **2019**, *198*, 11–27. [CrossRef]
34. Azar, E.; Menassa, C. Agent-based modeling of occupants and their impact on energy use in commercial buildings. *J. Comput. Civ. Eng.* **2012**, *26*, 506–518. [CrossRef]
35. Kashif, A.; Dugdale, J.; Ploix, S. Simulating Occupants' behavior For Energy Waste Reduction in Dwellings: A Multiagent Methodology. *Adv. Complex Syst.* **2013**, *16*, 1350022. [CrossRef]
36. Jia, M.; Srinivasan, R.; Ries, R.; Weyer, N.; Bharathy, G. A systematic development and validation approach to a novel agent-based modeling of occupant behaviors in commercial buildings. *Energy Build.* **2019**, *199*, 352–367. [CrossRef]
37. Jia, M.; Srinivasan, R.; Ries, R.J.; Bharathy, G.; Weyer, N. Investigating the Impact of Actual and Modeled Occupant Behavior Information Input to Building Performance Simulation. *Buildings* **2021**, *11*, 32. [CrossRef]
38. Ouf, M.; Brien, W.; Gunay, H. Optimization of electricity use in office buildings under occupant uncertainty. *J. Build. Perform. Simul.* **2020**, *13*, 13–25. [CrossRef]
39. Duarte, C.; Budwig, R.; Van, K. Energy and demand implication of using recommended practice occupancy diversity factors compared to real occupancy data in whole building energy simulation. *J. Build. Perform. Simul.* **2015**, *8*, 408–423. [CrossRef]
40. Ahmed, O.; Sezer, N.; Mohamed, O.; Wang, L.; Hassan, I. State-of-the-art review of occupant behavior modeling and implementation in building performance simulation. *Renew. Sustain. Energy Rev.* **2023**, *185*, 113558. [CrossRef]
41. Santangelo, A.; Yan, D.; Feng, X. Renovation strategies for the Italian public housing stock: Applying building energy simulation and occupant behaviour modelling to support decision-making process. *Energy Build.* **2018**, *167*, 269–280. [CrossRef]
42. Jiang, H.; Yao, R.; Han, S.; Du, C.; Li, B. How do urban residents use energy for winter heating at home? A large-scale survey in the hot summer and cold winter climate zone in the Yangtze River region. *Energy Build.* **2020**, *223*, 110131. [CrossRef]
43. Zheng, X.; Zhong, T.; Liu, M. Modeling crowd evacuation of a building based on seven methodological approaches. *Build. Environ.* **2009**, *44*, 437–445. [CrossRef]
44. Clevenger, C.; Haymaker, J. The impact of the occupant on building energy simulations. In Proceedings of the Joint International Conference on Computing and Decision Making in Civil and Building Engineering, Montreal, QC, Canada, 14–16 June 2006.
45. Hoes, P.; Hensen, J.; Loomans, M.; Vries, B.; Bourgeois, D. User behavior in whole building simulation. *Energy Build.* **2009**, *4*, 295–302. [CrossRef]
46. Chen, S.; Du, B.; Liu, S.; Xu, F.; Lu, Y.; Xing, W. Statistical Analysis on the Characteristics of Energy Use Behavior in Students'Dormitories in Universities. *Build. Sci.* **2015**, *31*, 133–139.
47. Xu, J.; Kang, X.; Chen, Z.; Yan, D.; Guo, S.; Jin, Y.; Hao, T.; Jia, R. Clustering-based probability distribution model for monthly residential building electricity consumption analysis. *Build. Simul.* **2021**, *14*, 149–164. [CrossRef]
48. Yao, R.; Costanzo, V.; Li, X.; Zhang, Q.; Li, B. The effect of passive measures on thermal comfort and energy conservation. A case study of the hot summer and cold winter climate in the Yangtze River region. *J. Build. Eng.* **2018**, *15*, 298–310. [CrossRef]
49. An, J.; Yan, D.; Hong, T. Clustering and statistical analyses of air-conditioning intensity and use patterns in residential buildings. *Energy Build.* **2018**, *174*, 214–227. [CrossRef]
50. *GB 50176-2016*; Code for Thermal Design of Civil Building. China Architecture & Building Press: Beijing, China, 2016.
51. NetLogo User Manual Version 6.4.0. 2023. Available online: https://ccl.northwestern.edu/netlogo/docs/ (accessed on 30 January 2024).
52. Hu, B.; Yu, Z.; LI, J. Investigation and analysis of standby energy consumption and usage behavior of electrical appliances in university buildings. *Build. Sci.* **2017**, *33*, 55–61.
53. Sun, W. Design and Implementation of University Energy Saving Control and Energy Consumption Monitoring and Management. Master's Thesis, Shandong University, Jinan, China, 2012.
54. Huang, C. High School Study of the Problem of Energy Conservation and Emission Reduction. Master's Thesis, Guangxi Normal University, Guilin, China, 2015.
55. Ecotect Software v.5.2 [Computer Software], Sq. One Res, PTY Ltd. 2014. Available online: http://www.squ1.com (accessed on 15 January 2014).
56. Amani, N. Energy efficiency using the simulation software of atrium thermal environment in residential building: A case study. *Adv. Build. Energy Res.* **2019**, *13*, 65–79. [CrossRef]

buildings

MDPI

Article

Indoor Air Temperature Distribution and Heat Transfer Coefficient for Evaluating Cold Storage of Phase-Change Materials during Night Ventilation

TaeCheol Lee [1], Rihito Sato [2], Takashi Asawa [3] and Seonghwan Yoon [1,*]

[1] Department of Architecture, Pusan National University, 2, Busandaehak-ro 63beon-gil, Geumjeong-gu, Busan 46241, Republic of Korea; leetc@pusan.ac.kr
[2] School of Systems Engineering, Kochi University of Technology, 185 Miyanokuchi, Tosayamada, Kami City 782-8502, Kochi, Japan; sato.rihito@kochi-tech.ac.jp
[3] School of Environment and Society, Tokyo Institute of Technology, 4259 Nagatsuda-cho, Midori-ku, Yokohama 226-8502, Kanagawa, Japan; asawa.t.aa@m.titech.ac.jp
* Correspondence: yoon@pusan.ac.kr; Tel.: +82-51-510-2355

Abstract: This paper focuses on clarifying the heat transfer coefficient necessary for determining the indoor temperature distribution during night ventilation using floor-level windows. Measurements were used to identify the factors that influence the vertical temperature distribution within a room wherein phase-change materials (PCMs) were installed at the floor level. The investigation revealed a temperature differential ranging from 1 °C to a maximum of 3 °C between the floor and the center of the room, attributable to external climatic conditions (outdoor temperature and wind speed). This variation was found to depend on the degree of mixing of indoor air currents. This deviation was critical because it significantly affected the phase-change temperature of PCMs, thereby impacting their thermal storage capabilities. Consequently, this study aimed to refine the predictive accuracy of indoor temperature distributions by proposing a modified vertical temperature distribution model that incorporated these findings. The results of this study are expected to provide better design strategies for building constructions that incorporate PCMs, and to optimize their functionality in passive cooling systems.

Keywords: passive house; indoor air temperature distribution; heat transfer coefficient; night ventilation; phase-change materials

1. Introduction

1.1. Background

The global climate crisis, which is characterized by unprecedented changes in temperature, precipitation patterns, and extreme weather events, is largely attributed to the increased concentration of greenhouse gases in the Earth's atmosphere. This has rendered the goal of net-zero carbon emissions a topic of global interest. According to the 6th comprehensive report by the Intergovernmental Panel on Climate Change (IPCC), the global average temperature is expected to rise by 1.5 °C by 2040 owing to greenhouse gas emissions, which is 10 years earlier than predicted in the previous report [1]. The building sector accounts for approximately 40% of the world's total energy consumption [2]. Santamouris reported that the average cooling energy demands of residential and commercial buildings are expected to increase by 750% and 275%, respectively, by 2050 [3]. These figures highlight the sector's carbon footprint as part of its carbon neutrality goals. Consequently, the architectural industry has been focused on zero-energy buildings and the incorporation of passive designs [4], active techniques, and renewable energy sources.

Passive design aims to enhance indoor thermal comfort and reduce energy consumption in buildings. The passive design principle facilitates the prevention of outdoor-to-indoor heat transfer in summer and indoor-to-outdoor heat transfer in winter. Passive design is targeted at the minimization of indoor heat gain and the maximization of the use of solar radiation for indoor heating without the use of active mechanical systems. Techniques such as shading, natural ventilation, thermal mass, and reflective materials can lower indoor temperatures, enhance occupant comfort, and reduce reliance on energy-intensive air-conditioning systems in the summers.

Research on night ventilation was conducted according to building uses in various climate zones [5]. The indoor temperature reduction effect by night ventilation is significantly affected by the heat capacity of the building, the amount of ventilation, and the change in outdoor temperature [6]. Exizidou et al. [7] found that thermal comfort levels improved by an average of 26% with the use of night ventilation in vernacular buildings. Geros et al. [8] presented the key parameters, which include the effective night-time airflow rate, as well as the indoor–outdoor temperature differential. There-fore, when planning the night ventilation in building, it is important to take the opening area, outdoor air temperature, as well as heat capacity of the building into account. In wooden buildings, PCMs are a good means to gain additional heat capacity.

On the other hand, studies on the application of PCMs to buildings have recently been conducted. Borderon et al. [9] conducted in-situ tests using PCMs combined with ventilation in the crawl space in office buildings. They concluded that overall thermal inertia in the crawl space was increased, leading to better pre-cooling of air in summer. Ling et al. [10] carried out a study using experimental and numerical methods to evaluate the effect of PCMs on the indoor thermal environment of greenhouses. This study demonstrated that sunny weather could help to promote the efficiency of PCMs. Wang et al. [11] evaluated the year-round applicability of a kind of composite PCMs wall through a full scale experiment. The PCMs wall was able to reduce cooling energy and nearly 10–30% in heating energy. Wonorahardjo et al. [12] implemented an experimental study on the influence of a PCM-based passive air conditioner on temperature distribution in the room under tropical climatic conditions. Researchers have attempted to maximize the performance of PCMs under various climactic conditions. Additionally, coupling PCMs with natural or mechanical ventilation, and incorporating them into building design, have been important considerations for achieving optimal effectiveness using PCMs. The surrounding environment of the building is also crucial to fully exploit the benefits of PCMs in buildings.

This study focused on the passive cooling of houses using night ventilation and phase-change materials (PCMs). Figure 1 visually depicts the proposed research methodology. Cold air entering a room through a floor-level window, which is commonly used to promote natural ventilation in Japan, solidifies the PCMs, thereby preventing an increase in the indoor temperature in the morning and reducing the cooling load. Typically, night ventilation occurs through buoyancy. Cold air enters the room from the outside and moves to the upper floors through stairwells [13,14]. Thermal buoyancy creates a vertical temperature gradient in an indoor space. The larger the temperature difference between the upper and lower levels of the indoor area, the greater the thermal discomfort for occupants [15]. Therefore, indoor spaces are typically designed to prevent variations in temperature. However, in the case of an unoccupied space, there is no thermal discomfort. Space usage in residential buildings can vary depending on the time of the day; for example, living rooms may not be used frequently at night. Therefore, living rooms in residential buildings are suitable for night ventilation without compromising the thermal comfort of occupants. Moreover, the incorporation of PCMs into the fabric of rooms ventilated at night promotes thermal mass cooling. Therefore, by utilizing the cold air accumulated in the lower part of the room, the PCM of the indoor floor structure can maximize the cooling effect of night ventilation using floor-level windows.

Considering that indoor thermal stratification occurs owing to night ventilation, and that the PCM is placed on the floor of the room, the air temperature near the floor must be predicted while designing the cold storage of a PCM. The existing methods for predicting the heating and cooling loads of a building, or the prediction of indoor air temperature, conventionally involve calculations using one node per room. However, night ventilation results in a vertical air temperature distribution that is difficult to interpret using a nodal model. Meanwhile, the computational fluid dynamics (CFD) model cannot calculate the annual cooling and heating loads or predict the indoor air temperature for the cold storage of PCMs because of the computational load. The zonal model is an intermediate model between the nodal and CFD models. Figure 2 shows the visual features of the numerical analysis methodologies for the indoor air temperature distribution. Therefore, in this study, we employed a type of zonal model that can analyze the indoor temperature distribution by considering the thermal characteristics.

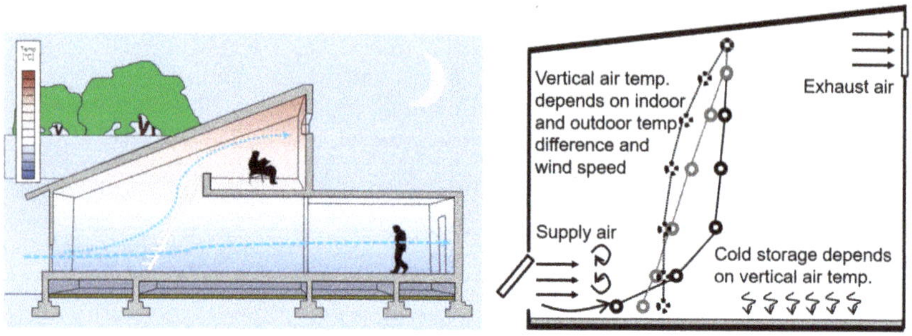

Figure 1. Visual representation of natural night ventilation with phase-change materials (PCMs).

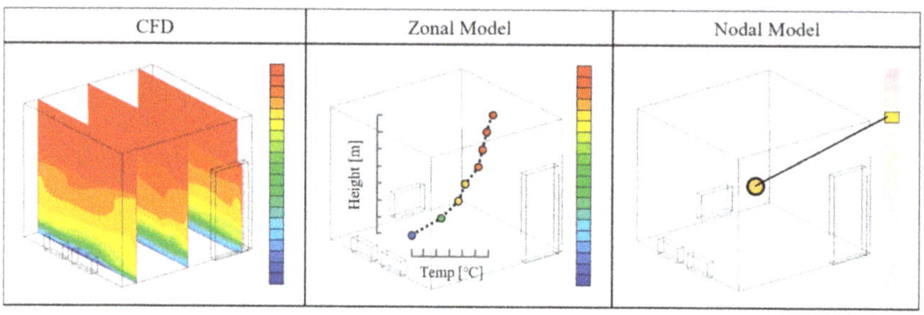

Figure 2. Numerical analysis methodologies for indoor air temperature distribution.

1.2. Previous Research on Indoor Air Temperature Distribution

Ito et al. [16] reported a correlation between the indoor temperature distribution and Archimedes numbers at the intake while heating an air-conditioned room with air intake at the top and exhaust from the floor of the room. The vertical air temperature distribution of the room was predicted using the correlation between the ratio of the height at which the indoor and outdoor air was perfectly mixed with the room height, and the Archimedes number. Togari et al. [17,18] proposed a simple prediction model for describing the vertical air temperature distribution generated by an air conditioner flow when the upper part of the space was warmed by solar radiation and cooled via downflow along the wall. Considering a room with temperature variations, Inard et al. [19] established a model for the determination of indoor air temperature and airflow distribution using mass and heat balance equations between each zone via vertical and horizontal zoning. Wurtz et al. [20] simplified the calculation of natural and mixed convection, and verified two-dimensional and three-dimensional zonal models to study the temperature and airflow distributions in spaces with temperature variations. Haghighat et al. [21] developed and proposed a simplified numerical model called the pressurized zonal model with an air diffuser (POMA) to predict indoor airflow patterns and heat distribution. The jet airflow equation model was based on the balance of the airflow rate, buoyancy force, and pressure loss because of fan and duct resistances. Musy et al. [22,23] improved Inard et al.'s zonal model by expanding it to multiple rooms to predict temperature and airflow distribution in the presence of a heat source. Higashimoto et al. [24] evaluated the predictability of vertical air temperature and pollutant distribution in a large room, based on displacement ventilation using a block model. Ren et al. [25] used Conjunction of Multlzone Infiltration Specialists(COMIS), a building airflow modeling program, to predict indoor temperature distribution and changes in airflow.

Wurtz et al. [26] used a mixed convection model to predict the indoor airflow and temperature distribution in residential buildings with an electric heater, and conducted an annual thermal comfort analysis. Some models [27] examined the effects of temperature stratification, natural convection, and wind-induced ventilation on indoor temperature distribution.

Various studies have proposed models for the analysis of temperature distribution based on the air age of a room where substitution ventilation is performed [28]. A review paper by Chen [29] highlighted the computational load problem of a zonal model.

Megri and Yu [30] developed a POMA model for predicting indoor air temperature distribution based on the differences in indoor surface temperature, and validated the prediction accuracy of the model by comparing its results with the measurement results.

Huang et al. [31] predicted indoor temperature distribution using the Block–Gebhart model (BG model), which combines the Block model with the Gebhart model. They predicted the indoor temperature distribution of a large space with a low-sidewall air supply system. This predictive model includes additional considerations for calculating radiant heat transfer loads, and its validity was verified through measurements. X. Wang et al. [32] used the BG model to predict and evaluate the temperature distribution in a large atrium building.

Considering the computational load and thermal characteristics of the target space in several existing studies, we have chosen the Block model by Togari et al. [17,18] as the analytical model for this research.

1.3. Purpose and Methodology

As previously mentioned, natural ventilation at night, facilitated by floor-level windows, contributes to the distribution of indoor temperatures. This ventilation primarily operates on the basis of temperature differentials, resulting in relatively stable indoor air. Considering the stability and nature of the ventilation process, this study posited that detailed airflow analysis models would not be required for predicting the temperature distribution under these specific conditions.

It was anticipated that the temperature distribution resulting from the disparity between indoor and outdoor temperatures would surpass the inflow wind speed. Thus, our study initially aimed to elucidate the factors influencing indoor vertical temperature distribution during night ventilation through floor-level windows by utilizing measurements. Subsequently, based on the measurements' findings, this study aimed to determine the heat transfer coefficient that is essential for accurately predicting indoor temperature distribution through the application of a vertical temperature distribution model.

2. Measurement

Measurements [33] were conducted to clarify the variables that affect the indoor temperature distribution during night ventilation through the floor-level windows. During night ventilation, the cold storage of the PCMs installed on the floor of the room was determined, and the variables affecting cold storage were analyzed.

2.1. Overview

The measurements were conducted in a single-story wooden public building in Kumagaya City, Japan. The measurement target was Room 1, shown in Figure 3a, with a floor area of 13.3 m^2 and a ceiling height of 2700 mm at the center of the room. Measurements were conducted from the end of September to the beginning of October 2014, and some portions of the data were used for analysis. The measured parameters (Table 1) were temperature, humidity, wind speed, and heat flow on the floor. The measurement points

(Figure 3c) were at the inflow side (MP1), center of the room (MP2), and inside the room (MP3). When night ventilation was performed using a floor-level window, it was assumed that indoor temperature stratification occurred along the vertical direction. Therefore, the measurements were conducted vertically ($T_{F.L+0.1m}$, $T_{F.L+0.5m}$, $T_{F.L+1.1m}$, $T_{F.L+1.9m}$, and $T_{F.L+2.6m}$). A heat flow meter was installed to measure the cold storage of the PCMs during night ventilation. The value obtained by adding up the differences between the values recorded by the upper and lower heat flow meters during ventilation was calculated as the daily accumulated cold storage.

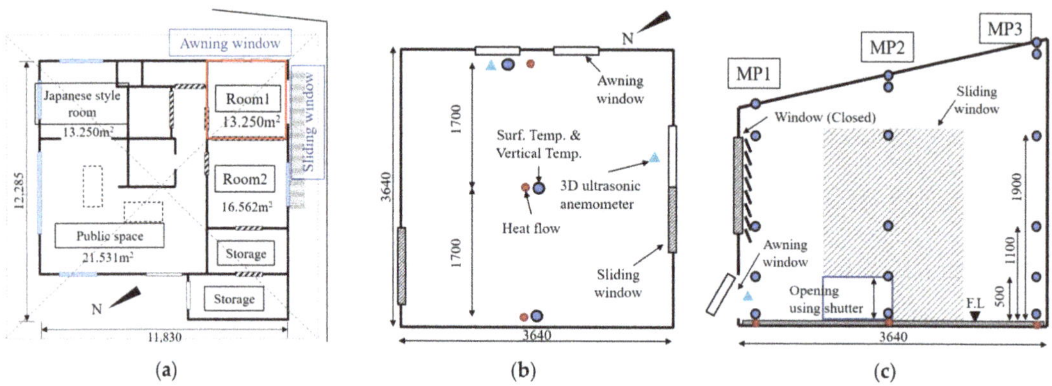

Figure 3. Floor plan and measurement points: (**a**) floor plan of the target building; (**b**) measurement points (floor plan); (**c**) measurement points (cross-section) [33].

Table 1. Measurement items.

Items	Instrument	Accuracy
Air temperature	0.1-mmΦ T-type thermocouple	−40 °C to 125 °C: ±0.5 °C
Relative humidity	Resistance change type (TDK (Tokyo, Japan), CHS-UPS)	0 °C to 50 °C: ±3% RH
Globe temperature	0.1-mmΦ T-type thermocouple inside a globe thermometer	−40 °C to 125 °C: ±0.5 °C
Surface temperature	0.3-mmΦ T-type thermocouple	−40 °C to 125 °C: ±0.5 °C
Wind speed in indoor spaces	Air velocity probe (KANOMAX (Tokyo, Japan), 0965-09)	±3% (KANOMAX)
Wind direction & speed	3D ultrasonic anemometers (Young-81000): Outdoor measurement (KAIJO (Tokyo, Japan), DA-600): Indoor measurement	±1%, ±0.05 m/s (Young-81000) ±2° (Young-81000) ±3% (KAIJO, DA-600) ±3° (KAIJO, DA-600)
Weather data	Air temp.: 0.1-mmΦ T-type thermocouple Humidity: Resistance change type (TDK, CHS-UPS) Wind: Wind vane anemometer (EKO (Tokyo, Japan), MA-110) Solar radiation: Thermopile pyranometer	±5% (EKO, MA-110) ±5° (EKO, MA-110) ±5% (Thermopile pyranometer)

Figure 4 shows a cross section of the floor, and a photograph of the target measurement room. The maximum cooling effect of a 6 mm resinous PCM mat installed on the floor was measured. The PCM mat contained microcapsules primarily composed of n-paraffin. The phase-change temperature of PCMs is within 23.8–24.2 °C during solidification and 26.6–27.6 °C during melting.

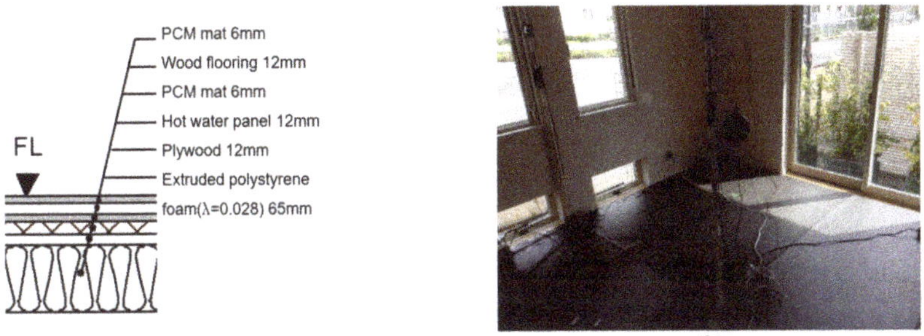

(**a**) (**b**)

Figure 4. (**a**) cross-section of the floor; (**b**) photo of target space [33].

The PCMs were ventilated at night for 10 or 11 h (8 p.m. to 6 a.m. or 7 a.m.) to maximize the cooling effect. Two lower windows (awning windows) measuring 818 mm × 362 mm were opened (Table 2). One side of the sliding window, with a height of 2118 mm, was opened to facilitate the lowering of the external shutters by 500 mm for maximum ventilation. The windows are shown in Figure 4b.

Table 2. Measurement information.

Case No.	Date & Time of Natural Ventilation	Additional PCM Mats on the Floor
Case 1	22 September—20 p.m.–23 September—7 a.m.	○
Case 2	25 September—20 p.m.–26 September—6 a.m.	○
Case 3	2 October—20 p.m.–3 October—7 a.m.	×
Case 4	9 October—20 p.m.–10 October—7 a.m.	×

2.2. *Weather Conditions*

Figures 5 and 6 present the meteorological conditions observed on the measurement days. Here, $T_{F.L+1.1m}$ indicates the indoor air temperature at a height of 1.1 m, which is shown in Figure 5 to demonstrate the difference between the indoor and outdoor air temperatures on the measurement day. Case 1 denotes a day when the outside temperature reduced to 14.8 °C and the wind speed was low (approximately 2.3 m/s). Case 2 denotes a day when the outside temperature was 19.4–24.8 °C and the wind speed was high during the night-ventilation period. Case 3 was a cloudy day, where the outside temperature remained stable at approximately 20 °C. Lastly, Case 4 was an almost windless day where the outside temperature was 15.7 °C and the wind speed was less than 1 m/s. Four measurements were performed to determine the distribution of the indoor temperature during night ventilation under various weather conditions.

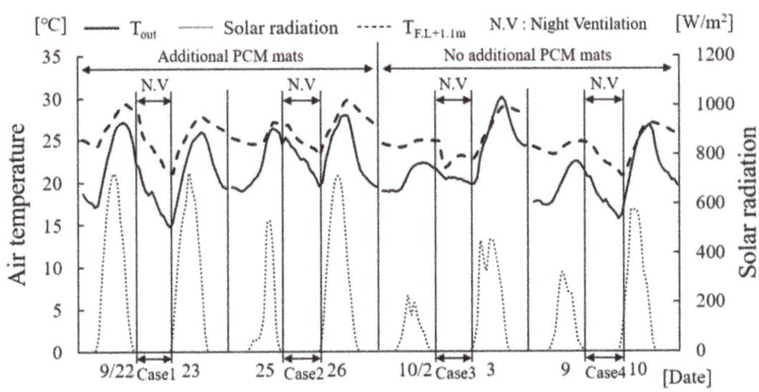

Figure 5. Weather conditions (air temperature and solar radiation) [33].

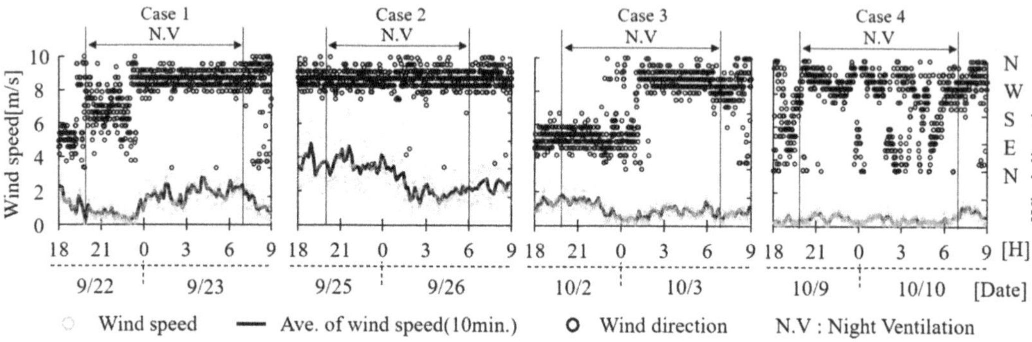

Figure 6. Weather conditions (windspeed and direction) [33].

2.3. Parameters Affecting the Vertical Indoor Temperature Distribution

For each measurement day, Figure 7 presents the relationship between the indoor temperature distribution, indoor and outdoor temperature differences, and inflow wind speed. In Cases 1 and 4, when the outside temperature decreased to approximately 15 °C during night ventilation, the most significant temperature drop was observed, particularly on the room floor. In Case 4, the external wind speed was negligible and the inflow wind speed was less than 0.1 m/s. Compared to that in Case 1, the temperature difference near the floor, at a height immediately above the opening, exceeded 2 °C, indicating less heat transfer and a stable temperature distribution. In Cases 2 and 3, with minor changes in the outside temperature, the indoor temperature changes were minimal. In Case 3, the indoor air was well-mixed because of the high inflow wind speed, which facilitated an insignificant change in temperature.

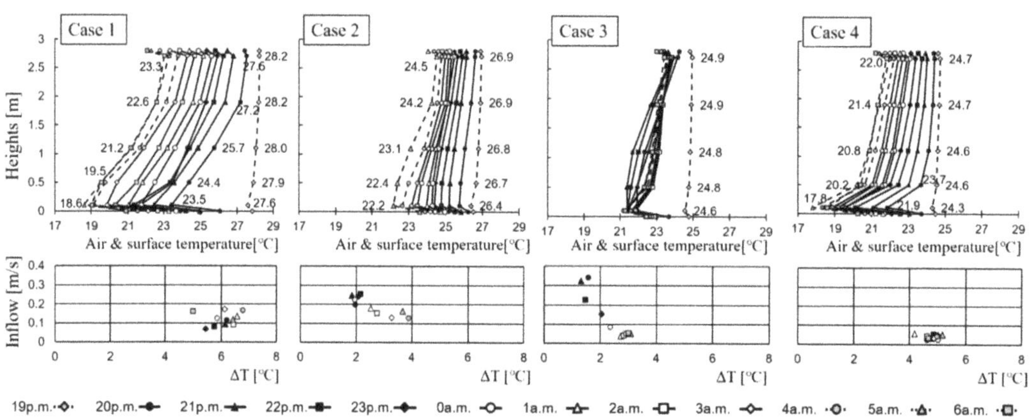

Figure 7. Measurement results (indoor air and surface temperature distribution, and inflow air) [33].

Figure 8 shows the temperature difference between the blocks based on the inflow wind speed for each block, and that between the indoor and outdoor temperatures (circle size). The temperature difference between the blocks increased with an increase in the indoor–outdoor temperature, while the inlet wind speed decreased.

Thus, the measurement findings indicate that cold air entered the room during night ventilation, thereby generating a vertical temperature distribution. As is evident, the greater the difference between the outdoor and indoor air temperatures, the smaller the inflow velocity, and the lower the air temperature near the floor.

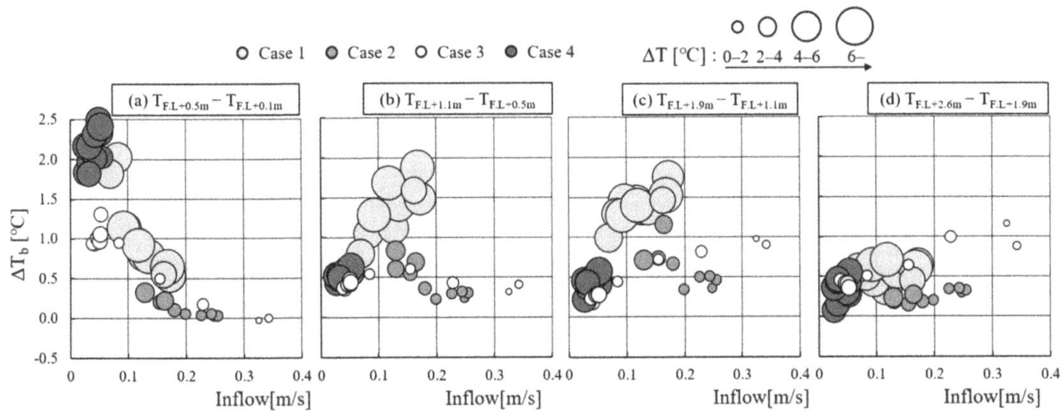

Figure 8. Correlation between inflow speed and difference in air temperature in each block (ΔT_b), and difference in indoor and outdoor air temperatures [33].

2.4. Thermal Characteristics of the Space with Night Ventilation

Figure 9 shows the results of the classification of the characteristics of the indoor vertical temperature distribution under each weather condition, obtained from the current measurement results. (a) At inflow wind speeds of 0.1 m/s or less, a temperature gradient that corresponded to the difference in the temperature between the inside and outside occurred up to the height of the opening; however, almost no temperature difference was observed in spaces exceeding the height of the opening. (b) Whereas, when the indoor inflow wind speed was 0.2 m/s or greater, a temperature gradient was observed even in the upper part of the room, depending on the difference in the temperature between the inside and outside, as well as on the magnitude of the inflow wind speed. (c) The size of the temperature gradient tended to be determined via the difference in temperature between the inside and outside of the room, and the speed of the inflowing wind. With an increase in the speed of the inflow air, the air in the lower part of the room was completely mixed, and the temperature difference tended to almost disappear.

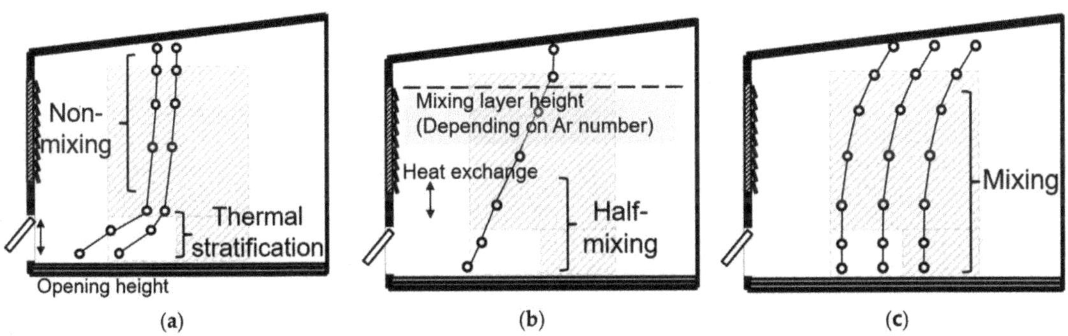

Figure 9. Classification of shape of vertical air temperature distribution by wind speed: (**a**) V_{in}: below 0.1 m/s; (**b**) V_{in}: approximately 0.2 m/s; (**c**) V_{in}: exceeding 0.3 m/s [33].

3. Block Model (Zonal Simulation or Numerical Analysis)

3.1. Model Description

Togari et al. [17,18] proposed a block model to predict the vertical temperature distribution in large spaces as a simplified theoretical framework that vertically divided a large space into multiple zones to assess and calculate the temperature gradients. This model comprises the following three types:

- Wall surface current model: This model evaluates the ascending and descending air currents along vertical surfaces due to convective heat transfer. It employs boundary layer analysis on a flat plate to determine the air currents formed along the wall owing to the temperature difference between the wall and air (Figure 10);
- Primary airstream evaluation model: This model considers the air streams discharged from the outlets as non-isothermal free jets. It assesses the influence of these jets on the vertical temperature distribution by entraining air from different zones, moving upward owing to buoyancy, and affecting the surrounding air temperatures;
- Heat transfer by temperature difference between adjacent zones: This component is used to calculate the heat transfer caused by the temperature difference between vertically adjacent zones. It employs a heat transfer factor to model the conduction and convection processes between zones.

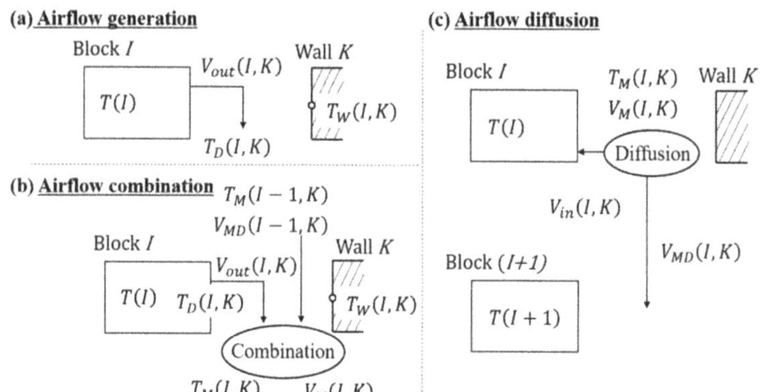

(d) Airflow direction

Air temp. condition	$V_{in}(I,K)$	$V_{MD}(I,K)$
$T_M(I,K) \geq T(I)$	$= V_M(I,K)$	$= 0$
$T(I) > T_M(I,K) > T(I+1)$	$= V_M(I,K) \cdot \dfrac{T_M(I,K) - T(I+1)}{T(I) - T(I+1)}$	$= V_M(I,K) - V_{IN}(I,K)$
$T_M(I,K) \leq T(I+1)$	$= 0$	$= V_M(I,K)$

Figure 10. Togari et al.'s wall surface current model (descending flow) [17,18].

These combined models aid the prediction and analysis of the vertical temperature distribution in large spaces.

In this study, the indoor air temperature distribution was predicted using a block model owing to its similarity to the indoor thermal environment caused by natural ventilation at night and temperature stratification. The primary airstream evaluation model was not considered in this study owing to the low air inflow.

3.2. Indoor Air Temperature Distribution

The block model, which predicts the vertical temperature distribution through the division of the space into multiple blocks, wherein non-uniform temperature distributions are known to occur, was used as the indoor temperature distribution model in this study. The room was divided into 27 upper and lower blocks, as well as blocks adjacent to the wall. The vertical temperature distribution was predicted by solving the mass balance equation between each adjacent block and the heat balance equation, considering advection, diffusion, and convective heat transfer with the wall (Figure 11). This analysis was performed in the same room as that used for the measurements described in Section 2. To facilitate easier calculations, a room with a flat rather than sloped roof, set at a height of 2.7 m, was used. Figure 12 shows the block model for the numerical simulations used in this study.

— Mass conservation

$$0 = \sum_{K=1}^{m} \{\gamma V_{IN}(I,K) - \gamma V_{OUT}(I,K)\} + \gamma V_{SI}(I) - \gamma V_{SO}(I) + \gamma V_C(I+1) - \gamma V_C(I)$$

— Heat balance

$$\begin{aligned} 0 = & \sum_{K=1}^{m} C\gamma V_{IN}(I,K)\{T_M(I,K) - T(I)\} + C\gamma V_{SI}(I)\{T_{SI} - T(I)\} \\ & + C\gamma V_C(I+1)\{T(I+1) - T(I)\} \quad [V_C(I+1) > 0] \\ & - C\gamma V_C(I)\{T(I-1) - T(I)\} \quad [V_C(I) < 0] \\ & + C_B(I)A_B\{T(I-1) - T(I)\} + C_B(I+1)A_B\{T(I+1) - T(I)\} \end{aligned}$$

— Nomenclature

V_{IN} : Inlet air flow from wall down flow $[m^3/s]$
V_{OUT} : Outlet air flow to wall down flow $[m^3/s]$
V_{SI} : Supply air flow rate $[m^3/s]$
V_{OUT} : Exhaust air flow rate $[m^3/s]$
V_C : Vertical flow rate between blocks $[m^3/s]$
T_M : Temperature of wall down flow [°C]
T_{SI} : Supply air temperature [°C]
T : Room block temperature [°C]
C_B : Heat transfer coefficient $[J/K \cdot m^3]$
$C \cdot \gamma$: Heat capacity
A_B : Area of boundary surface of block $[m^2]$

Figure 11. Mass conservation and heat balance equations of the Togari model [17,18].

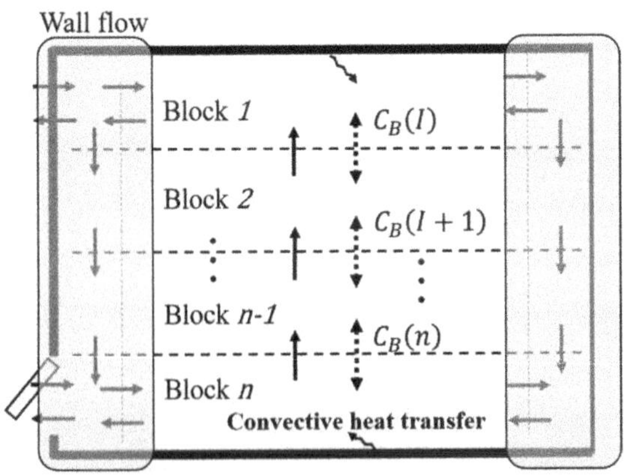

Figure 12. Schematic of numerical model of vertical air temperature distribution.

The block model employed a non-isothermal jet model to calculate the mass and heat movement attributable to the vortex generated by the inflow air. However, owing to the inflow not being sufficiently strong in this study, the non-isothermal jet model was not used. This is because it did not significantly affect the analysis results. The air movement was calculated by considering the flow rate entering through the floor-level window, and the cooling downflow along the wall at night. Heat transfer between the blocks in the room was determined using the heat transfer coefficient (C_B). Herein, C_B is generally assumed to be 2.3 W/(m^2·°C) for large-space buildings [17,18]. However, the C_B was determined to be 14.8 W/(m^2·°C) in the DOE-2 software (version DOE-2) [34], and set at 10 W/(m^2·°C) in an atrium space [35]. This coefficient varies according to the different regions of an indoor space, and its value can be calibrated using CFD [36]. The maximum C_B reached 28 W/(m^2·°C) in the case of an extremely turbulent flow [37]. Furthermore, the indoor temperature distribution was predicted considering the measured values of the heat transfer coefficient.

3.3. Heat Transfer in Each Block

The indoor temperature and airflow during night ventilation were proportional to the indoor–outdoor temperature difference, and inversely proportional to the inflow wind speed. The Archimedes number (Ar) was introduced to represent the relationship between the indoor–outdoor temperature difference and wind speed, based on the measurement results. Consequently, Ar can be used to determine the turbulent diffusion coefficient in each block, and to predict the indoor temperature distribution in a room with night ventilation.

The relationship between the heat transfer and turbulent diffusion coefficients, and the formula for Ar, are as follows:

$$C_B(I) = a_t \cdot C \cdot r / H_b, \tag{1}$$

$C_B(I)$: Heat transfer coefficient [J/K·m^3]
a_t: Turbulent diffusion coefficient [m^2/s]
C·r: Heat capacity [J/K]
H_b: Height [m]

$$Ar = \{g \times \beta \times \sqrt{A_{in}} \times (T - T_{SA})\}/V_{SA}^2, \tag{2}$$

g: Gravitational acceleration [m^2/s]
V_{SA}: Inflow wind speed [m/s]
T_{SA}: Inflow air temperature [°C]
A_{in}: Opening area [m^2]
β: Coefficient of expansion [1/K]
T: Air temperature [°C]

The relationship between Ar and the turbulent diffusion coefficient in each block is shown in Figure 13. The lowest turbulent diffusion coefficient (a_{t_lower}) was highly correlated with Ar; the correlation decreased with increased distance from the opening (upper block). However, the cold storage capacity of PCMs was closely related to the floor surface temperature. The air temperature of the lowest block strongly influenced the surface temperature of the PCMs on the floor of a detached house. These results can be used to predict the air temperature distribution.

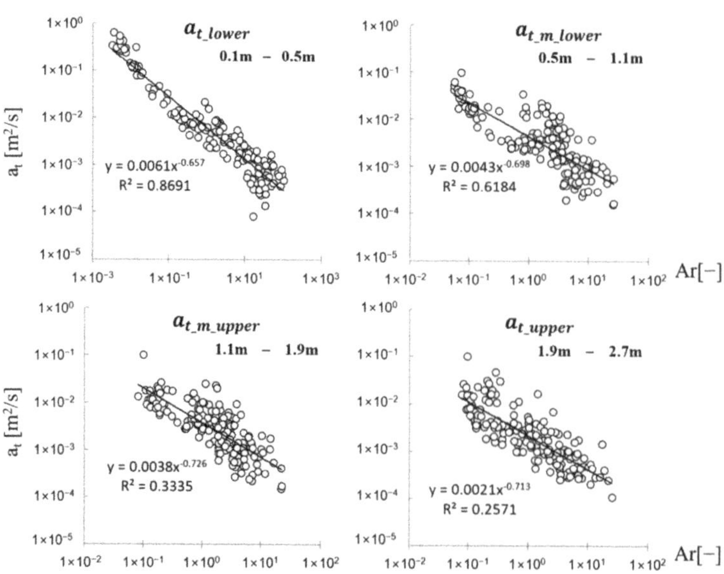

Figure 13. Correlation between Ar number and turbulent diffusion coefficient.

3.4. Verification

In the verification phase of this study, the indoor air temperature distribution was systematically predicted by employing the heat transfer coefficient derived from the empirical correlations, as depicted in Figure 14. For each day of measurement, predictions were based on weather data specific to the study's target area. The wall surface temperatures, floor and ceiling temperatures, inflow air temperature, and wind speed were obtained as input data from the weather data of the target area.

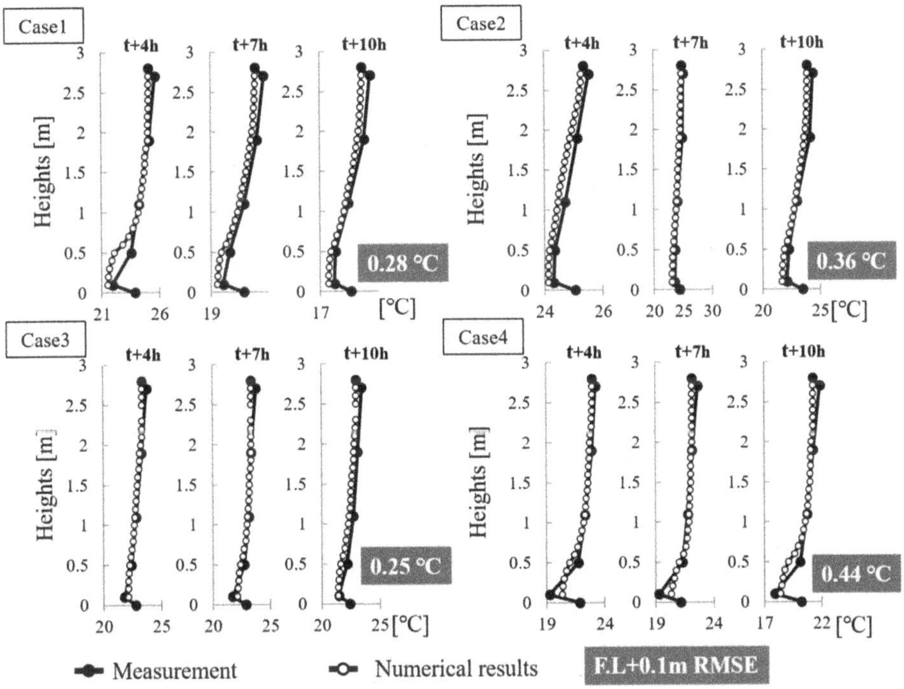

Figure 14. Comparison of verification of air temperature distribution measurement, and numerical results.

Additionally, the convective heat transfer coefficient was determined to be 5 kcal/hm^{2}K. The initial temperatures of the building blocks were estimated using assumed baseline values. The simulation's predictive accuracy was validated by comparing the predicted results with measurements taken during the study. These comparisons are graphically represented in Figure 14. The alignment between measured and predicted temperature distributions was quantitatively assessed at several intervals—specifically 4, 7, and 10 h following the initiation of night ventilation. The root–mean–square error (RMSE) was calculated for temperatures near the floor (0.1 m above floor level), where the thermal effects of the PCM were most pronounced. The RMSE values for Cases 1 through 4 were found to be 0.28, 0.36, 0.25, and 0.44 °C, respectively.

These findings not only confirm the reliability of our predictive model but also underscore the significance of the heat transfer coefficient in modeling indoor temperature dynamics during night ventilation with PCM.

3.5. Discussion

A typical nodal model predicts the temperatures based on the assumption of homogeneous thermal conditions within a room. However, in this study, which utilized night ventilation combined with floor-installed PCMs as a passive cooling strategy, some temperature differences were observed. Specifically, temperatures near the floor were found to be approximately 3 °C lower than those at the center of the room. This is attributable to the outdoor climatic conditions. Such a disparity is critical, as a deviation of 3 °C can significantly influence the phase-change temperature of the PCMs, resulting in inaccuracies in the assessment of their cold storage. Therefore, this study underscores the need to reconsider the placement and utilization of PCMs in building designs, particularly when they are employed in night-time floor-level installations.

4. Conclusions

This study conducted in-situ measurements to clarify the indoor temperature distribution during night ventilation through a floor-level window, and consequently confirm the synergistic effect. Subsequently, the heat transfer coefficient was clarified for use in a numerical analysis model to design an optimal building with night ventilation and PCMs. The indoor air temperature was distributed during night ventilation with a floor-level window. The temperature distribution was proportional to the difference between the indoor and outdoor temperatures, and inversely proportional to the inflow wind speed. Further, the numerical analysis model predicted indoor air temperature distributions using blocks along the vertical direction. In such analyses, the numerical heat transfer coefficient must be specified. The heat transfer coefficient can be calculated from the measurement findings using Ar, which is a dimensionless number based on the temperature difference and wind speed. The predicted and measured indoor air temperature distributions were compared using the calculated heat transfer coefficients. The predicted air temperature near the floor yielded an RMSE of 0.25–0.44 °C.

This study focused primarily on thermal management strategies during the summer season. However, for comprehensive year-round efficiency, special consideration is required for the selection of phase-change materials (PCMs) during the winter. The phase-change temperature of these materials is critical, particularly in ensuring effective indoor heating. Moreover, strategic placement of PCM, such as in the flooring where direct heat gain is more prevalent during winter months, can facilitate the melting of PCM. This placement leverages the latent heat release of PCM to enhance the indoor heating environment effectively in winter, while still contributing to cooling in the summer.

In future research, we will focus on the cold storage of PCM with night ventilation, and explore models for evaluating its performance. This research will be expanded to encompass various general climatic conditions and different types and scales of buildings.

Author Contributions: Conceptualization, T.L., R.S. and T.A.; methodology, T.L. and R.S.; software, T.L.; validation, T.L.; formal analysis, T.L.; investigation, T.L. and R.S.; resources, T.L., R.S. and T.A.; data curation, T.L., R.S. and T.A.; writing—original draft preparation, T.L. and R.S.; writing—review and editing, T.L., R.S., T.A. and S.Y.; visualization, T.L.; supervision, T.A. and S.Y.; project administration, R.S. and T.A.; funding acquisition, S.Y. All authors have read and agreed to the published version of the manuscript.

Funding: This research was funded by 2023 BK21 FOUR Graduate School Innovation Support funded by Pusan National University (PNU-Fellowship program), Korea Ministry of Environment (MOE) as "Graduate School specialized in Climate Change" and National Research Foundation of Korea (NRF) grant funded by the Korea government (MSIT) (No. RS-2023-00218875).

Data Availability Statement: The authors do not have permission to share data.

Acknowledgments: The authors would like to thank the Misawa Homes Institute of Research and Development Co., Ltd. for their support, especially for providing the measurement facilities.

Conflicts of Interest: The authors declare no conflicts of interest.

References

1. IPCC AR6 SYR, Summary for Policemakers. Available online: https://www.ipcc.ch/report/ar6/syr/ (accessed on 1 April 2023).
2. International Energy Agency. *World Energy Outlook 2018*; International Energy Agency: Paris, France, 2018. [CrossRef]
3. Santamouris, M. Cooling the buildings—past, present and future. *Energy Build.* **2016**, *128*, 617–638. [CrossRef]
4. Santamouris, M.; Asimakopoulos, D. *Passive Cooling of Buildings*; James and James Science Publishers: London, UK, 1996.

5. Solgi, E.; Hamedani, Z.; Fernando, R.; Skates, H.; Orji, N.E. A literature review of night ventilation strategies in buildings. *Energy Build.* **2018**, *173*, 337–352. [CrossRef]
6. Shaviv, E.; Yezioro, A.; Capeluto, I.G. Thermal mass and night ventilation as passive cooling design strategy. *Renew. Energy* **2001**, *24*, 445–452. [CrossRef]
7. Exizidou, P.; Christoforou, E.; Fokaides, P.A. Numerical assessment of night ventilation impact on the thermal comfort of vernacular buildings. *J. Build. Pathol. Rehabil.* **2017**, *2*, 2. [CrossRef]
8. Blondeau, P.; Spérandio, M.; Allard, F. Night ventilation for building cooling in summer. *Sol. Energy* **1997**, *61*, 327–335. [CrossRef]
9. Borderon, J.; Virgone, J.; Cantin, R.; Kuznik, F. Full-scale study of a building equipped with a multi-layer rack latent heat thermal energy storage system. *HVACR Res.* **2011**, *17*, 566–576. [CrossRef]
10. Ling, H.; Chen, C.; Wei, S.; Guan, Y.; Ma, C.; Xie, G.; Li, N.; Chen, Z. Effect of phase change materials on indoor thermal environment under different weather conditions and over a long time. *Appl. Energy* **2015**, *140*, 329–337. [CrossRef]
11. Wang, X.; Yu, H.; Li, L.; Zhao, M. Experimental assessment on the use of phase change materials (PCMs)-bricks in the exterior wall of a full-scale room. *Energy Convers. Manag.* **2016**, *120*, 81–89. [CrossRef]
12. Wonorahardjo, S.; Sutjahja, I.M.; Tunçbilek, E.; Achsani, R.A.; Arıcı, M.; Rahmah, N. PCM-based passive air conditioner in urban houses for the tropical climates: An experimental analysis on the stratum air circulation. *Build. Environ.* **2021**, *192*, 107632. [CrossRef]
13. Andersen, K.T. Theory for natural ventilation by thermal buoyancy in one zone with uniform temperature. *Build. Environ.* **2003**, *38*, 1281–1289. [CrossRef]
14. Fanger, P.O. Local discomfort to the human body caused by non-uniform thermal environments. *Ann. Occup. Hyg.* **1977**, *20*, 285–291. [CrossRef] [PubMed]
15. Santamouris, M.; Sfakianaki, A.; Pavlou, K. On the efficiency of night ventilation techniques applied to residential buildings. *Energy Build.* **2010**, *42*, 1309–1313. [CrossRef]
16. Ito, H.; Yokoi, M.; Nakahara, N. Simplified calculation model of vertical air temperature distribution during heating in air-conditioned room—Studies on thermal properties in air-conditioned space Part 2. *J. Archit. Plann. Environ. Engng AIJ* **1989**, *398*, 59–67. [CrossRef] [PubMed]
17. Togari, S.; Arai, Y.; Miura, K. Simplified prediction model of vertical air temperature distribution in a large space—Study on a thermal environment design system for large spaces, Part 1. *J. Archit. Plan. Environ. Eng. AIJ* **1991**, *427*, 9–19. [CrossRef] [PubMed]
18. Togari, S.; Arai, Y.; Miura, K. A simplified model for predicting vertical temperature distribution in a large space. *ASHRAE Trans.* **1993**, *99*, 84–99. Available online: https://www.scopus.com/inward/record.uri?eid=2-s2.0-0027269039&partnerID=40&md5=8a73bb2ccc4ca8ee55fa2759b0ced9b3 (accessed on 2 April 2024).
19. Inard, C.; Bouia, H.; Dalicieux, P. Prediction of air temperature distribution in buildings with a zonal model. *Energy Build.* **1996**, *24*, 125–132. [CrossRef]
20. Wurtz, E.; Nataf, J.M.; Winkelmann, F. Two- and three-dimensional natural and mixed convection simulation using modular zonal models in buildings. *Int. J. Heat Mass Transf.* **1999**, *42*, 923–940. [CrossRef]
21. Haghighat, F.; Li, Y.; Megri, A.C. Development and validation of a zonal model—POMA. *Build. Environ.* **2001**, *36*, 1039–1047. [CrossRef]
22. Musy, M.; Wurtz, E.; Winkelmann, F.; Allard, F. Generation of a zonal model to simulate natural convection in a room with a radiative/convective heater. *Build. Environ.* **2001**, *36*, 589–596. [CrossRef]
23. Musy, M.; Winkelmann, F.; Wurtz, E.; Sergent, A. Automatically generated zonal models for building air flow simulation: Principles and applications. *Build. Environ.* **2002**, *37*, 873–881. [CrossRef]
24. Higashimoto, T.; Yamanaka, T.; Kotani, H.; Hanano, H. Temperature and contaminant concentration distribution of displacement ventilated rooms with cooled wall. *J. Environ. Eng. AIJ* **2003**, *68*, 47–53. [CrossRef] [PubMed]
25. Ren, Z.; Stewart, J. Simulating air flow and temperature distribution inside buildings using a modified version of COMIS with sub-zonal divisions. *Energy Build.* **2003**, *35*, 257–271. [CrossRef]
26. Wurtz, E.; Mora, L.; Inard, C. An equation-based simulation environment to investigate fast building simulation. *Build. Environ.* **2006**, *41*, 1571–1583. [CrossRef]
27. Gao, J.; Zhao, J.; Xiaodong, L.; Gao, F. A Zonal Model for Large Enclosures With Combined Stratification Cooling and Natural Ventilation Part 1—Model Generation and its Procedure. *J. Sol. Energy Eng.* **2006**, *128*, 367–375. [CrossRef]
28. Zhao, B.; Yang, X.; Jiang, Y.; Gopal, V.; Dobbs, G.; Sahm, M. A new approach on zonal modeling of indoor environment with mechanical ventilation. *Build. Environ.* **2008**, *43*, 278–286. [CrossRef]
29. Chen, Q. Ventilation performance prediction for buildings A method overview and recent applications. *Build. Environ.* **2009**, *44*, 848–858. [CrossRef]
30. Megri, A.C.; Yu, Y. New calibrated zonal model (POMA+) for temperature and airflow predictions. *Build. Environ.* **2015**, *94*, 109–121. [CrossRef]
31. Huang, C.; Li, R.; Liu, Y.; Liu, J.; Wang, X. Study of indoor thermal environment and stratified air-conditioning load with low-sidewall air supply for large space based on Block-Gebhart model. *Build. Environ.* **2019**, *147*, 495–505. [CrossRef]
32. Wang, X.; Yang, Y.; Xu, Y.; Wang, F.; Zhang, Q.; Huang, C.; Shi, C. Prediction of vertical thermal stratification of large space buildings based on Block-Gebhart model: Case studies of three typical hybrid ventilation scenarios. *J. Build. Eng.* **2021**, *41*, 102452. [CrossRef]

33. Lee, T.; Sato, R.; Asawa, T.; Kawai, H.; Hirayama, Y.; Ohya, I.; Sato, Y.; Hayashi, Y. Study on cold storage of phase change material on the floor of a house with night natural ventilation. *J. Environ. Eng. (Trans. AIJ)* **2017**, *82*, 1025–1034. (In Japanese) [CrossRef]
34. Landsberg, D.R.; Misuriello, H.P.; Moreno, S. Design strategies for energy-efficient atrium spaces. *ASHRAE Trans.* **1986**, *92 Pt 2*, 310–328. Available online: https://www.scopus.com/inward/record.uri?eid=2-s2.0-0022940710&partnerID=40&md5=88e2d69119e7811b3c330f27ac78b0d2 (accessed on 15 January 2023).
35. Chow, W.K. Assessment of thermal environment in an atrium with air-conditioning. *J. Environ. Syst.* **1996**, *25*, 409–420. [CrossRef]
36. Gao, J.; Zhang, X.; Zhao, J.; Gao, F. A heat transfer parameter at air interfaces in the BLOCK model for building thermal environment. *Int. J. Therm. Sci.* **2010**, *49*, 463–470. [CrossRef]
37. Wang, H.; Zhou, P.; Guo, C.; Tang, X.; Xue, Y.; Huang, C. On the calculation of heat migration in thermally stratified environment of large space building with sidewall nozzle air-supply. *Build. Environ.* **2019**, *147*, 221–230. [CrossRef]

buildings

MDPI

Article

An Investigation of Occupants' Thermal Requirements in Indoor Transitional Space in Entertainment Buildings

Guoying Hou [1,2,*], Chris Tweed [2] and Shen Wei [3]

[1] School of Architecture and Art Design, Hebei University of Technology, Tianjin 300400, China
[2] Architectural Science Group, Welsh School of Architecture, Cardiff University, Wales CF10 3NB, UK; tweedac@cardiff.ac.uk
[3] The Bartlett School of Sustainable Construction, University College London (UCL), 1–19 Torrington Place, London WC1E 7HB, UK; shen.wei@ucl.ac.uk
* Correspondence: 2019052@hebut.edu.cn

Abstract: Indoor transitional space is a popular buffer space between buildings' interior and exterior environments nowadays. Maintaining a comfortable indoor thermal comfort for transitional spaces often poses challenges to building designers and engineers. Some existing studies have already explored this topic, but they are mainly carried out in academic buildings. There are, however, still many other types of buildings containing transitional space, including entertainment buildings such as theaters and tourist centers. To provide useful information about people's thermal requirements in the transitional space of entertainment buildings, this study has adopted both field measurement and questionnaire methods. Additionally, the same method has been repeated in an academic setting as well, so the results can be compared with existing studies. By comparing participants' thermal requirements, it indicates that people's thermal requirements are significantly impacted by operative temperature, which can give architects suggestions to improve the thermal environment in transitional spaces. In addition, in transitional spaces, people had a high tolerance for their thermal environment, especially participants in entertainment buildings, who showed a fairly high thermal satisfaction rate of 96% in winter and 94% in summer, far beyond the rates of 89% and 73% in academic buildings. Combined with the analysis of participants' thermal preferences and the reason people stay in transitional spaces, it implies a close relationship between participants' thermal comfort differences and the function that transitional spaces provide.

Keywords: indoor transitional space; thermal requirement; field studies; thermal adaptation

1. Introduction

In current society, transitional spaces are a special type of room in buildings, as they provide a link between indoor and outdoor environments [1]. Transitional space commonly includes outdoor transitional space (e.g., bus stations and courtyards), semi-outdoor space (e.g., train stations and porches), and indoor transitional space (e.g., foyers and lift lobbies) [2]. Within these types, indoor transitional space has been developed to work as a buffer space between indoor and outdoor environments. In modern non-residential buildings, indoor transitional space has become very common, providing functional areas like resting areas, performance areas, and meeting areas [3]. Although indoor transitional space often takes up no more than 40% of buildings' total floor area, their energy demand could be up to three times that from the remaining parts of buildings [4]. Therefore, reducing the energy consumption of this special space type becomes very important for reducing overall building energy consumption. To achieve this intention, a good understanding of occupants' thermal requirements is essential for guiding architectural design, system selection, and building operation [2,5–7].

Existing studies in terms of the thermal requirements in indoor transitional spaces were mainly carried out in academic buildings. In Bangkok, Thailand, Jitkhajornwanich et al. [8] studied the thermal environment of indoor transitional space in two academic buildings, using calculated neutral temperatures and preferred temperatures for both cool and warm seasons. In a study carried out by Pitts et al. [9] in one academic building and one office building, a wider thermal comfort range in transitional space was observed than was predicted by Fanger's PMV model. In a laboratory building, Wu and Mahdavi [10] assessed people's thermal requirements in transitional states and suggested that the change in thermal sensation was consistent with the respective difference in temperature. In Malaysia, Kwong and Adam [11] conducted a field study in the enclosed lift lobby of an educational institution, and their result showed a lower thermal expectation than conventional space. In another study by Kwang and Adam [12], it was found that occupants' thermal perception was directly related to the level of human occupancy in transitional space. Additionally, a sudden change in temperature may lead to discomfort for occupants. To investigate the thermal performances of both semi-opened and fully enclosed transitional spaces in different seasons, Hui and Jiang [13] conducted a study in one academic building in Hong Kong. In this study, they observed that people accept a wider thermal environment in transitional spaces, and their thermal response varies with dressing, activity level, past thermal experience, and prior thermal preference. Therefore, they believed that the current comfort standards and criteria are not suitable for designing transitional spaces.

In addition to the studies carried out in academic buildings, there are some relevant studies carried out in other types of buildings as well. For example, Hwang et al. [14] studied occupants' thermal requirements in the transitional space of one commercial building in Bangkok, Thailand. From the study, they observed that the thermal requirements of occupants in the transitional space were similar to those in an office environment. Kotopouleas and Nikolopoulou [15] conducted a study in airport terminals, evaluating the thermal perceptions of both passengers and staff. Their study revealed the complications of thermal requirements in airport terminals. Avantaggiato [16] conducted a study in three Italian shopping centers, and the study demonstrated the inappropriateness of using both Fanger's PMV model and the adaptive comfort model for transitional spaces, as customers were found to have a much wider range of indoor comfort feelings. However, research data regarding the thermal environment of transitional areas are still lacking [13,17,18]. The majority of studies [4,5,13,19–23] came to the conclusion that a greater variety of comfort conditions than those suggested by Fanger's model can be accepted by individuals in transitional settings. They contend that Fanger's model is limited by the steady-state settings in which it was created, but transitional regions include dynamic characteristics that render Fanger's model inappropriate. According to certain research [8–11], users in transitional spaces are also more adaptive. This observation may be influenced by the amount of time spent in the transitional area as well as prior thermal experience. All these studies concluded that further investigation on thermal comfort in transitional spaces at different times of the year is required to expand the database of evidence. All of these studies came to the conclusion that in order to increase the body of evidence, more research on thermal comfort in transitional spaces at various times of the year is needed.

The above review work reflects that most existing studies in terms of indoor transitional space focused on academic buildings, with some investigations carried out in other building types, including airport terminals and lobbies. To obtain a comprehensive understanding, more data are still highly required, especially from non-academic buildings. In current society, entertainment buildings have started to play an important role, as they can provide functions like cinemas, leisure centers, cafés, amusement arcades, et al. This building type may consume a high level of energy, but relevant studies seem to be rare, especially for their transitional space. To fill this gap, this study has selected one entertainment building located in the UK, and data relevant to occupants' thermal requirements were collected by both subjective and objective methods. To compare occupants' thermal requirements between this building type and academic buildings, which have been better understood, data were collected using the same methods from one academic building close by. We hope that the results from this study will provide field evidence about the thermal requirements of the transitional space in entertainment buildings to help control this space more efficiently.

2. Methodology

The methodology conducted in this research included on-site questionnaire surveys and physical measurements, one existing entertainment building and one academic building with indoor transitional spaces were chosen as the case study building. The monitoring of both interior and outdoor environmental conditions occurred concurrently with the field studies and the administration of questionnaire surveys.

2.1. Case Study Buildings

In this study, two buildings with different operational functions were investigated. One is an academic building, and another is an entertainment building. Both buildings are in the city of Cardiff, UK (51°29′0″ N, 3°11′0″ W, alt. 65 m) and have transitional space inside. Figure 1 depicts the two buildings and the layouts of the transitional spaces in them. Both transitional spaces connect with the main entrances of the two buildings, and they contain communal spaces like cafés, bars, and shops. For the transitional spaces in the two buildings under investigation, there are some common functional areas, such as corridors for passing through, resting areas, eating and drinking areas, and informal meeting areas. The TSAB (Transitional Space in the Academic Building), however, has one more function than the TSEB (Transitional Space in the Entertainment Building), which is the study area. To maintain a comfortable indoor thermal environment, both spaces were centrally controlled by dedicated HVAC systems, so the users have no control over the indoor thermal environment. Table 1 lists some key information about the two indoor transitional spaces investigated in this study.

Figure 1. Investigated buildings and transitional spaces. (**a**) Entertainment building, (**b**) academic building.

Table 1. Key characteristics of the investigated transitional spaces.

		TSEB	TSAB
Year of building		2004	2007
Business type		Entertainment	Academic
Building floor area		19,020 m^2	2490 m^2
Measured Transitional space area		1298 m^2	810 m^2
Main orientation		West	North
Window type		Double glazing	Double glazing
Openable windows		No	No
Type of building services	Winter	Air conditioning	Air conditioning
	Summer	Air conditioning	Air conditioning

2.2. Physical Measurements

The survey was conducted in both winter (the last week of January for the TSAB and the first week of February for the TSEB) and summer (the first week of August for the TSAB and the second week of August for the TSEB) to reflect seasonal variations. The measurement was performed in two consecutive weeks to ensure a similar outdoor environment. The data were collected daily, from the opening of the buildings (9 a.m. for both TSEB and TSAB) to the closure of the buildings (7 p.m. for the TSEB and 5 p.m. for the TSAB).

In this study, indoor environmental parameters, namely, air temperature, mean radiant temperature, relative humidity, and air velocity, were measured and recorded. Operative temperature and relative humidity were measured by an AREXX TSN-TH70E Wireless Temperature and Humidity Sensor (accuracy: ±0.2 °C for temperature and ±5% for humidity), and air velocity was measured by a Testo RS327-0640 hot-wired anemometer (accuracy: ±0.1 m/s). All measurements were carried out at the center of different thermal zones within the indoor transitional space, 1.1 m above the floor (the abdomen level recommended in ISO 7726) [24], with the avoidance of direct sunlight and local heating and cooling sources. The measurement intervals were set at 2 min. The outdoor temperature was measured by a DELTA-T WS-GP1 weather station (accuracy: ±0.2 °C), located on the roof of one building on the campus of Cardiff University, at an interval of 2 min as well. The weather station was installed three meters above the roof level to avoid the heat generated by the building.

2.3. Questionnaire Surveys

In addition to the above physical parameters, relevant subjective information was also collected using questionnaires to reflect occupants' thermal sensations and preferences. The questions were developed according to ASHRAE Standard 55 [25]. The questionnaire consists of three main sections. Section 1 collects relevant personal information, such as participants' gender, age, and occupation. Section 2 investigates participants' historical interactions with the building and the indoor transitional space under investigation, such as their purposes for using the space, activity level, previous visit frequency, time spent at the interviewed location, and clothing insulation. Section 3 focuses on participants' thermal perceptions of the indoor transitional space. All participants were randomly selected from the users of the investigated transitional spaces. All participants required a minimum setup time of 5 min to ensure sufficient time to experience the local thermal environment. To determine participants' metabolic rate, their activities were recorded by the researcher of this study through observation. Their clothing insulation was estimated by themselves using the checklist given in ASHRAE Standard 55. Finally, a total of 494 interviews were conducted throughout the survey period, with 243 (49%) performed in the winter and 251 (51%) performed in the summer. To balance gender, there were 232 (47%) females and 262 (53%) males who participated. All participants were selected randomly from the targeted transitional spaces.

2.4. Data Analysis

The data collected from the field studies were first compiled into spreadsheets and then analyzed using the Statistical Package for Social Science (SPSS) version 20. The data were prepared separately for both buildings, and the results were then compared. To statistically estimate the potential effect of various factors on participants' thermal sensation, such as operative temperature and cloth insulation, linear regression was adopted [26], with the outcomes analyzed based on two significance levels, namely, average statistical significance ($p < 0.05$) and high statistical significance ($p < 0.01$). For the analysis of participants' thermal preference and preferred temperature, probit analysis was applied, as in previous studies [14].

3. Results and Discussion

3.1. Results from the Academic Building (with Comparisons to Existing Studies)

The data analysis of this study started in the academic building, so its results can be compared with existing studies. Table 2 has included the main results from both this study and previous studies. The results indicate that in a tropical climate country, the cloth value of people using the indoor transitional spaces is close to the cloth value in the summer UK. In terms of the metabolic rate, it indicates that except for the quite large metabolic rate range of 0.7–3.8 in research on transient thermal comfort in the UK, the metabolic rate in other research both in the UK and tropical countries has a little difference of 1.0–1.9. The results indicate that people in cooler climates favor cooler conditions than people in warmer climates. Therefore, it can be assumed that people in the UK should favor a cooler temperature range than those in tropical climate countries. As approved by the survey results in this research, the neutral temperature in Cardiff (UK) transitional space is about 4 °C lower than that in those tropical climate countries, while preferred temperature is 3 °C lower and the comfort temperature range is 4–5 °C lower. It can be seen that the comfort range in this study is wider than both these two surveys in Greece and the UK.

Table 2. Comparison of results from other field research related to indoor transitional space.

Location	Cardiff (Current study)	Malaysia	Bangkok, Thailand	Taichung, Taiwan	Greece	Sheffield, UK	Hong Kong, China	Jaipur, India
Space type	Foyer, café, corridor	Enclosed lift lobby	Lobby, foyer, atrium	Foyer	Atrium	Lobby	Fully enclosed lift lobbies and corridors	Corridor, entrance canopy, stairwell
Samples	245	113	1143	587	300	1794	84	50
clo value (clo)	1.24 (winter) 0.46 (summer)	0.62	0.53–0.65	0.54	-	0.72 (spring) 0.57 (summer) 1.01 (autumn) 1.06 (winter)	-	0.45–0.51
Metabolic rate (met)	1.0–1.5	1.2	1.0–1.9	1.0–1.2	-	0.7–3.8	-	1.3–1.4
Surveyed temperature range (°C)	20.2 (winter) 24.3 (summer)	23–32	23–32	20–30	10.2–16.6 (winter) 19.0–29.1 (summer)	21.9 (spring) 23.5 (summer) 21.2 (autumn) 20.0 (winter)	18.20–21.31 °C (winter)/ 25.62–27.32 °C (summer)	28.1 °C
Neutral temperature (°C)	20.8 (winter) 22.5 (summer)	-	26.5	26.3	14.98 (winter) 24.22 (summer)	-	-	-
Preferred temperature (°C)	21.4 (winter) 22.3 (summer)	-	25.5	24.5	-	-	-	-
Comfortable range (°C)	14.0–27.0 (winter) 14.5–27.8 (summer)	26.8	25.5–31.5	24.0–27.8	13.47–16.49 (winter) 22.71–25.73 (summer)	21–22 (spring) 23 (summer) 21.0 (Autumn) 19–20 (winter)	-	30.06–30.39

The previous studies also showed multiple space types in the transitional spaces of academic buildings, such as lift lobby, foyer, atrium, café, and corridor. Furthermore, occupants' activity level was different as well while in different space types, such as walking and standing occurring mainly in corridors or lift lobby, resting and meeting occurring mainly in resting areas, and eating and drinking occurring mainly in cafés). This may explain why people showed different neutral temperatures and acceptable temperature ranges in different transitional spaces. To further investigate the relationship between thermal comfort and the architecture design of indoor transitional spaces, a comparison between the research on thermal comfort in indoor transitional spaces of academic buildings and entertainment buildings is conducted.

3.2. Results from the Entertainment Building (with Comparisons to the Academic Building)

3.2.1. Descriptive Analysis

The total number of responses collected from the questionnaire surveys was 494, with detailed statistics listed in Table 3. The results reflect a good distribution in both winter and summer, as well as between males and females. It includes the measured values of both environmental and non-environmental parameters that affect occupant thermal sensations. In terms of age, the results indicate that people who visit TSAB are younger than those who visit TSEB. Due to the academic function of TSAB, quite a lot of visitors are students. In terms of clothing insulation, the results indicate that visitors wear more clothing in TSAB than in TSEB in winter, but TSAB's visitors wear less in summer. This is affected by the time participants stay in the spaces, especially in winter. If people stay in the transitional spaces for a very short time or just pass through the spaces, they will not take off their coat. It is indicated that participants in the transitional spaces of TSEB and TSAB have approximate cloth insulation. In summer, the transitional space occupants had very similar clothing insulation values, while in winter, the effect of outdoor weather resulted in distinct variations in clothing levels between the different individuals.

Table 3. Summary of the surveyed and monitored results.

		TSEB		**TSAB**	
		Winter	**Summer**	**Winter**	**Summer**
Total response (N)		120	129	123	122
Male respondents (N)		56	64	57	61
Female respondents (N)		64	65	66	51
Age	Mean	41	39	26	26
	SD	16	15	13	11
Clothing value (clo)	Mean	1.11	0.53	1.24	0.46
	SD	0.34	0.16	0.33	0.13
Activity level (met)	Mean	1.51	1.50	1.34	1.32
	SD	0.29	0.24	0.26	0.17
Indoor temperature (°C)	Mean	19.4	20.1	22.1	25.6
	SD	1.4	3.3	1.3	2.1
Relative humidity (°C)	Mean	41.0	64.1	45.0	56.0
	SD	5.6	8.9	8.4	8.2
Air velocity	Mean	0.20	0.15	0.12	0.19
	SD	0.10	0.22	0.13	0.18

Temperatures shown were the record taken during the time when the questionnaire survey was conducted.

Participants' metabolic range is indicated similarly in the transitional spaces of TSEB and TSAB. In TSEB, the metabolic rate of participants is similar to that in TSAB, even though the activity level is a little different: participants rarely use entertainment as a transitional space for study and working. The participants' activity in the entertainment transitional space was predominantly eating and drinking (1.5 met), social, watching performances (1.2 met), or sitting (1.0 met). In TSAB, this was predominantly reading and writing, social (meeting and talking), watching performances and working on computers (1.2 met) or sitting (1.0 met), but a small number of them were eating or drinking (1.5 met).

As mentioned in the research methodology, three major environmental parameters, namely, operative temperature (T_o), air velocity (AV), and relative humidity (RH), have been monitored in this study. It could be observed that in terms of average values, most parameters are similar between the two types of buildings, except the indoor operative temperature in summer (25.6 °C in the TSAB vs. 22.1 °C in the TSEB).

3.2.2. Comparisons of Occupant Thermal Comfort

The best that people can realistically hope to achieve is a thermal environment that satisfies the majority of people in the space, or, more simply, 'reasonable comfort'. Fanger claimed that, in general, a comfort zone is an environment situation in which 80% of the occupants feel satisfied with their environment, and the HSE (Health & Safety Executive) in the UK also considers 80% of occupants as a reasonable limit for the minimum number of people who should be thermally comfortable in an environment [27,28].

Thermal sensation: The statistical distributions of the survey participants' perceptions of the thermal environment are summarized in Figure 2. In winter, 55% of participants in TSEB expressed their thermal sensation as "neutral", while in TSAB, the percentage is 29%. In summer, 54% of participants in TSEB and 34% of TSAB expressed their thermal sensation as "neutral". In winter TSAB, almost 50% of the votes fell in the "warmer than neutral" region of the scale, i.e., including "slightly warm" (34%), "warm" (7%), and "hot" (3%) and 50% of the votes "cooler than neutral", i.e., including "slightly cool" (19%), "cool" (9%), and "cold" (2%). In entertainment, this ratio is less and relatively equal to 18% and 27% separately. In summer TSAB, more than ten times as many votes fell in the "warmer than neutral" region of the scale, i.e., including "slightly warm" (43%), "warm" (12%), and "hot" (6%), compared to the votes "cooler than neutral", i.e., including "slightly cool" (5%), "cool" (1%), and "cold"(0%), which shows a relatively equal ration in TSEB, as 24% and 23% separately, even the votes on "slightly cool" are 2% higher than "slightly warm".

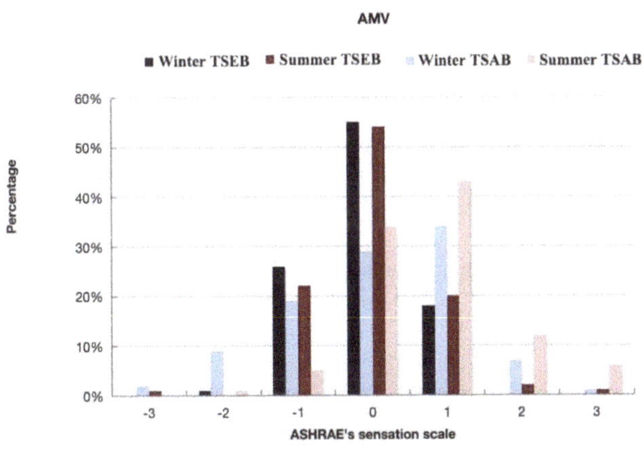

Figure 2. Distributions of thermal sensation votes in TSEB and TSAB.

The PPD thermal comfort index is based on the assumption that people voting in the middle three categories (i.e., "slightly cool" −1, "neutral" 0, and "slightly warm" +1) of the 7-point thermal sensation scale are satisfied with their thermal environment. Extending the assumption to the AMV in the survey of TSAB, 82% of the participants in the winter and summer were satisfied with their transitional space thermal conditions. But in TSEB, it is as high as 99% in winter and 98% in summer. By logical extension, votes on +2 (warm), +3 (hot), −2 (cool), and −3 (cold) can be regarded as an expression of thermal dissatisfaction, which in TSAB amounted to 18% both in winter and summer when in entertainment it was only 1% and 2%. It indicates that in both buildings, a quite high percentage of participants are satisfied with their thermal environment compared to the industry-accepted minimum standard of 80% acceptability, as recommended in regulatory documents such as ASHRAE's Standard 55 [29].

Thermal satisfaction: The data analysis indicated that the thermal conditions in the two surveyed transitional spaces in buildings regularly meet people's thermal requirements. In both cases, the satisfaction with the thermal environment according to the ASHRAE scale central categories (−1, 0, +1) meets the ASHRAE standards requirement of 80% [30]. Actually, it is far more than 80% in TSEB; the satisfactory rate is close to 100% both in winter and summer. In terms of direct acceptability, higher acceptability also happens at TSEB for 96% in winter and 94% in summer, whereas at TSAB it is 89% and 73%. It indicates that, according to the thermal sensation scale or direct acceptability, people in TSEB indoor transitional spaces have a much higher thermal satisfaction rate than in TSAB, even though the measured operative temperature in them is quite beyond the comfort temperature boundary.

Figure 3 indicates the operative temperature during survey time, together with the 80% and 90% acceptability temperature ranges. During the winter surveys, the temperature lies within the acceptable range of 80% in TSEB, and 85% in TSAB during the monitoring time. However, the summer operative temperature in all of these two cases was not 100% within the 80% acceptable range. Especially in TSAB, the percentage of operative temperature remains within that range for only 73% of the time, which highlights the periods of overheating. Additionally, the acceptable temperature range in winter TSEB is far wider than the actual operative temperature range, which also illustrates a high thermal tolerance of the environment for participants in TSEB.

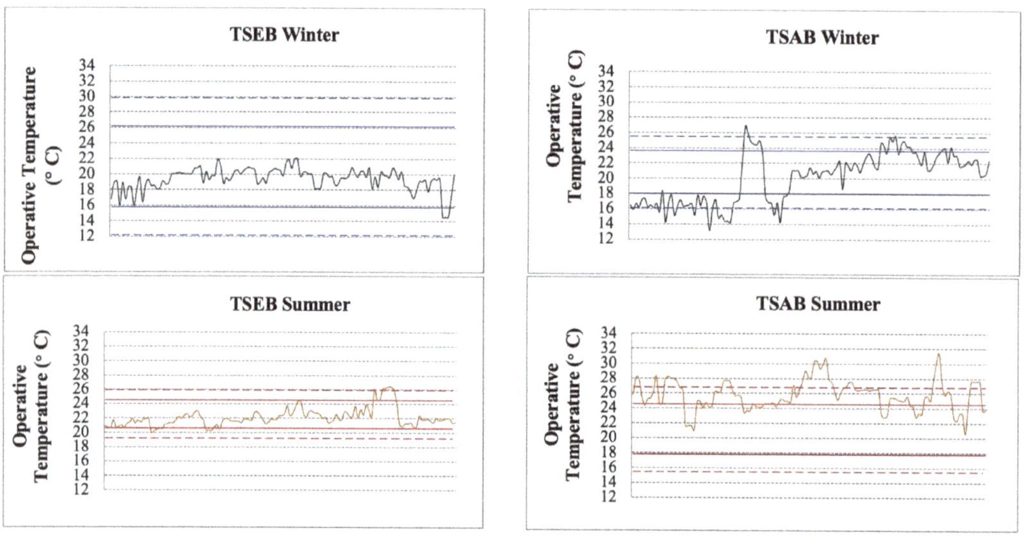

Figure 3. Operative temperature, 80% (dotted lines) and 90% (continuous lines) acceptability temperature ranges in winter and summer of two cases.

Thermal perception: This study defines the relative contribution of environmental and personal parameters (heat-balance parameters) to the thermal perception of subjects in indoor transitional spaces. To find which environmental parameters had a strong influence on thermal comfort in two indoor transitional spaces, two steps of statistical analysis were conducted. Firstly, a correlation analysis was carried out between AMV and the evaluation parameters (environmental and personal parameters). Secondly, a further ordinal regression analysis was carried out on the correlated environmental and the personal parameters with AMV.

Table 4 shows the results of the correlation analysis between AMV and the evaluated parameters in TSEB and TSAB, respectively. The results indicate that AMV is more correlated with operative temperature than with any other physical variable, with the associated coefficients being 0.256 for TSEB and 0.492 for TSAB (all significant at $p < 0.01$). In terms of personal parameters, AMV correlated to clothing insulation rather than activity, as indicated by the results of −0.019 for TSEB (significant at $p > 0.05$) and −0.327 for TSAB (significant at $p < 0.01$). Therefore, these variables are analyzed by using ordinal regression analysis (because the AMV value is an ordinal variable).

Table 4. Correlation analysis between AMV and evaluate parameters.

		T_o	AV	Rh	Clo	Met
TSEB	Pearson Correlation	0.256 **	0.008	0.050	0.010	−0.019
	Sig. (2-tailed)	0.000	0.899	0.431	0.875	0.766
TSAB	Pearson Correlation	0.492 **	0.069	0.067	−0.327 **	0.076
	Sig. (2-tailed)	0.000	0.279	0.296	0.000	0.235

* Correlation is significant at the 0.05 level (2-tailed). ** Correlation is significant at the 0.01 level (2-tailed).

Table 5 shows the results of the ordinal regression to predict the criterion variable (AMV). For the data collected in TSEB, operative temperature ($p < 0.001$) and clothing insulation ($p = 0.001$) are the significant variables and account for almost 10% of the variation in the actual sensation vote (AMV). However, $r^2 = 0.095$ indicates weakness in the ability of the predicted model to fit that data. By comparing it to clothing insulation, the Wald value also indicates that operative temperature (Wald = 26.801 $p < 0.001$) is the most important predictor that influences the actual thermal sensation votes of TSEB's participants. Clothing insulation for participants in TSEB was found to be significant in predicting the AMV; however, it had less influence on the AMV compared to operative temperature. The Wald value of clothing insulation is 10.715 and $p = 0.001$. In TSAB, explanatory variables include operative temperature and clothing insulation, which screen out correlation analysis. As can be seen, operative temperature is the significant variable ($p < 0.001$) and accounts for almost 24% (Cox and Snell $r^2 = 0.235$) of the variation in the actual sensation vote (AMV). However, $r^2 = 0.235$ indicates weakness in the ability of the predicted model to fit that data. The Wald value also indicates that operative temperature (Wald = 39.792 $p < 0.001$) is the most important predictor that influences the actual thermal sensation votes of TSAB's participants. This result suggests the importance of solar radiation intensity together with air temperature. Thus, the mitigation of solar and air temperatures is significant for the design of indoor transitional spaces, and these two parameters could have a great impact on the use of the indoor transitional spaces and may determine the number of people and activities in them.

Table 5. Ordinal regression statistics and best fit model to predict AMV using environmental and personal data from TSEB and TSAB.

	R Squared (Dependent Variable: AMV)	Wald		Chi-Square
TSEB	0.095	T_o	12.664	20.287
		Clo	0.375	
TSAB	0.235	T_o	39.792	15.974
		Clo	0.306	

Thermal preference: The preferred sensation is the sensation people actually expected, compared to the neutral sensation, which indicates people feel comfortable. The comparison of preferred sensations between different groups could help to explore differences or similarities in their thermal perception. The smaller difference between the neutral and preferred sensations of people relates to their good adaptation to the thermal environment. Figure 4 shows the distribution of survey participants' thermal preference votes in relation to their thermal sensation votes using probit analysis. In TSEB, it shows participants' preference did not coincide with the thermal neutral condition, but it was shifted slightly toward a positive value both in winter and summer on the sensation scale. In TSAB, this preference did not coincide with the thermal neutral condition but was shifted slightly toward a positive value in winter and a negative value in summer on the sensation scale. As thermal sensation increased (i.e., from cold to hot), the percentage of subjects voting for 'want cooler' generally increased. As one might expect, the percentage of those preferring to be warmer (i.e., 'want warmer' responses) tended to increase as thermal sensation decreased from warm to cool. The preferred thermal sensation indicates that in winter, both groups were a little warmer than neutral, with 0.21 and 0.22 sensation levels in TSAB and TSEB, while in summer, people prefer cooler with −0.18 and 0.02 sensation levels in TSAB and TSEB. The higher sensitivity level of 0.02 in summer proves the operative temperature in summer TSEB is lower than TSAB. The analysis of preference votes demonstrated an asymmetrical correlation between thermal sensation and thermal preference in TSAB both in winter and summer, and an asymmetrical correlation between thermal sensation and thermal preference in winter and a symmetrical correlation in summer in TSEB.

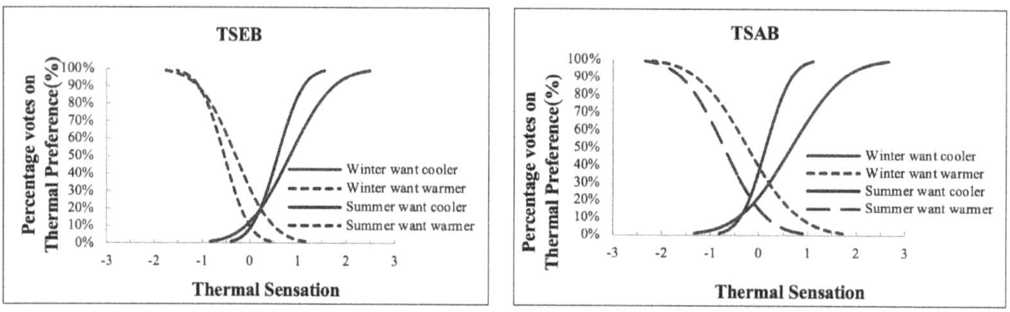

Figure 4. Percentage of thermal preference against thermal sensation in two cases.

3.2.3. Investigation of the Thermal Conflict

This study raised the question of how the thermal comfort requirements in indoor transitional spaces in both entertainment and academic buildings in UK climate, such as the thermal sensitivity and the neutral temperature. Additionally, if the different thermal comfort requirements in different indoor transitional spaces depend on the different functions (different ways of subjects using them). The analysis of thermal neutral temperature, preferred temperature, and comfort temperature range was explained in detail.

Neutral temperature: Neutral temperature is the thermal index value corresponding with a maximum number of building occupants voting neutral on a thermal sensation scale [31,32]. When a neutral temperature can be achieved, most people will feel thermally comfortable and accept the thermal environmental condition [25]. The average neutral temperature has been used in thermal comfort research to study the effects of experience on respondents' thermal perception [33]. A binned method was adopted by setting the increments of indoor operative temperature at half-degree Celsius in order to eliminate the outliers [34–37]. This study adopted linear regression to evaluate neutral temperatures, which has been used to investigate thermal comfort datasets since the 1930s [38,39].

Data collected from TSAB and TSEB were used to calculate the neutral temperatures to examine how people from different indoor transitional spaces adjust to their thermal perceptions. The "bin mean thermal sensation vote" rather than the individual actual votes was used to reduce individual differences [14,40]. This can be achieved by gathering several votes that correspond with half or more T_o degrees, depending on the highest value of r^2 obtained.

Because of the significance of the operative temperature as a predictor of the thermal sensation in the context of this study, it was therefore used as a thermal index to calculate the neutral temperature and examine the thermal sensitivity. The sensitivity of subjects' thermal sensation to operative temperature was evaluated by examining the mean thermal sensation vote response for each half-degree interval. The plotted data are in Figure 5, and the fitted regression lines for subjects' sensation prediction versus operative temperature in winter and summer are as follows:

Winter TSEB:

$$AMV = 0.162\ T_o - 3.264, \quad r^2 = 0.716 \tag{1}$$

Summer TSEB:

$$AMV = 0.254\ T_o - 5.730, \quad r^2 = 0.867 \tag{2}$$

Winter TSAB:

$$AMV = 0.180\ T_o - 3.740, \quad r^2 = 0.600 \tag{3}$$

Summer TSAB:

$$AMV = 0.150\ T_o - 3.173, \quad r^2 = 0.586 \tag{4}$$

In adaptive thermal comfort theory, it regards the gradient of this regression equation as being inversely proportional to the adaptability of the building occupants under analysis. A very shallow gradient indicates the subjects were able to adapt very effectively to changes in temperature (instead of feeling over- or under-heated and shifting their thermal sensation accordingly), whereas a steep regression line suggests the subjects were not successful in adapting because they quickly felt warm (or cool) as the room temperature shifted away from their neutrality. At more than five and six degrees per thermal sensation unit, the regression equation shows this sample to be remarkably successful at adapting to changes in indoor temperature.

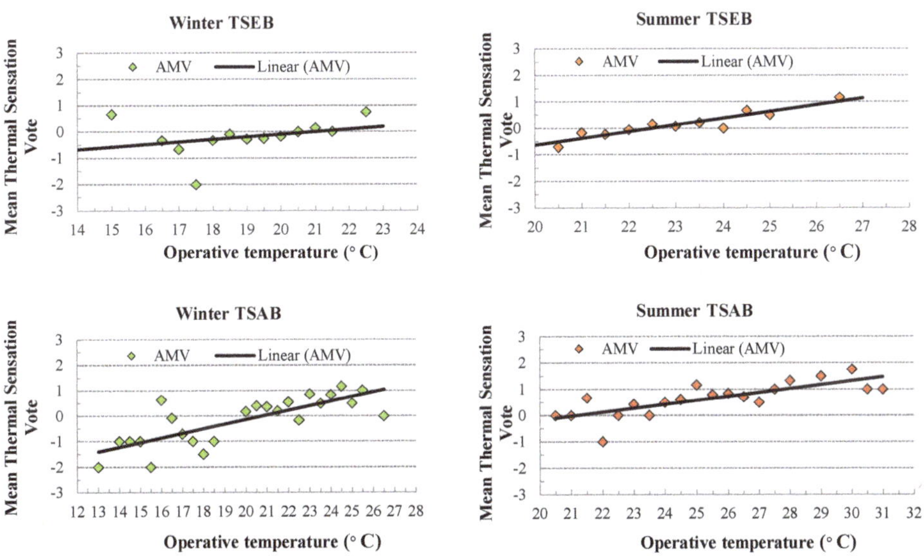

Figure 5. Mean observed sensation in indoor transitional spaces in winter and summer.

The neutral temperatures (Tn) can be calculated by using Equations (1) to (4) when AMV = 0. The actual neutral temperature in winter is 20.8 °C and 20.1 °C in TSAB and TSEB, while in summer it is 21.1 °C and 22.6 °C, respectively. As can be seen, the neutral temperature of the TSAB group is lower in winter but higher in summer. The result shows differences in thermal sensitivity and neutrality between these two indoor transitional spaces. This finding shows that the thermal requirements of people in indoor transitional spaces must be considered separately according to the function of the space. This is mainly due to the difference in the prevailing thermal environment in both of these two locations and the influence of thermal adaptive methods.

Preferred temperature: The preferred sensation and preferred temperature are the sensation and temperature people actually expected, compared to the neutral sensation and temperature in which people feel comfortable. The comparison of temperatures between different groups could help in exploring differences in their thermal perceptions or similarities. The smaller difference between the neutral and preferred choice of sensation and temperature for a group of people relates to their good adaptation to the thermal environment. Probit analysis is employed for advanced analysis in thermal studies to survey thermal preference sensations and calculate the preferred temperature. This method is used for thermal sensation assessments by Ballantyne, Hill and Spencer [41], which were conducted separately on the preferences of the participants in winter and summer for warmer and cooler conditions.

The cumulative frequency distributions for the "wanting warmer" and "wanting cooler" inclinations were plotted against the operative temperature scale of the environment in winter and summer (Figure 6). The point located at the intersection of the two cumulative curves corresponds to the participants' preference in terms of sensation. To investigate preferred temperatures, participants' preference votes were binned into half degree intervals of indoor operative temperature. The point of intersection between the "want cooler" and "want warmer" probit models is taken to represent the group's preferred temperature. According to the regression model, the preferred temperature in winter is 20.2 °C in TSAB and 21.6 °C in TSEB; in summer, the temperature is 21.5 °C and 21.9 °C separately.

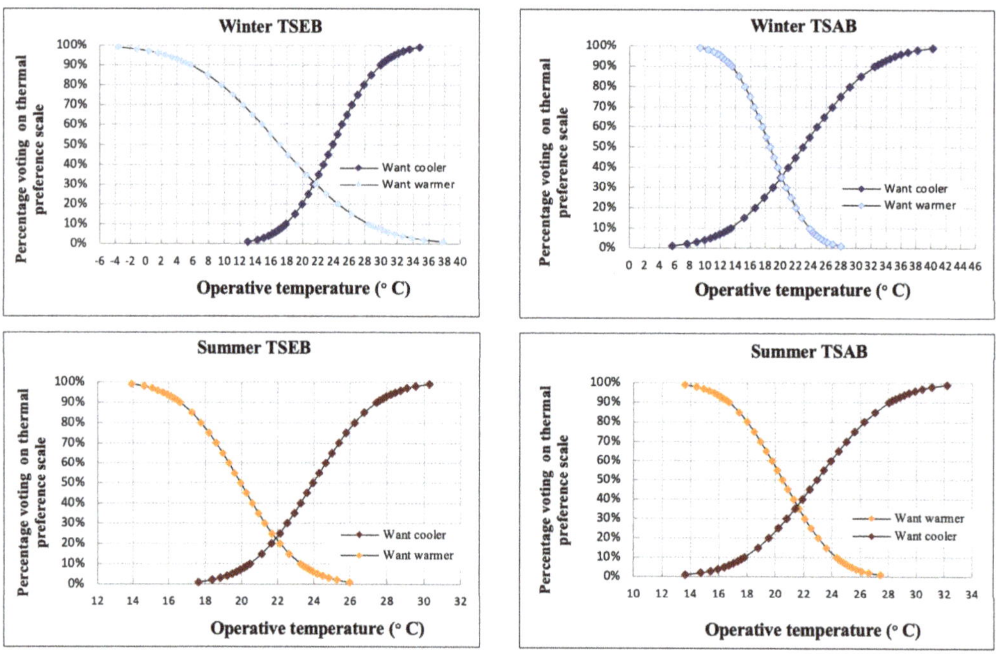

Figure 6. Probit regression models fitted to thermal preference percentages.

Thermal comfort range: regression equations describing the dependence of sample mean thermal sensation on mean indoor operative temperature are often used to define acceptable temperature limits for a particular sample. In the case of ASHRAE 55-2013, the so-called "comfort zone," as expressed on a temperature-humidity graph has its boundaries defined as −0.5 PMV on the cool side and +0.5 PMV on the warm side. Different from the fundamental logic adopted in ASHRAE and also ISO to define their comfort zones [25,30,42], some scholars research in the thermal comfort of transitional spaces and expand the comfort zone in these spaces as $-1 < \text{PMV} < +1$ [7], which can be applied to the results obtained in this thermal comfort survey of participants in transitional spaces in the present study. But with key differences, rather than using predicted mean thermal sensations (PMV), this survey has the advantage of actual mean thermal sensations. The mean indoor operative temperatures corresponding to mean thermal sensations of +1 and −1 stretch from 14 °C to 27.0 °C in winter and from 14.5 °C to 27.8 °C in summer (marked region on Figure 7).

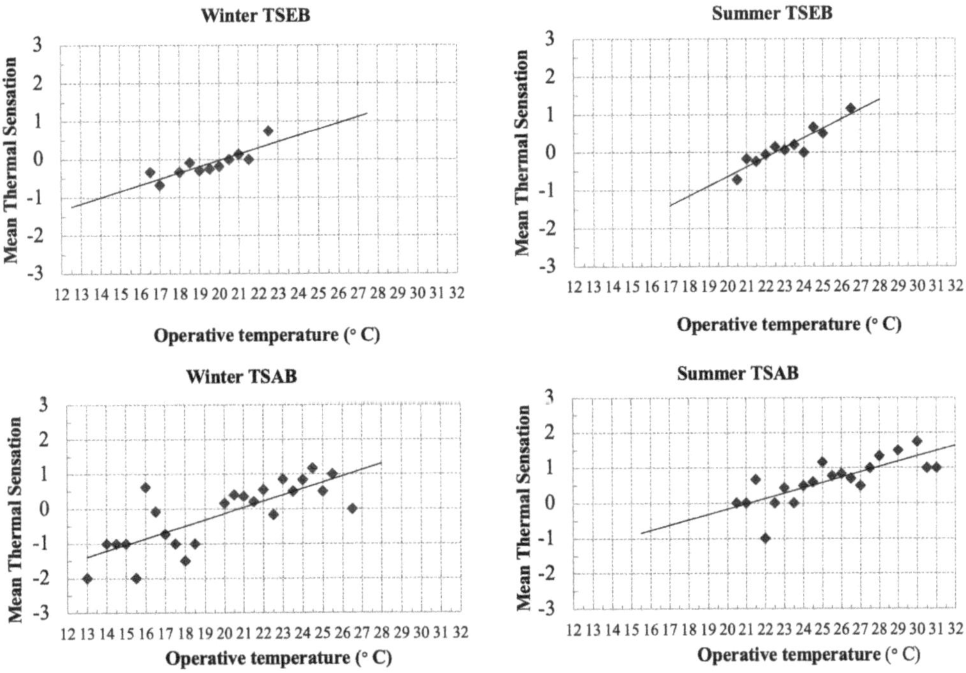

Figure 7. Mean thermal sensation votes (−3 = cold, through 0 = neutral to +3 = hot) related to indoor operative temperature in summer.

In TSEB, the mean indoor operative temperatures corresponding to group mean thermal sensations of +1 and −1 stretch from 14.0 °C to 26.3 °C in winter and from 18.6 °C to 26.5 °C in summer. In TSAB, the mean indoor operative temperatures corresponding to group mean thermal sensations of +1 and −1 stretch from 15.2 °C to 26.3 °C in winter and from 14.5 °C to 27.8 °C in summer. The widest comfort temperature range in winter is TSEB at 12.3 °C, and in summer it is TSAB at 13.3 °C. The results prove again that in winter TSEB participants and in summer TSAB participants had the highest tolerance for their thermal environment. This result is in line with previous findings about acceptable temperatures in transitional spaces [4,5,9,13,19].

3.3. Thermal Preference and the Use of the Spaces

Table 6 shows the most popular areas the participants chose to stay in and the popular reason for it in these two cases. It indicates that in both of these two indoor transitional spaces, thermal consideration truly is a reason with quiet proportion when people choose a favorite space to stay. But this reason is always not the most important one; people frequently put their use requirements as the most significant reason to choose a space to stay.

Table 6. Most popular area and popular reasons.

Transitional Space		TSEB	TSAB
Most Popular Area		**South Foyer Seat Area**	**South Foyer Seat Area**
Popular reason	For warm/cool	13%	25%
	For quite	6%	4%
	For light	19%	4%
	For facilities	13%	35%
	Good view		13%
	For performance	44%	
	For social	6%	5%
	Fresh air		9%
	For refreshment		9%
	No draft		

Figure 8 indicates that people in TSEB have a lower thermal preference rate for all of the stay reasons, but in TSAB, this rate is higher. It can be seen that the reason for passing through has the highest thermal non-preference rate, which means people passing through the indoor transitional space care less or less sensitivity about the thermal environment. The stay reason for working or study shows the highest thermal preference rate, which means people stay in the indoor transitional space because working or study cares more and more sensitivity to the thermal environment in the indoor transitional space. People staying in the indoor transitional spaces for thermal reasons (warm or cool) show a high preference rate, as expected.

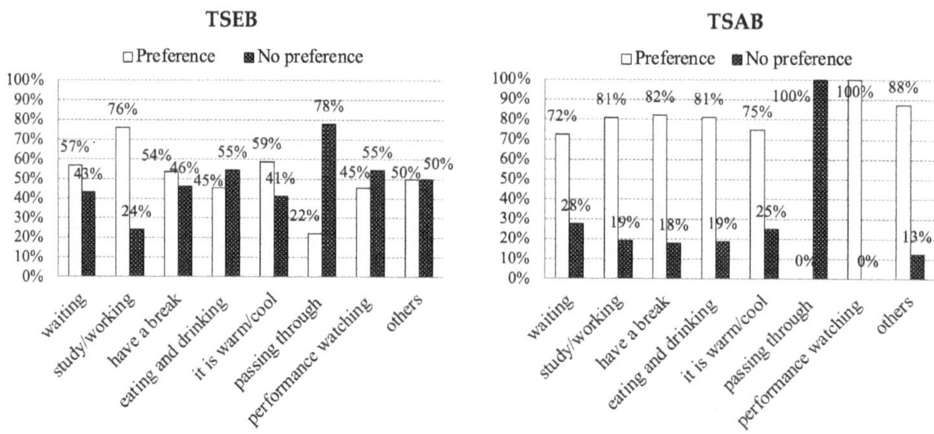

Figure 8. Thermal preference of participants depends on the stay reason.

The data analysis showed a high thermal comfort acceptable level of 96% and 94% of in winter and summer in TSEB, which is quite higher than in TSAB. Correlation analysis was applied in this study to investigate the relationship between the environmental variables and the actual thermal sensation votes of participants. Operative temperature appears to be the most important predictor of thermal sensation in two indoor transitional spaces, which could have an excessive impact on the use of the indoor transitional spaces in the UK climate and may determine the number of people and activities in them. Strong correlations were identified for the influences of the operative temperature on people's thermal sensations.

Similar trends were identified for the two surveyed buildings, where the gradients were all positive. In other words, the higher the indoor operative temperature, the warmer the thermal sensation the building occupants would have [36,39]. Both the design and management of indoor transitional spaces can influence the operative temperature in them, for instance, moderate the size of the glass facade, change the seat area to avoid direct sun light, and use a moderate heating–cooling system to mitigate the air temperature in indoor transitional spaces [43]. Additionally, the cloth insulation also significantly affects subjects' evaluation of their thermal sensation.

The investigation results of thermal sensation showed that people in TSAB were more thermally sensitive than people in TSEB in winter, as the slope value corresponds to 5.5 °C per sensation unit in TSAB and 6.1 °C in TSEB. However, in summer, people in TSEB were more thermally sensitive than people in TSAB; the corresponding degrees were 3.6 °C and 3.9 °C, respectively. This also means that people in winter in TSEB and people in summer in TSAB were thermally comfortable at a wider range of operative temperatures. The mean operative temperature in winter TSEB is lower when summer TSAB has a higher mean operative temperature. A possible explanation for why people in winter TSEB and summer TSAB were found to be less thermally sensitive might be due to the difference in clothing insulation and the way people use indoor transitional spaces (stay area, visit frequency, visit time, and stay time). A further analysis was carried out to quantify the acceptable temperatures in terms of neutral temperature, preferred temperature, and comfort temperature range. It shows the differences in acceptable temperature between the two different groups of participants in indoor transitional spaces. Excluding in summer TSAB, all other samples' results reveal that neutral temperatures lie above or are the same with the mean operative temperature occupants experienced. In summer TSAB, the mean operative temperature is as high as 4.5 °C than the neutral temperature. The comparison revealed that the summer group of participants in TSAB was more comforted by the cooler environment than the other three groups of participants. Additionally, the difference between the preferred temperature and the neutral temperature of the two groups demonstrates the occurrence of thermal adaptation. In addition, the comparison of a neutral temperature with a preferred temperature could explain which group is better adapted to its thermal environment. The differences between neutral temperature and preferred temperature in winter in academic and TSEB are 0.6 and 1.5 °C separately; in summer, they are 0.3 and 0.7 °C separately. The results prove that TSAB's participants have a better adaptation ability (smaller difference between neutral temperature and preferred temperature) to their thermal environment. The profile of the preferred temperature follows that of the neutral temperatures, and the two almost do not coincide. In TSEB, it demonstrates the preference for cooler temperatures in summer and warmer temperatures in winter. However, even in winter TSAB, participants still prefer a cooler environment, which explains the overheated thermal environment in it. This phenomenon is deduced due to the higher operative temperature in the sitting area in TSAB; even in winter, some participants complain it is too warm in this area. The high temperature of this area is gained through the big facade of glazing, and the north–south layout also results in longer sunshine time.

4. Conclusions

In this research, the field studies carried out in the indoor transitional spaces in different types of buildings show that people use the transitional spaces in different ways. There were 494 customers who attended the interview survey, and two surveyed buildings produced an evaluation of the thermal environmental performance and people's adaptability. The detailed calculation of thermal comfort requirement parameters includes neutral temperature, preference temperature, and comfort temperature range. The comfort temperature range is quite wide as in TSEB from 14.0 °C to 26.3 °C in winter and 18.6 °C to 26.5 °C in summer; and from 15.2 °C to 26.3 °C in winter and 14.5 °C to 27.8 °C in summer in TSAB. The results indicate that participants in TSEB's indoor transitional space have a higher thermal tolerance than in TSAB due to the different way of using the space, which also means the possibility of energy savings and specific design depends on the different functions of indoor transitional spaces.

The ordinal regression analysis was carried out to examine how the environmental variable related to the actual thermal sensation votes of participants. Operative temperature appeared to be the most important predictor of thermal sensation in two cases. In both entertainment and academic indoor transitional spaces, people had a wider thermal comfort temperature range than the CIBSE guide and fully occupied spaces such as offices in a moderate climate. It indicates participants in indoor transitional spaces had a higher tolerance for their thermal environment, which evidently suggests the possibility of useful energy savings through a modest (and realistic) relaxation of comfort standards regulation in transitional spaces. Additionally, this study has found that the way people use indoor transitional spaces has an important influence on participants' thermal requirements. The design of indoor transitional spaces has a significant influence on the thermal environment and the way people use indoor transitional spaces.

Environmental variables could have a great impact on the use of indoor transitional spaces in the UK's moderate climate. The thermal environment in the indoor transitional space can be determined by the design and setting of the cooling and heating system inside it. Design-related environmental improvements are necessary, but they may not be sufficient for a successful design of indoor transitional spaces. On the other hand, physical features appear to play an important role in attracting people to indoor transitional space, but thermal comfort plays a significant influence on how people use the indoor transitional spaces. Access to good indoor transitional spaces is a luxury or a need for people, which is determined by the function of the building and each area of indoor transitional spaces. Indoor transitional spaces that offer good design and allow social activities are likely to influence their users to stay longer.

The surveyed buildings served different purposes, and the people's activity inside the buildings was different. People visited TSEB for public events such as building tours and watching the show; this led to a relatively higher respondent's activity level when compared to TSAB because a greater number of people were walking or standing before taking the questionnaire surveys. On the other hand, in TSAB, more people used the transitional spaces for resting, dining, and discussion. This may explain why TSEB had a lower neutral temperature and a wider acceptable temperature range when compared to TSAB. From another perspective, different architectural designs of transitional spaces could influence thermal comfort [43–45]. In this study, it was explained that this may be due to the different people's usage and activity levels within the spaces as a result of architectural designs.

This statement was supported by the investigations of thermal preference and stay reasons. It indicated that thermal consideration truly is a reason with quiet proportion when people choose a favorite space to stay. But this reason is always not the most important one; people frequently put their use requirements as the most significant reason to choose a space to stay. However, the results of the analysis of thermal preference depending on the stay reason indicate that the participants' thermal perception has a close relationship with the participants' stay reasons in the indoor transitional spaces. Currently, transitional spaces are quite common in many types of buildings. In this study, clear evidence has been provided to justify the different thermal comfort requirements in academic buildings and entertainment buildings. From this conclusion, it would be reasonable to say that data need to be collected from other types of buildings with transitional spaces to control their thermal environment with higher energy efficiency.

This study focuses on human thermal comfort and the use of indoor transitional spaces in the UK. The studied sites are only located in Cardiff, UK, and the research focuses only on transitional spaces in academic and entertainment buildings. There are benefits and drawbacks to the field study methodology for research. The inability to directly regulate environmental factors and the challenge of accurately determining human physiological parameters are sources of limitations. However, field research is crucial to understanding thermal perception in an actual setting with regular operations. Moreover, only guests were considered in this study; staff were excluded from analysis to ensure that the sample presented participants who have self-adaptive freedom (staff were limited to moving their location and opening windows to modify thermal comfort conditions). Based on this research, further work needs to be conducted to cover more geographical areas within the UK climate since this study covers only Cardiff. Such an expansion may generalize the findings of this study or explain any particularities associated with the sites of the current study. Moreover, this investigation of the thermal comfort in indoor transitional space and the way people use the space should be further investigated. It has emphasized the important influence of improving the thermal environment in indoor transitional space and the possibility of energy saving in indoor transitional space.

Author Contributions: Conceptualization, G.H., C.T. and S.W.; Methodology, G.H. and S.W.; Software, G.H.; Validation, G.H.; Formal analysis, G.H.; Investigation, G.H.; Resources, G.H.; Data curation, G.H.; Writing—original draft, G.H.; Writing—review & editing, G.H., C.T. and S.W.; Visualization, G.H.; Supervision, C.T.; Funding acquisition, G.H. All authors have read and agreed to the published version of the manuscript.

Funding: This research was funded by Hebei University of Technology.

Data Availability Statement: Data is contained within the article.

Acknowledgments: The authors would like to express their gratitude to staff of the Glamorgan University and Millennium Center of Wales for their support to this study. Not only for providing the basic information about the buildings, but also lending the spaces for carrying out the field study data collections and physical measurements.

Conflicts of Interest: The authors declare no conflict of interest.

Abbreviations

ASHRAE	American Society of Heating, Refrigerating and Air-Conditioning Engineers
HVAC	Heating, Ventilation and Air-conditioning
AMV	Actual Mean Vote
ISO	International Organization for Standardization
PMV	Predicted Mean Vote
PPD	Predicted Percentage of Dissatisfied
CIBSE	Chartered Institution of Building Services Engineers
UK	United Kingdom
TSAB	Transitional Space in the Academic Building
TSEB	Transitional Space in the Entertainment Building

References

1. Pitts, A.; Saleh, J.B. Transition spaces and thermal comfort—Opportunities for optimising energy use. In Proceedings of the PLEA 2006—23rd International Conference on Passive and Low Energy Architecture, Geneva, Switzerland, 6–8 September 2006.
2. Adrian, P. Thermal Comfort in Transition Spaces. *Buildings* **2013**, *3*, 122–142. [CrossRef]
3. Ilham, S. *Thermal Comfort in Transitional Spaces in Desert Communities: The Study of Cases in Tucson*; The University of Arizona: Tucson, AZ, USA, 2006.
4. Chun, C.; Kwok, A.; Tamura, A. Thermal comfort in transitional spaces—Basic concepts: Literature review and trial measurement. *Build. Environ.* **2004**, *39*, 1187–1192. [CrossRef]
5. Pitts, A.; Saleh, J.B.; Sharples, S. Building Transition Spaces, Comfort and Energy Use. In Proceedings of the PLEA 2008—25th Conference on Passive and Low Energy Architecture, Dublin, Ireland, 22–24 October 2008.
6. Pitts, A.; Saleh, J.B. Potential for energy saving in building transition spaces. *Energy Build.* **2007**, *39*, 815–822. [CrossRef]
7. Pitts, A.; Douvlou-Beggiora, E. Post-occupancy analysis of comfort in glazed atrium spaces. In Proceedings of the Closing the Loop: Ways forward for Post-Occupancy Evaluation, Cumberland Lodge, Windsor, UK, 29 April–2 May 2004.
8. Jitkhajornwanich, K.; Pitts, A.C.; Malama, A.; Sharples, S. Thermal comfort in transitional spaces in the cool season of Bangkok. *ASHRAE Trans.* **1998**, *104*, 1181–1193.

9. Jitkhajornwanich, K.; Pitts, A. Interpretation of thermal responses of four subject groups in transitional spaces of buildings in Bangkok. *Build. Environ.* **2002**, *37*, 1193–1204. [CrossRef]
10. Wu, Y.-C.; Mahdavi, A. Assessment of thermal comfort under transitional conditions. *Build. Environ.* **2014**, *76*, 30–36. [CrossRef]
11. Kwong, Q.J.; Tang, S.H.; Adam, N.M. Thermal comfort evaluation of the enclosed transitional spaces in tropical building: Subjective response and computational fluid dynamics simulation. *J. Appl. Sci.* **2009**, *9*, 3480–3490. [CrossRef]
12. Jie Kwong, Q.; Adam, N.M. Perception of thermal comfort in the enclosed transitional spaces in tropical building. *Indoor Built Environ.* **2011**, *2011*, 524–533. [CrossRef]
13. Hui, S.C.M.; Jiang, J. Assessment of thermal comfort in transitional spaces. In Proceedings of the Joint Symposium 2014: Change in Building Services for Future, Hong Kong, China, 25 November 2014.
14. Hwang, R.L.; Yang, K.H.; Chen, C.P.; Wang, S.T. Subjective responses and comfort reception in transitional spaces for guests versus staff. *Build. Environ.* **2008**, *43*, 2013–2021. [CrossRef]
15. Kotopouleas, A.; Nikolopoulou, M. Understanding thermal comfort conditions in airport terminal buildings. In Proceedings of the 8th Windsor Conference: Counting the Cost of Comfort in a Changing World Cumberland Lodge, Windsor, UK, 10–13 April 2014; Network for Comfort and Energy Use in Buildings: London, UK, 2014. Available online: http://nceub.org.uk (accessed on 25 October 2014).
16. Avantaggiato, M.; Belleri, A.; Oberegger, U.F.; Pasut, W. Unlocking thermal comfort in transitional spaces: A field study in three Italian shopping centres. *Build. Environ.* **2021**, *188*, 107428. [CrossRef]
17. Monterio, L.M.; Alucci, M.P. Transitional spaces in São Paulo, Brazil: Mathematical modeling and empirical calubration for thermal comfort assessment. In Proceedings of the Building Performance Simulation Association Conference Exhibition, Beijing, China, 3–6 September 2007.
18. Rupp, R.F.; Vásquez, N.G.; Lamberts, R. A Review of Human Thermal Comfort in the Built Environment. *Energy Build.* **2015**, *105*, 178–205. [CrossRef]
19. Kotopouleas, A.; Nikolopoulou, M. Thermal comfort conditions in airport terminals: Indoor or transition spaces? *Build. Environ.* **2016**, *99*, 184–199. [CrossRef]
20. Hou, G. An Investigation of Thermal Comfort and the Use of Indoor Transitional Spaces. Ph.D. Thesis, Cardiff University, Cardiff, Wales, 2016.
21. Li, Y.; Geng, S.; Chen, F.; Li, C.; Zhang, X.; Dong, X. Evaluation of thermal sensation among customers: Results from field investigations in underground malls during summer in Nanjing, China. *Build. Environ.* **2018**, *136*, 28–37. [CrossRef]
22. Du, X.; Zhang, Y.; Lv, Z. Investigations and analysis of indoor environment quality of green and conventional shopping mall buildings based on customers' perception. *Build. Environ.* **2020**, *177*, 106851. [CrossRef]
23. Lu, P.; Li, J. Acceptable Temperature Steps for Occupants Moving between Air Conditioned Main Space and Naturally Ventilated Transitional Space of Building. *Build. Environ.* **2020**, *182*, 107150. [CrossRef]
24. *ISO 7726*; Ergonomics of the Thermal Environment. ISO: Geneva, Switzerland, 1998.
25. *ASHRAE Standard 55-2010*; Thermal Environmental Conditions for Human Occupancy. ASHRAE: Atlanta, GA, USA, 2010; ISSN 1041-2336.
26. Pallant, J. *SPSS Survival Manual: A Step by Step Guide to Data Analysis Using the SPSS Program*, 4th ed.; McGraw Hill: New York, NY, USA, 2010.
27. Fanger, P.O. *Thermal Comfort: Analysis and Applications in Environmental Engineering*; Danish Technical Press: Copenhagen, Denmark, 1970.
28. Health and Safety Statistics. 2007. Available online: https://www.hse.gov.uk/statistics/ (accessed on 21 June 2007).
29. *ANSI/ASHRAE Standard 55-2013*; Thermal Environmental Conditions for Human Occupancy, Atlaa. ASHRAE: Atlanta, GA, USA, 2013.
30. *ANSI/ASHRAE Standard 55-2004*; Thermal Environmental Conditions for Human Occupancy. American Society of Heating: Atlanta, GA, USA, 2004.
31. Humphreys, M. Field Studies of Thermal Comfort Compared and Applied. *Build. Serv. Eng.* **1976**, *44*, 5–27. [CrossRef]
32. Brager, G.S.; de Dear, R.J. Thermal adaptation in the built environment: A literature review. *Energy Build.* **1998**, *27*, 83–96. [CrossRef]
33. Lin, Z.; Deng, S. A study on the thermal comfort in sleeping environments in subtropics—Developing a thermal comfort model for sleeping environments. *Build. Environ.* **2008**, *43*, 70–80. [CrossRef]
34. Palme, M.; Gaona, G.; Lobato-Cordero, A.; Beltrán, R.; Gallardo, A. Evaluating thermal comfort in a naturally conditioned office in a temperate climate zone. *Buildings* **2016**, *6*, 27. [CrossRef]
35. Luo, M.; Ke, Z.; Ji, W.; Wang, Z.; Cao, B.; Zhou, X.; Zhu, Y. The time-scale of thermal comfort adaptation in heated and unheated buildings. *Build. Environ.* **2019**, *151*, 175–186. [CrossRef]
36. Khalid, W.; Zaki, S.A.; Rijal, H.B.; Yakub, F. Investigation of comfort temperature and thermal adaptation for patients and visitors in Malaysian hospitals. *Energy Build.* **2019**, *183*, 484–499. [CrossRef]
37. Wu, Z.; Li, N.; Wargocki, P.; Peng, J.; Li, J.; Cui, H. Adaptive Thermal Comfort in Naturally Ventilated Dormitory Buildings in Changsha. *Energy Build.* **2019**, *186*, 56–70. [CrossRef]
38. Nicol, F.; Humphreys, M.; Roaf, S. *Adaptive Thermal Comfort: Principles and Practice*; Taylor & Francis Group: London, UK, 2012. [CrossRef]

39. Zaki, A.; Damiati, S.A.; Rijal, H.B.; Hagishima, A.; Razak, A.A. Adaptive thermal comfort in university classrooms in Malaysia and Japan. *Build. Environ.* **2017**, *122*, 294–306. [CrossRef]
40. De Dear, R.J.; Brager, G.S. Thermal comfort in naturally ventilated buildings: Revisions to ASHRAE Standard 55. *Energy Build.* **2002**, *34*, 549–561. [CrossRef]
41. Ballantyne, E.R.; Hill, R.K.; Spencer, J.W. Probit analysis of thermal sensation assessments. *Int. J. Biometeorol.* **1977**, *21*, 29–43. [CrossRef] [PubMed]
42. *ISO 7730*; Moderate Thermal Environments—Determination of the PMV and PPD Indices and Specification of the Conditions for Thermal Comfort. International Organization for Standardization: Geneva, Switzerland, 2005.
43. Bodach, S.; Lang, W.; Hamhaber, J. Climate Responsive Building Design Strategies of Vernacular Architecture in Nepal. *Energy Build.* **2014**, *81*, 227–242. [CrossRef]
44. Dili, A.S.; Naseer, M.A.; Varghese, T.Z. Passive Control Methods for a Comfortable Indoor Environment: Comparative Investigation of Traditional and Modern Architecture of Kerala in Summer. *Energy Build.* **2011**, *43*, 653–664. [CrossRef]
45. Chandel, S.S.; Sharma, V.; Marwah, B.M. Review of energy efficient features in vernacular architecture for improving indoor thermal comfort conditions. *Renew. Sustain. Energy Rev.* **2016**, *65*, 459–477. [CrossRef]

buildings

MDPI

Article

Research on Temporal–Spatial Partition Control Strategies for Luminous and Thermal Environment in High Space of Gymnasiums

Yanpeng Wu *, **Kaikai Guo**, **Shaoxiong Li** and **Meitong Jin**

School of Civil and Resource Engineering, University of Science and Technology Beijing, Beijing 100083, China; m202310068@xs.ustb.edu.cn (K.G.); m202110070@xs.ustb.edu.cn (S.L.); m202310072@xs.ustb.edu.cn (M.J.)
* Correspondence: wuyanpeng@ustb.edu.cn

Abstract: The lighting design of large-space buildings in gymnasiums can impact the indoor luminous and thermal environment, resulting in an uneven light and thermal distribution. This paper investigates the luminous and thermal environment control strategies for high spaces in gymnasiums, by simulating the luminous and thermal environment under different lighting forms and establishing a comprehensive evaluation model. The results show that the weights of the indoor luminous environment, thermal environment, and comprehensive energy consumption change with season and time under different lighting forms, which provides a basis for developing a temporal–spatial partition control strategy. The temporal–spatial partition control strategy is proposed for summer and winter, including the shading angle control under the lighting forms of south-facing side windows, west-facing side windows, and top skylights. Under summer conditions, the south-facing side windows have no shading from 8:00 to 10:00 and 14:00 to 16:00, and the shading angle is 0° from 10:00 to 14:00; the west-facing side windows have no shading from 8:00 to 14:00, the shading angle is 0° from 14:00 to 15:00, and the shading angle is 15° from 15:00 to 16:00; and the top skylight has a shade angle of 15° from 8:00 to 9:00, 30° from 9:00 to 11:00, 45° from 11:00 to 13:00, and no shade from 13:00 to 16:00. Under winter conditions, the south-facing side windows have no shading all day; the west-facing side windows have no shading from 8:00 to 14:00, and the shading angle is 30° from 14:00 to 16:00; and the top skylight has no shading from 8:00 to 13:00, a shading angle of 45° from 13:00 to 15:00, and a shading angle of 75° from 15:00 to 16:00. This paper provides a set of scientific and reasonable luminous and thermal environment regulation strategies for large-space buildings, which can help optimize the building energy consumption and improve the indoor environment quality.

Keywords: large-space building; luminous and thermal environment; temporal–spatial partition; control strategy

1. Introduction

With the continuous advancement of urbanization in China, the share of energy use in the building sector in the total energy consumption of society has been increasing, which also brings about an increase in carbon emissions [1]. The use of large light windows in the form of high spaces in sports stadiums provides sufficient natural light and, at the same time, brings the challenge of controlling the luminous and thermal environment. Under direct sunlight conditions, drastic changes in indoor illuminance and thermal radiation intensity are likely to cause inhomogeneity in the distribution of the luminous and thermal environment [2,3], which not only affects the comfort of the athletes and spectators but also increases the energy consumption of the building. According to statistics, in the energy consumption structure, cooling and heating energy consumption account for as much as 40% to 50%, while the lighting component is about 28% of the total energy consumption [4]. Therefore, optimizing the luminous and thermal environment of stadiums is of great significance in reducing energy consumption and improving the quality of the indoor environment.

The quality of the building's luminous and thermal environment directly affects the indoor lighting as well as the building energy consumption [5,6], and, with the continuous improvement in the quality of life, people have higher requirements for the indoor light environment and thermal comfort. Due to the potentially contradictory relationship between indoor light and heat gain [7,8], continuous research on regulating and optimizing the luminous and thermal environment is needed to balance light and building energy consumption, seeking a relative balance of the luminous and thermal environment. A study on the effect of building shading on cooling energy consumption showed that the use of exterior shading devices, particularly exterior blinds that are 50% closed, significantly reduces seasonal cooling loads, reduces energy consumption, and lowers the operating costs of air conditioning systems [9]. Examining the interaction between light and temperature perception in indoor environments, research has shown that improved visual comfort through lighting can partially compensate for thermal discomfort, which may help to extend the acceptable range of indoor temperatures, resulting in reduced building energy consumption and possible health benefits [10]. A study on the effects of residential building design variables on daylighting and energy performance showed that the daylighting performance of the building improved with an increase in the window-to-wall ratio and the use of tinted double-glazing, but that these design choices could lead to overheating of the residential building, and, therefore, it is important to find the appropriate design variables to balance the daylighting and energy performance of the residential building in hot climates [11]. The optimization of windows and shading systems in building design is studied to achieve the best design of window systems for different orientations [12].

Energy consumption and the indoor environment have become more and more emphasized in building design [13], with artificial lighting as the main energy-consuming items in the building, accounting for 20–30% of the total electricity consumption of the building [14], and, at the same time, in some time periods, there are some indoor areas of the building over-illuminated, resulting in a waste of energy. Therefore, it is important to make full use of the natural outdoor light and cost-effectively manage the electric lighting. Relevant scholars have studied the indoor luminous and thermal environment under the lighting control method and explored the visual performance and electric lighting energy use of atrium buildings with high-frequency dimming control. The results showed that dimming control applies to sites with similar architectural layouts and lighting schemes, and can effectively reduce electric energy consumption and improve the lighting effects [15]. However, the study mainly focuses on the lighting control of atrium corridors, without fully considering the specific needs of other areas or different functional spaces, lacking an in-depth analysis of time-dynamic dimming strategies, and less research on the regulation of the thermal environment. The application of intelligent lighting control systems in large sports stadiums can significantly improve the energy efficiency and the comfort of the light and heat environment. By using the PID incremental control model and Karatsuba multiplication model, combined with the multi-level fuzzy comprehensive evaluation model, precise lighting control for different sports activities can be realized [16]. In addition, considering the utilization of natural light, by installing strip light windows and vertical shading devices, it is possible to increase the intensity of natural light on the east and west elevations while preventing glare [17]. In addition, some scholars have explored the energy and environmental performance of atrium spaces controlled using daylight-linked lighting, as well as its potential for energy savings [18].

The presence of large transparent envelopes in large-space buildings leads to high indoor energy consumption as well as visual and thermal discomfort, and shading devices are often used as passive regulation measures to shield transparent envelopes from excessive solar radiation and natural light [19]. For shading system regulation strategies, cooling load control strategies for non-insulated shading systems are not applicable during the heating season [20,21]. In some cases, shading systems implemented during the heating season can instead increase the building's energy consumption [22]. This phenomenon is possible for both manual and automated shading systems due to the opposing energy-saving mechanisms in summer and winter. Therefore, for shading regulation, the time-phased regulation should be optimized for different seasons. Conducting actual measurements and comfort surveys on the luminous and thermal environment of high-speed rail stations, the study found that the station is over-illuminated during certain periods, which easily causes glare, and, due to the extensive use of glass curtain walls, the indoor area is exposed to too much solar radiation, which affects the thermal comfort level, and, for the differences in the demand of different indoor areas, the dynamic design of shading regulation by a temporal–spatial partition is proposed [23]. However, the study used a questionnaire survey method, and the data had the bias of the subjective feelings of the respondents, which affected its accuracy and objectivity. A study on the lighting level and visual comfort of indoor sports facilities at the Institute of Physical Education (IPE) was conducted to evaluate the lighting performance of five representative indoor sports facilities in a comprehensive manner through field measurements and computer simulations, and the results showed that the proper maintenance, cleaning, and possible replacement of lamps and lanterns can significantly improve the lighting performance; in addition, the use of natural light plays a positive role in improving the level of indoor illumination but also poses a glare problem, which needs to be addressed through optimized control schemes and anti-glare measures [24].

Although previous researchers have made some contributions to the study of luminous and thermal environments in buildings, there are still some shortcomings. First, most of the existing studies have focused on a single performance of buildings, such as daylighting or energy consumption, and lacked the assessment of the comprehensive performance of the luminous and thermal environment. Second, the research on daylighting regulation strategies for large-space buildings fails to fully consider seasonal changes and light characteristics at different times of the day, resulting in the limited applicability and flexibility of regulation strategies. Finally, the existing studies rely on experts' experience and subjective judgment in the selection and weight distribution of evaluation indices, which lack objectivity and universality.

Aiming at the above problems, this paper aims to develop a temporal–spatial partition control strategy for large-space buildings such as gymnasiums by simulating the light and thermal environments under different lighting forms and establishing a comprehensive evaluation model. The study proposes a dynamic temporal–spatial partition control strategy that can be adjusted according to changes in natural light during the seasons and different times of the day by simulating the light and thermal environments under different lighting forms and establishing a comprehensive evaluation model that takes into account the indoor light environment, thermal environment, and energy consumption. It is found that the weights of the indoor light environment, thermal environment, and comprehensive energy consumption change during different seasons and time periods, which provides a scientific basis for the development of the corresponding control strategies. The recommendations for shading measures also differed between summer and winter to accommodate changes in natural light and to improve the comfort of the indoor environment. This study not only provides a new method for evaluating building performance, but also offers practical guidance for the sustainable design and operation of large buildings such as stadiums.

2. Methodology

2.1. Benchmark Modeling

Considering the comprehensive study of the photothermal environment, it is necessary to take into account both static and dynamic lighting simulation, and, at the same time, it is convenient to study the impact of different building design parameters on the indoor photothermal performance; therefore, we choose to use Honeybee and Ladybug to carry out the experimental simulation validation and subsequent simulation and analysis research.

Honeybee and Ladybug are plug-ins developed by the University of Pennsylvania in 2014 that are architected on the Rhino and Grasshopper parametric platforms [25]. The plug-ins are used to simulate lighting in building models by invoking the Daysim and Radiance computational engines. The joint use of this plug-in and the parametric modeling platform enables parametric performance simulation and analysis of buildings, realizing an efficient and integrated workflow of building model construction, performance simulation, and visualization and analysis of the results.

Considering that the study area is located in Beijing, a building model of a university gymnasium in Beijing is chosen as the benchmark model for the subsequent simulation study. Typical lighting methods in gymnasiums mainly include side window lighting and top skylight lighting. Due to the difference in indoor light and heat environments under different orientations of side window lighting, especially for south- and west-oriented lighting, south-oriented lighting has a better heat gain in winter, while west-oriented lighting is prone to causing obvious "west sun" phenomenon in summer, so the subsequent regulation simulation study focuses on the three lighting models of top skylight, and south-oriented and west-oriented side windows. Therefore, the subsequent modulation simulation study focuses on the top skylight, and south-facing and west-facing side window lighting models.

Three gymnasium lighting benchmark models of the top skylight, and southward and westward side windows were established on the modeling software Rhino (version 7.4) and Grasshopper (version 1.0) parametric platform, respectively, as shown in Figure 1. The dimensions of the stadium lighting benchmark model are 70 m × 40 m × 13 m, the roof form is horizontal, the side window form is a high side window, and the skylight form is a flat skylight [26]. The stadium is divided into the game area (game) and the two sides of the viewing area (watch1 and watch2); the size of the middle game area is 38 m × 40 m, and the size of the two sides of the viewing area is 32 m × 40 m. The division of the benchmark model area of the stadium is used as the basis for the subsequent zoning regulation.

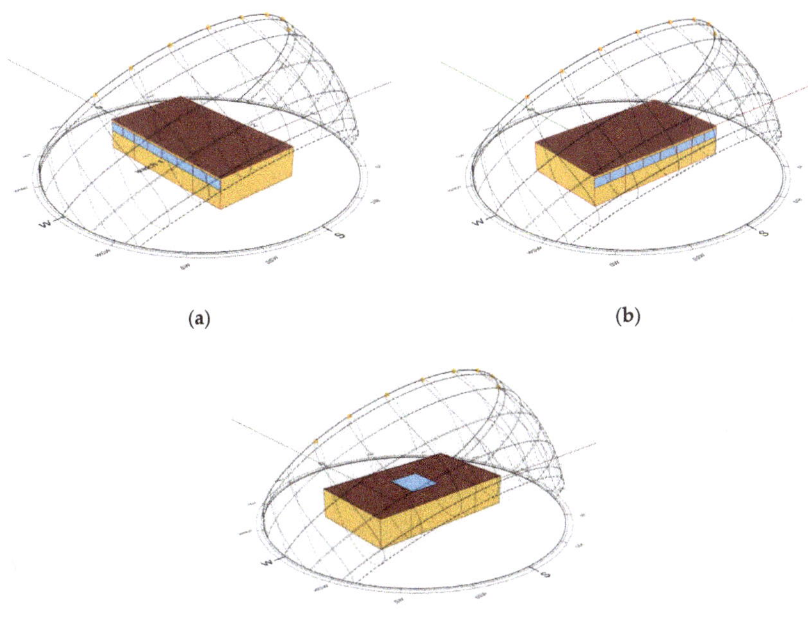

(a) (b) (c)

Figure 1. Benchmark model of high space gymnasium: (**a**) west-facing side window daylighting; (**b**) south-facing side window daylighting; and (**c**) skylight daylighting.

The window-to-wall ratio of the side window of the model is 0.3, and the ratio of the skylight to the roof area is 0.08. The optical and thermophysical parameters of the baseline model are shown in Table 1; the internal heat source parameters in the model, such as indoor personnel, lighting, and air conditioning, are shown in Table 2. Among them, the personnel, lighting, and equipment operation schedules and air-conditioning turn-on time are set according to the school building type in the General code for energy efficiency and renewable energy application in buildings. The target illuminance for the baseline model of the stadium was set to 300 lx [27].

Table 1. Optical and thermal property parameters of the enclosure.

Enclosure	Heat Transfer Coefficient [W/(m^2·K)]	Solar Heat Gain Coefficient (SHGC)	Visible Light Transmittance (VT)	Reflectance
External wall	0.35	-	-	0.75
Roofing	0.18	-	-	0.58
Floor	0.85	-	-	0.2
Side window	1.8	0.37	0.62	-
Skylights	1.3	0.34	0.39	-

Table 2. Heat source parameters in the model.

Indoor Parameters	Setting Values
Density of personnel (Person/m^2)	0.167
Lighting power density (W/m^2)	8
Equipment power (W/m^2)	5
Heating design temperature (°C)	20
Refrigeration design temperature (°C)	26
Illumination standard value (lx)	300

The coefficient of performance (COP) of the air conditioner is set to 3. The lighting system adopts continuous dimming control, which is based on the actual indoor natural light level and the set target light level for continuous dimming. When the indoor natural light exceeds the target illuminance, the artificial lighting will be automatically dimmed to reduce lighting energy consumption.

2.2. Temporal–Spatial Partition Regulation Simulation Scheme

In order to facilitate the simulation and analysis of building performance based on the preliminary model, Ladybug Tools, an indoor lighting and energy simulation software based on the Grasshopper (version 1.0) parametric platform, was selected. Ladybug Tools integrates Radiance, Energyplus, and Daysim simulation kernels for evaluating the performance of lighting, thermal comfort, and energy consumption of buildings. Ladybug Tools includes Ladybug and Honeybee plug-ins, of which Ladybug is mainly used for climate and thermal comfort analysis, with which you can import meteorological data from different regions and directly input the meteorological files for the analysis of the building's thermal performance. Honeybee builds on this by invoking the Radiance and Energyplus simulation engines for daylighting and energy simulation.

The simulated weather files are in epw format, which contain the typical meteorological year-by-year data of Beijing in China Weather Database (CWSD). In the lighting simulation calculation, the height of the reference plane is set as the height of the human standing point of view, 1.5 m, in order to be closer to the actual illumination of the human eye, and, considering the simulation accuracy and time cost, the quality of the lighting simulation is selected to be medium-quality, and Radiance is set for the specific relevant environmental parameters and analysis: the number of ambient reflections is 5, the ring ambient subsampling value is 2048, the ambient oversampling value is 512, the ambient resolution is 512, the ambient accuracy is 0.08, the simulation grid is 2 m × 2 m (700 sensor points), and the daylighting simulation time is from 8:00 to 17:00 on weekdays.

Sub-area regulation adopts sub-area lighting dimming. Due to the existence of two different types of functional areas within the stadium building, this results in different lighting needs for the different areas of the lighting level division. Lighting is mainly used for competition training and recreational activities, while the spectator area has relatively low lighting requirements, so the competition area and the spectator area are divided into lighting levels when carrying out lighting zoning.

As shown in Figure 2, the center of the stadium game area (game) is divided into lighting level 1 area, while the long side of the two sides of the viewing area (watch1 and watch2) is used for personnel to watch the game, which is divided into lighting level 2 area. The target illuminance of the first three areas of the lighting partition is 300 lx; after the lighting level division, the target illuminance value of the game area remains unchanged, and the target illuminance value of the viewing area is reduced to 150 lx, to better meet the needs of different functional areas.

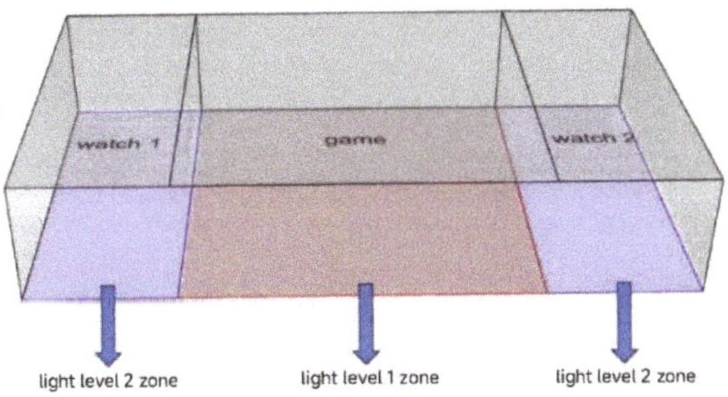

Figure 2. Lighting zoning diagram of a typical model of high-space gymnasium.

As shown in Figure 3, each regional lighting simulation in the grid is placed in the lighting dimming sensors; different regional lighting dimming is still using continuous dimming control, each region based on its actual illuminance value and the size of the target illuminance value of the automatic lighting dimming control, to achieve the actual illumination needs by the region for sub-regional lighting control.

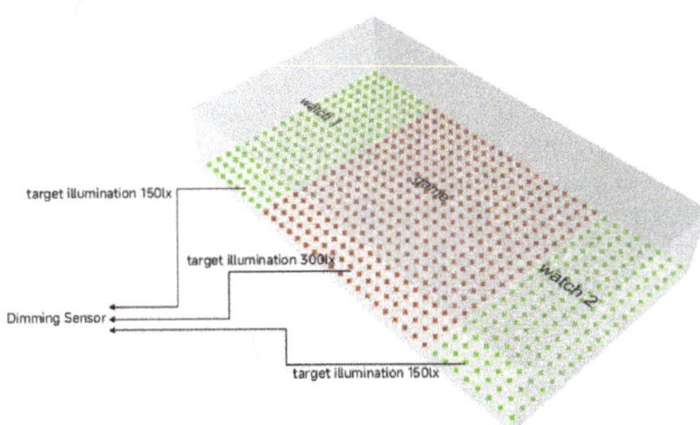

Figure 3. Gymnasium lighting zoning control diagram.

From the perspective of energy saving, external shading is more effective than internal shading [28], and, because the louvered shading set outside the windows has the ability to control solar heat gain and glare, its flexible and adjustable characteristics can be adjusted according to the outdoor weather changes, which is more suitable for time-phased regulation in this study [29]. Therefore, for the three stadium benchmark models, the time-phased regulation means of external louvered shading was selected. The time-sharing control starts from the building operation process, considering the influence of outdoor natural light on indoor light and the thermal environment under different time periods, and adopts the time-sharing control of exterior louver shading. The settings of exterior louver shading for different lighting forms are shown in Figure 4. To determine the key parameters of louver shading regulation, three exterior louver shading parameters are selected, which are shading angle (SA), sunshade length (SL), and shading distance from the window (SD). Through the sensitivity analysis of design parameters, it is concluded that SA has a more significant impact on the indoor performance parameters under different lighting forms, and, at the same time, taking into account the convenience of regulation, SA is selected as the sunshade regulation parameter for the study of sunshade regulation at different time periods, with a change range of 0°~75° and a change step of 15°. The length of the exterior louver shading boards and the distance of the shading from the window were determined mainly in consideration of being able to completely block the outdoor natural light under a fixed number of shading boards, so the length of the shading boards was set to be equal to the spacing of the shading boards, and the distance of the shading from the window was set to 0; i.e., the louver shading boards were close to the windows. The length of the exterior louver shading panels is set to 0.6 m for the west- and south-facing side window shading models and the length of the exterior louver shading panels is set to 0.8 m for the skylight model.

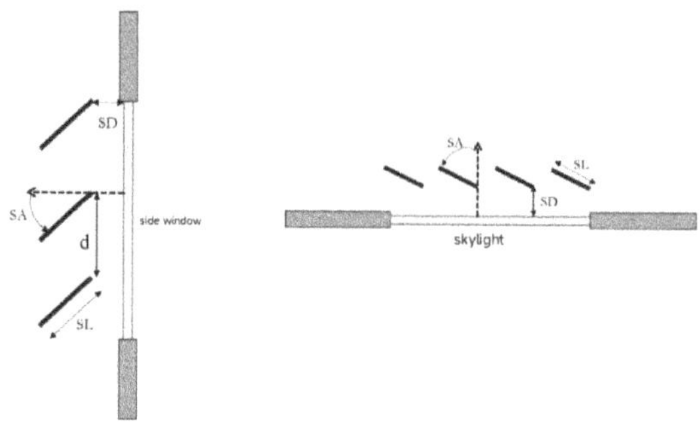

(**a**) (**b**)

Figure 4. Exterior louver shading diagram of different lighting forms: (**a**) side window sunshade; and (**b**) skylight sunshade.

The temporal–spatial partition regulation simulation study time was chosen to be 21 June, a typical day in summer, and 21 December, a typical day in winter, from 9:00 to 16:00. To analyze the comprehensive dynamic change in indoor natural light in time and space, the dynamic light environment indices and year-round energy consumption indices were used for the evaluation. Daylight Autonomy (DA), Spatial Daylight Autonomy (sDA), and Useful Daylight Illuminance (UDI) were selected as the dynamic light evaluation indices, and the lighting, cooling, and heating energy consumption were used as the year-round energy consumption evaluation indices. DA is the percentage of natural lighting throughout the year, which refers to the percentage of time, expressed as a percentage, that each lighting measurement point on the building's reference surface is able to achieve the target illuminance value by relying on natural lighting alone, during working hours throughout the year. sDA is the lighting threshold occupancy ratio, which can also be called the autonomous lighting area ratio, is a lighting evaluation index derived based on the DA value, which refers to the ratio of the area of the area in the building where the DA value exceeds the target time to the area of the total lighting area, and can be used to evaluate the spatial distribution characteristics of indoor lighting. sDA was calculated in this study by setting the target time to 50%. UDI is the effective natural lighting illuminance, which refers to the percentage of time that the natural lighting illuminance value is in the effective range during the working hours of the whole year, and UDI is divided into three indices, which are the percentage of time with insufficient illuminance, the percentage of time with comfortable illuminance, and the percentage of time with too much illuminance, and are expressed by UDI-l, UDI-a, and UDI-u, respectively, in this study, and the range of the illuminance in UDI before and after the lighting zoning is shown in Table 3. On this basis, sUDI is adopted as the lighting evaluation index for time regulation. sUDI is defined in this study as the percentage of the area within a specific illuminance range over the total area at a certain moment. The sUDI was divided into three types, sUDI-l, sUDI-a, and sUDI-u, in which sUDI-l indicates the percentage of the area that is under illuminated, sUDI-a indicates the percentage of the area that is in the effective illuminance range, and sUDI-u indicates the percentage of the area that is under excessive illuminance. In the simulation evaluation of shading regulation in the time period, the illuminance thresholds of the three indicators in each region were selected as the values after daylighting partitioning in Table 3.

Table 3. Illuminance threshold range of lighting index before and after lighting partition.

Zone	Functionality	sUDI-l		sUDI-a		sUDI-u	
		Pre-District	Post-District	Pre-District	Post-District	Pre-District	Post-District
watch1	Spectator	<300 lx	<150 lx	300 lx~2000 lx	150 lx~2000 lx	>2000 lx	>2000 lx
game	Campaigns	<300 lx	<300 lx	300 lx~2000 lx	300 lx~2000 lx	>2000 lx	>2000 lx
watch2	Spectator	<300 lx	<150 lx	300 lx~2000 lx	150 lx~2000 lx	>2000 lx	>2000 lx

Since the air-conditioning system is used in the stadium model, Predicted Mean Vote (PMV) is selected as the thermal environment evaluation index for time-sharing control. PMV is the Predicted Mean Voting Value, an evaluation index describing indoor thermal comfort, reflecting a subjective feeling of the human body towards the thermal environment; when PMV is close to 0, it is considered to be more comfortable, and the size of the PMV index is related to the indoor related parameters. In addition to the athletes' competition area, the overall situation is considered because the environmental control parameters of the spectator stand are the same. For the thermal comfort calculation in different seasons, the thermal comfort calculation conditions in summer and winter are as follows: clothing thermal resistance (clo) is 1 for winter conditions and 0.5 for summer conditions; metabolic rate (met) is 1.2 for winter conditions and 1.2 for summer conditions; and air flow rate (m/s) is 0.1 for winter conditions and 0.1 for summer conditions. The building energy consumption is mainly considered to be lighting, cooling, and heating energy consumption, and the sum of the three energy consumption types under each moment of a typical day is selected as the comprehensive energy consumption evaluation index for time-sharing control.

2.3. *Establishment of a Comprehensive Evaluation Model for Luminous and Thermal Environment by Entropy Weight Method*

To carry out an objective and comprehensive analysis of each performance indicator, this paper adopts the entropy weight method to establish a comprehensive evaluation model for the overall indoor light and thermal and comprehensive energy consumption indicators of the gymnasium. The entropy weighting method gives comprehensive consideration to all kinds of influencing factors, and the method does not introduce the subjective ideas of the decision maker and is objectively assigned through the entropy value of the indices, and the resulting weights are characterized by strong objectivity and high reliability [30–32]. The steps of applying the entropy weight method to calculate the weights are as follows:

Firstly, the overall lighting, thermal comfort, and comprehensive energy consumption of the stadium are selected as the three evaluation indices; due to the different scales of each index, it is necessary to standardize the raw data of each evaluation index of the samples, and there are m samples and n evaluation indices set, which results in the raw data matrix

$\{x_{ij}\}_{mn}$ (i = 1, 2, 3, . . ., m; j = 1, 2, 3, . . ., n). Under different working conditions, different indicators represent different meanings; for positive indicators (the larger the value of the indicator, the better) and negative indicators (the smaller the value of the indicator, the better), treatments are shown in Equations (1) and (2), respectively.

$$m_{ij} = \frac{x_{ij} - \min(x_{ij})}{\max(x_{ij}) - \min(x_{ij})} \tag{1}$$

$$m_{ij} = \frac{\max(x_{ij}) - x_{ij}}{\max(x_{ij}) - \min(x_{ij})} \tag{2}$$

where m_{ij} is the standardized data; x_{ij} is the original data of the sample; and max(x_{ij}) and min(x_{ij}) are the maximum and minimum values of the original data of the sample.

From the above normalization calculation, a normalization matrix M can be obtained as in Equation (3):

$$\mathrm{M} = \{m_{ij}\}_{mn} \tag{3}$$

Then, the information entropy value e_j of the j-th indicator is calculated as shown in Equation (4), where $k = \frac{1}{\ln m}$, $p_{ij} = \frac{m_{ij}}{\sum_{i=1}^{m} m_{ij}}$.

$$e_j = -k\sum_{i=1}^{m} p_{ij} \ln p_{ij} \tag{4}$$

Next, the weight w_j of each indicator is calculated by calculating the information entropy value obtained for each indicator, as shown in Equation (5), where $h_j = 1 - e_j$.

$$w_j = \frac{h_j}{\sum\limits_{j=1}^{n} h_j} \tag{5}$$

The weights of light environment, thermal environment, and comprehensive energy consumption are finally obtained as w_1, w_2, and w_3, respectively, which are brought into Equation (6) to obtain the comprehensive evaluation index Z.

$$Z = m_{i1} \times w_1 + m_{i2} \times w_2 + m_{i3} \times w_3 \tag{6}$$

According to the simulation results, sUDI, PMV, and comprehensive energy consumption are used to establish the comprehensive evaluation index, of which the sUDI index is defined as the percentage of the area in the effective illumination range of the gymnasium. In the calculation of the comprehensive evaluation index, the weight of each index is multiplied by the standardized raw data, so that the final evaluation index is unified in the range of 0 to 1. The index is used to carry out a comprehensive evaluation of the indoor luminous and thermal environment and energy consumption and to find a reasonable temporal–spatial partition control strategy.

3. Results and Discussion

3.1. Influence of Daylighting Zoning Regulation on Indoor Light and Heat Environments

3.1.1. Dynamic Daylighting Evaluation Analysis

The dynamic lighting simulation before and after lighting zoning was carried out for different lighting forms of gymnasium models, and the changes in the distribution of lighting indices of the gymnasium before and after lighting zoning were compared and analyzed, so as to analyze the impact of lighting zoning from the perspective of year-round dynamic lighting evaluation.

Table 4 shows the dynamic lighting indicators before and after the lighting partition of the gymnasium model; it can be seen that, after the lighting partition, different lighting forms of the gymnasium model of the dynamic lighting indicators have been improved to a certain extent, indicating that, after the lighting partition, the gymnasium is more in line with the actual demand for lighting of the overall evaluation of lighting. At the same time, the lighting partition takes into account the different indoor lighting needs of different areas, and sets a more reasonable illumination threshold for each area of the stadium, which provides a basis for the subsequent partition lighting dimming control, and then reasonably controls the indoor lighting on state, and realizes the purpose of rationally utilizing natural light to save energy.

Table 4. Dynamic lighting index before and after lighting partition of gymnasium model.

Lighting Model	Daylighting before Zoning			After Light Zoning		
	DA (%)	UDI-a (%)	sDA	DA (%)	UDI-a (%)	sDA
West-facing side window	57.2	52.87	0.56	69	64.65	0.76
South-facing side window	72.98	63.34	1	80.14	70.47	1
Top skylights	46.58	38.23	0.5	62.29	53.95	0.7

3.1.2. Annual Energy Consumption Analysis

On the basis of the above lighting zoning, each lighting area of the gymnasium is controlled by zoned lighting dimming, summarizing and analyzing the energy consumption indices of the gymnasium before and after the zoned lighting dimming control in different lighting forms for the whole year. Before the partitioned lighting control, the gymnasium adopts a uniform target illuminance of 300 lx for lighting dimming, and, after the partitioned lighting control, the lighting dimming control is carried out according to the standard value of illuminance after the lighting partition, and separate lighting dimming control is carried out for each area.

The results of the annual energy consumption indicators are shown in Table 5, which shows that, under the three lighting forms, although there is a certain increase in heating energy consumption after the zoned lighting control, the cooling and lighting energy consumption is less, and the total energy consumption can be reduced. The results show that zoned lighting control results in the overall energy efficiency of the stadium being improved, with a more efficient use of natural light, especially in the form of skylight lighting, so that the energy saving effect of natural light is more significant, and the total energy consumption is reduced by 4.85%. In the end, zoned lighting control makes the indoor lighting more reasonable, and achieves the purpose of lighting on demand and energy saving, which proves the necessity of zoned lighting control.

Table 5. Energy consumption before and after zoning lighting regulation.

Lighting Model	Energy Consumption before Zonal Regulation (kWh/m^2)				Energy Consumption after Zoning Control (kWh/m^2)			
	Illumination	Heating	Refrigeration	Subtotal	Illumination	Heating	Refrigeration	Subtotal
West-facing side window	8.5	12.01	7.68	28.19	7.99	12.07	7.61	27.67
South-facing side window	8.4	10.56	6.4	25.36	7.96	10.59	6.33	24.88
Top skylights	10.43	11.52	8.16	30.11	9.04	11.63	7.98	28.65

3.2. Simulation Results of Temporal–Spatial Partition Control

The light environment simulation, thermal environment simulation, and comprehensive energy consumption simulation are carried out for the three lighting forms, respectively, and the simulation results of the top skylight lighting form are selected for analysis.

3.2.1. Light Environment Simulation Results

The simulation results of a typical daylight environment in summer for top skylight lighting are shown in Figure 5, and the changes of sUDI indices in different lighting areas in the sunless room are shown in Figure 5a, and there are obvious opposite trends in the watch1 and watch2 areas. In the morning time, the sunlight is concentrated in the west side of the room, which makes the light quality of the watch1 area better, while the sUDI-l value of the watch2 area located in the east side reaches the highest value of 98%, and there is an obvious illuminance shortage phenomenon. As the time goes towards the afternoon, the change in the sun's orientation makes the daylight gradually move to the east side, so that the lighting quality of the watch2 area is greatly improved, while the lighting quality of the watch1 area is gradually deteriorating, and there is a serious lighting deficiency phenomenon. Due to the high solar altitude during most of the summer, daylight moves around the central game area, which makes the game area over-illuminated at all times (sUDI-u), while the fluctuation of its lighting quality is less than that of the remaining two areas. From Figure 5b–e, we can see the effect of shading angle on indoor lighting, at 9:00 and 14:00, the increase in shading angle makes the sUDI-a of the watch1 and game regions continue to decrease, and the quality of lighting gradually deteriorates, and the quality of lighting for the watch2 region increases slightly and then continues to decrease. At 12:00, with the increase in shading angle, the sUDI-a of the game and watch2 regions increased slightly and then decreased, and the sUDI-a of the watch1 region continued to decrease, and the lighting quality gradually became worse. At 16:00, the lighting quality of each region increased and then decreased with the increase in shading angle overall; especially for the watch2 region, the sUDI-a increased significantly with the increase in shading angle. Compared with the no-shading condition, moderate shading at all moments improves the excessive illumination in the game area and improves the overall lighting quality. At 9:00 and 16:00, moderate shading improves the illuminance deficit in the watch2 and watch1 areas, respectively, and improves the lighting quality.

Under a typical day in winter, the light environment simulation results are shown in Figure 6. From Figure 6a, it can be seen that, unlike the side window lighting, under the top skylight lighting, the lower sun altitude in winter makes the indoor lighting conditions worse compared to the summer. The percentage of under-illumination (sUDI-l) in the watch1 and watch2 areas reached 100% in the afternoon and morning hours, respectively, resulting in severe under-illumination. sUDI-l in the watch1 area reached a maximum of 76% in the morning hours due to the low sun altitude, resulting in poorer lighting quality overall. The game area saw a decrease in the percentage of over-illumination (sUDI-U) compared to the summer months, but an increase in the percentage of under-illumination (sUDI-l) in all moments compared to the summer months, resulting in poorer lighting quality overall compared to the summer months. The overall lighting quality was worse compared to summer. The effect of shading angle on the lighting in each area of the room is shown in Figure 6b–e; at 9:00 and 12:00, the increase in shading angle makes the sUDI-a in the watch1 and game areas decrease continuously, and the quality of the lighting becomes worse, and the sUDI-a in the watch2 area rises and then decreases with the increase in the shading angle. At 14:00, the sUDI-a of all regions increased with the increase in shading angle and then decreased. At 16:00, with the increase in shading angle, the light quality of the watch1 and watch2 regions did not change significantly as a whole, and the insufficient light phenomenon of the game region was improved, and the light quality first increased and then decreased. Compared with no-shading conditions, moderate shading can improve the lighting quality of the game area at all moments, and, at 14:00, moderate shading can reduce the illuminance insufficiency phenomenon in the watch1 area, and then improve its lighting quality.

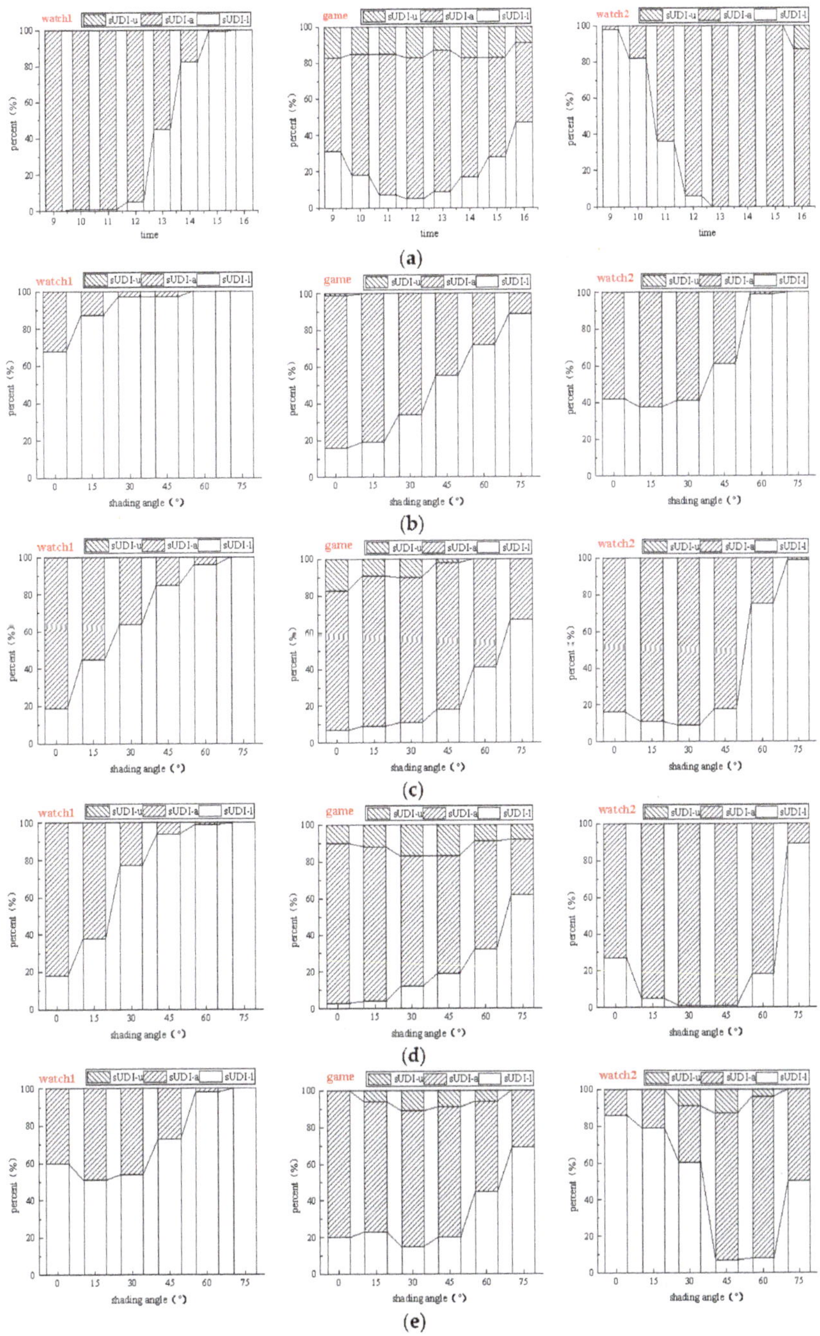

Figure 5. Skylight lighting typical summer daylight environment simulation results: (**a**) typical summer day sUDI values for skylight lighting without shading; (**b**) change in sUDI at 9:00 sUDI on a typical summer day for skylight shading; (**c**) change in sUDI at 12:00 sUDI on a typical summer day for skylight shading; (**d**) change in sUDI at 14:00 sUDI on a typical summer day for skylight shading; and (**e**) change in sUDI at 16:00 sUDI on a typical summer day for skylight shading.

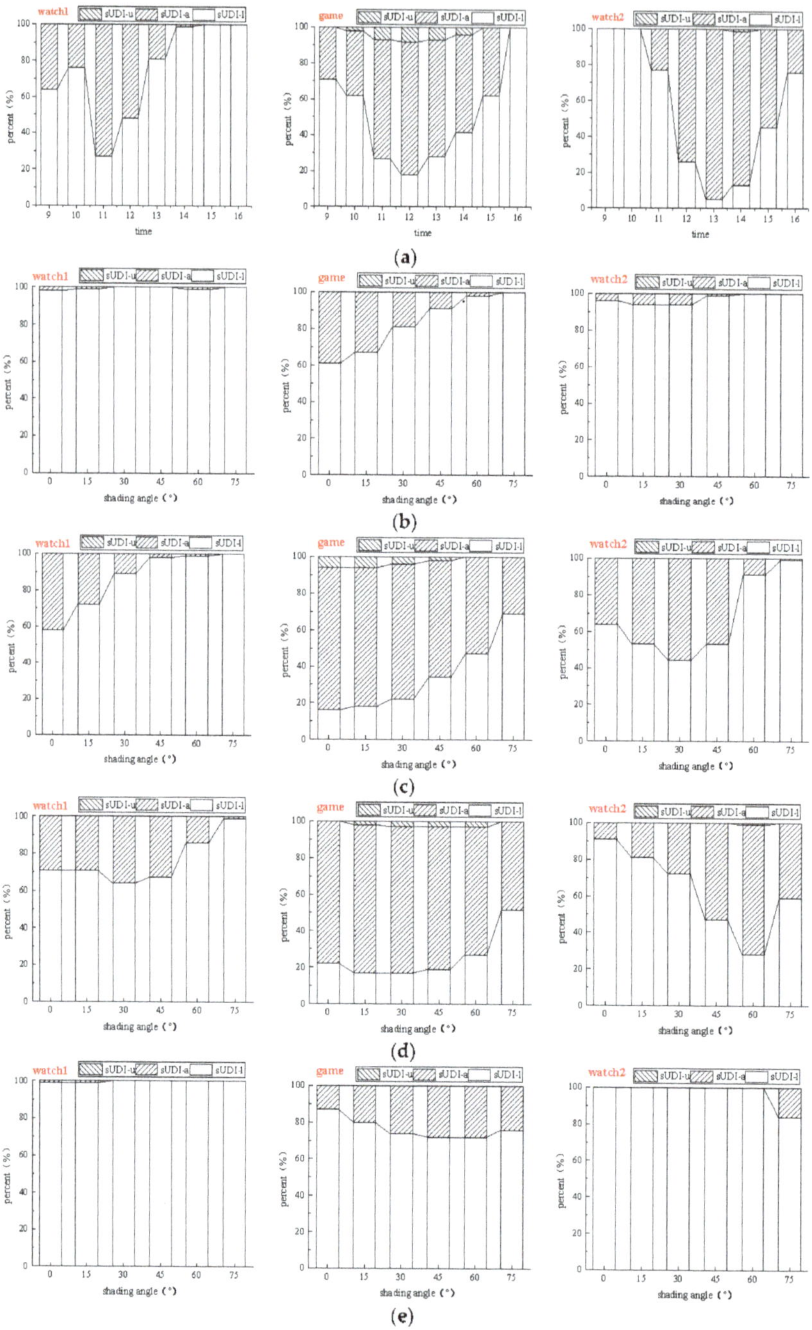

Figure 6. Skylight lighting typical winter daylight environment simulation results: (**a**) typical winter day sUDI values for skylight lighting without shading; (**b**) change in sUDI at 9:00 sUDI on a typical winter day for skylight shading; (**c**) change in sUDI at 12:00 sUDI on a typical winter day for skylight shading; (**d**) change in sUDI at 14:00 sUDI on a typical winter day for skylight shading; and (**e**) change in sUDI at 16:00 sUDI on a typical winter day for skylight shading.

3.2.2. Thermal Environment Simulation Results

The PMV changes on a typical summer day in the form of top skylight lighting are shown in Figure 7. From Figure 7a, it can be seen that the PMV in each area of the room increases and then decreases from the morning to the afternoon, and the effect of the PMV increase and decrease in the game area are more significant than that in the other areas. The game area has a larger PMV value compared to the other two areas at all moments in time, and the PMV value is close to 1 at 13:00, which is the most important time for indoor thermal comfort. The indoor thermal comfort is poor. From Figure 7b–d, it can be seen that the PMV of different zones decreases with the increase in shading angle from 9:00 to 12:00, and the increase in shading angle is beneficial for improving indoor thermal comfort. From 13:00 to 16:00, the increase in the shading angle caused the PMV of each area to increase and then decrease, which also shows that the continuous increase in the shading angle is favorable to indoor thermal comfort. Compared with no shading, PMV in all indoor areas decreased at different shading angles at all times, and indoor thermal comfort was improved to some extent.

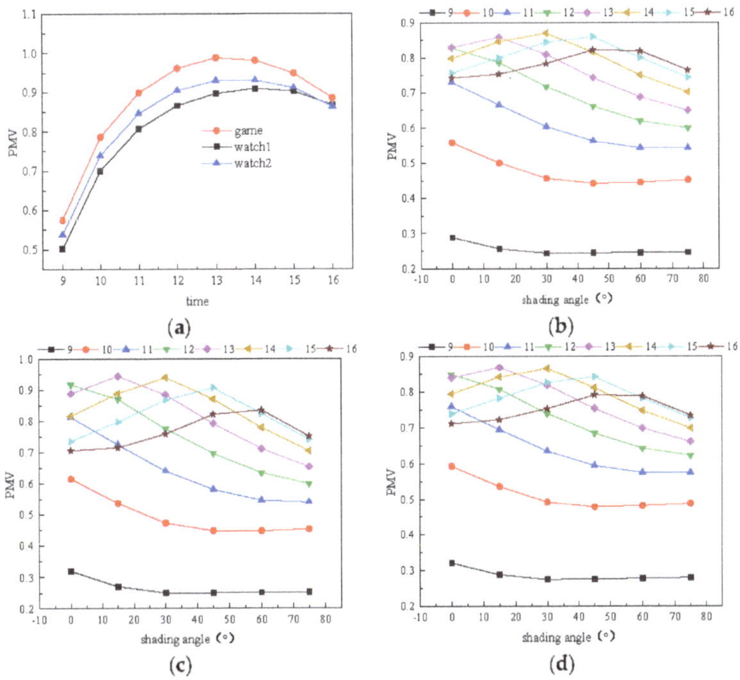

Figure 7. PMV changes on a typical summer day for skylight: (**a**) changes in light unshaded PMV in each area; (**b**) change in light shading PMV in the watch1 region; (**c**) change in light shading PMV in the game region; and (**d**) change in light shading PMV in the watch2 region.

The PMV changes on typical winter days for skylight daylighting are shown in Figure 8. From Figure 8a, it can be seen that the trends of PMV changes in each area of the interior on typical winter and summer days without sunshade are more consistent, but the thermal comfort in each area is better overall under typical winter days. The effect of the shading angle on the PMV in different regions of the room is shown in Figure 8b–d, and the overall trend of the PMV in each region is similar to that of summer in most moments, but, compared with summer, the PMV in each region does not change significantly with increasing shading angle in different moments under winter conditions due to less sunlight entering the room. Compared with the no-shading condition, the PMV of all indoor areas decreased slightly at different shading angles at all moments, and shading has a negative effect on indoor thermal comfort in the form of skylight lighting in winter.

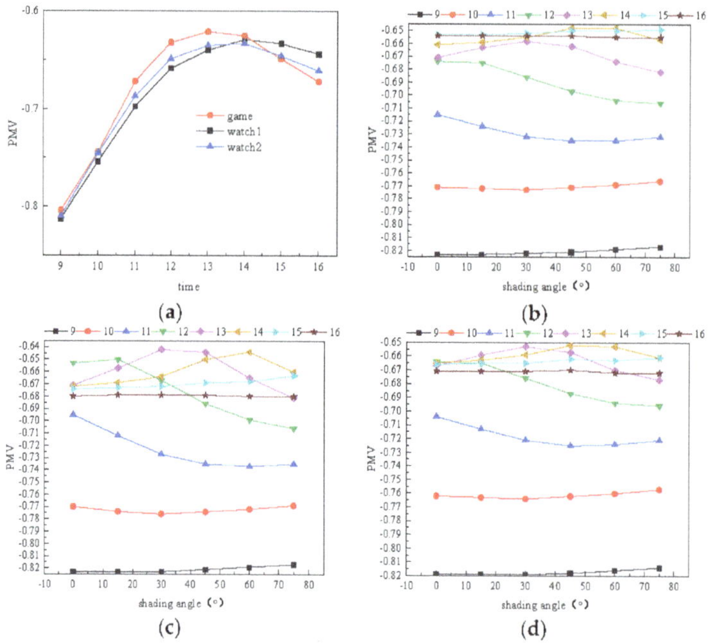

Figure 8. PMV changes on a typical winter day for skylight: (**a**) changes in light unshaded PMV in each area; (**b**) change in light shading PMV in the watch1 region; (**c**) change in light shading PMV in the game region; and (**d**) change in light shading PMV in the watch2 region.

3.2.3. Comprehensive Energy Consumption Simulation Results

The sum of the energy consumption for the lighting and cooling of the gymnasium is taken as the combined summer energy consumption, and the sum of the energy consumption for the lighting and heating of the gymnasium is taken as the combined winter energy consumption.

In the form of top skylight lighting, the integrated energy consumption of the gymnasium on a typical summer day without shading and under different shading angles is shown in Table 6; at 9:00, the integrated indoor energy consumption increases continuously with the increase in shading angle; at each moment from 10:00 to 12:00, the integrated energy consumption decreases firstly and then rises with the increase in shading angle; while, from 13:00 to 16:00, the integrated energy consumption increases firstly and then decreases with the increase in shading angle. Compared with no sunshade, different sunshade angles at each time reduce the comprehensive energy consumption to different degrees.

The integrated energy consumption of the gymnasium without shading and under different shading angles at each moment of a typical day in winter is shown in Table 7, and the influence of an increasing shading angle on the integrated energy consumption has a big difference in different time periods; the integrated energy consumption continues to rise with increasing shading angle in each moment from 9:00 to 12:00, and the integrated energy consumption firstly decreases and then rises with increasing shading angle in each moment from 13:00 to 15:00. At 16:00, the integrated energy consumption continues to decrease with the increase in shading angle. Compared with no sunshade, different sunshade angles at each moment make the comprehensive energy consumption increase to different degrees.

Table 6. Comprehensive energy consumption (kWh) of skylights on a typical summer day in a gymnasium with different shading angles.

Angle	9:00	10:00	11:00	12:00	13:00	14:00	15:00	16:00
unshaded	13.87	35.48	43.28	39.15	42.92	53.23	57.77	60.22
0°	7.81	21.45	35.03	35.2	37.01	43.27	45.23	47.92
15°	8.6	20.26	30.67	32.57	38.96	46.8	47.04	47.94
30°	9.63	20.27	28.84	29.74	36.18	49.45	51.07	49.66
45°	11.39	21.74	29.27	28.56	33.65	45.94	53.44	52.52
60°	14.27	25.41	32.11	29.53	33.17	43.47	50.01	54.51
75°	17.3	29.31	36.16	33.01	35.92	44.78	49.76	53.3

Table 7. Comprehensive energy consumption (kWh) of skylights on a typical winter day in a gymnasium with different shading angles.

Angle	9:00	10:00	11:00	12:00	13:00	14:00	15:00	16:00
unshaded	56.03	49.09	40.04	38	35.5	32.33	34.21	39.41
0°	57.75	51.16	41.56	39.12	37.75	35.13	35.68	39.46
15°	58.47	51.86	42.59	39.26	36.86	34.5	35.41	39.33
30°	59.19	52.65	44.14	40.28	36.34	33.59	34.8	39.12
45°	60.18	53.77	45.5	41.74	36.83	32.58	34.17	38.75
60°	61.53	55.18	47.76	43.81	38.85	32.92	33.86	38.38
75°	62.87	56.72	50.25	46.49	41.87	35.99	34.89	37.99

3.3. South-Facing Side Window Daylighting Temporal–Spatial Partition Control Strategy

Calculated by the entropy weighting method, the weights of the indoor light environment, thermal environment, and comprehensive energy consumption on typical days in summer and winter under the lighting conditions of south-facing side windows are shown in Figure 9, and it can be seen that the light environment indices have the largest weights in the summer conditions, followed by the thermal environment indices and the comprehensive energy consumption indices, which indicates that the indoor light environment should be prioritized under the lighting conditions of the south-facing side windows in the summer and that the energy consumption should not be taken as a key consideration. In winter conditions, the weight of the thermal environment indicator is significantly higher than the remaining two indicators, indicating that more attention should be paid to ensuring the comfort of the indoor thermal environment in winter conditions. The weights of the calculated indicators and the standardized raw data are brought into Equation (6) to obtain the comprehensive score under different working conditions, which is used as a comprehensive evaluation index for the formulation of control strategies.

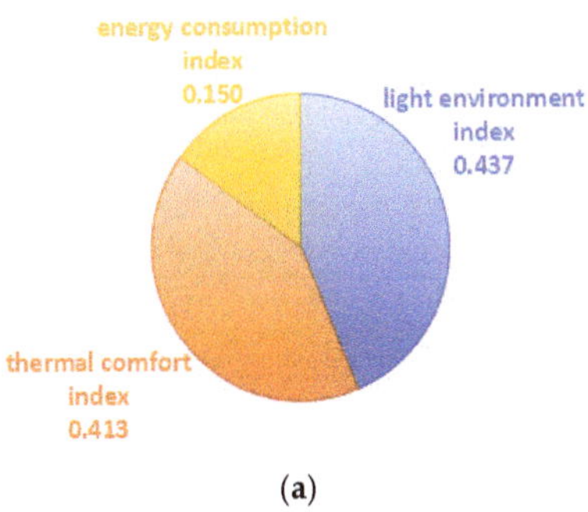

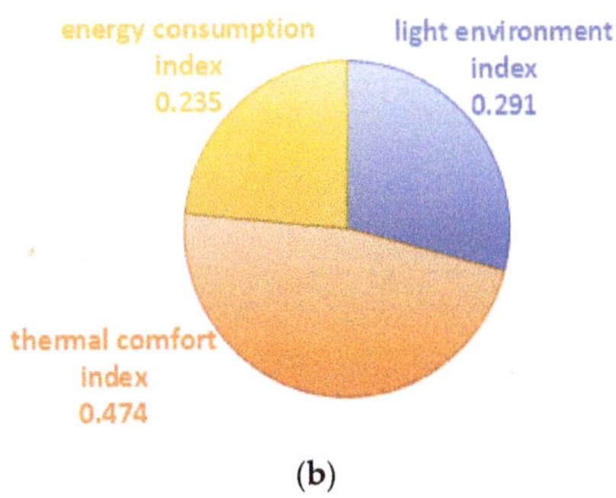

Figure 9. South-facing side window light indicator weights: (**a**) summer conditions; and (**b**) winter conditions.

From Figure 10a, it can be seen that, under the summer condition of south-facing side window lighting, the highest indoor composite scores were achieved under the no-shading condition at 9:00 and 10:00, the highest indoor composite scores were achieved under 0° shading at all moments from 11:00 to 14:00, and the highest indoor composite scores were achieved under the no-shading condition at 15:00 and 16:00. It can be seen that, due to the maximum weight of the light environment index in summer, this results in weak light in the morning and afternoon, to ensure that the indoor lighting conditions, to take measures without shading can better meet the comprehensive needs of the indoor, and, in the midday period when the light is strong, to take the appropriate shading angle to better meet the comprehensive needs of the indoor environment. According to the highest composite score of each time period, the regulation strategy for south-facing side window lighting under typical summer days is formulated as follows: 8:00~10:00, no sunshade; 10:00~14:00, 0° sunshade; and 14:00~16:00, no sunshade.

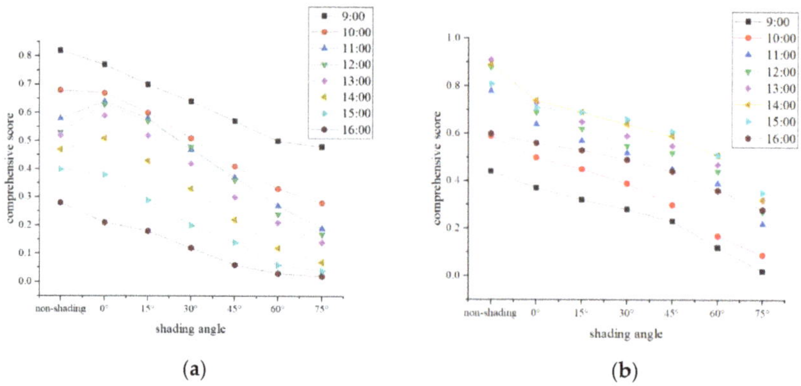

Figure 10. South-facing side window light composite score: (**a**) summer conditions; and (**b**) winter conditions.

South-facing side window daylighting in winter conditions under the integrated score is shown in Figure 10b; it can be seen that all moments in winter are indoors without shading under the highest integrated score, which is mainly due to the fact that thermal environment indicators in winter accounted for a larger weight; to prioritize the satisfaction of the indoor thermal comfort conditions, in the winter, as much as possible should be allowed to give more sunlight into the room to increase the indoor radiant heat gain, which, in turn, improves the indoor thermal comfort conditions. According to the highest composite score of each time period, the regulation strategy for south-facing side window lighting under typical winter days is formulated as follows: 8:00~16:00, no sunshade.

3.4. West-Facing Side Window Daylighting Temporal–Spatial Partition Control Strategy

Under the lighting form of west-facing side windows, the weights of each index in different working conditions are shown in Figure 11. It can be seen that, in summer and winter conditions, the weight of indoor light environment indicators is higher than that of thermal environment and energy consumption indicators; especially in summer, the weight of light environment indicators reaches 0.702, which plays a more significant role in the comprehensive indoor evaluation. Therefore, under the west-oriented lighting form, more attention should be paid to considering the quality of the indoor light environment, to achieve a more reasonable comprehensive evaluation.

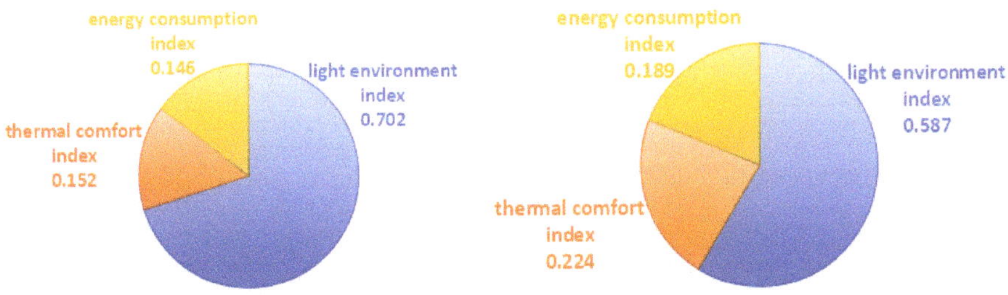

(a) (b)

Figure 11. West-facing side window light indicator weights: (**a**) summer conditions; and (**b**) winter conditions.

The comprehensive score under the summer condition of west-facing side window lighting is shown in Figure 12a, and it can be seen that the comprehensive score is the highest in the period from 9:00 to 14:00 under no shading, which is mainly because the indoor lighting quality is better in the condition of no shading in this period, and, at the same time, the weight of the light environment index is larger, and the comprehensive demand of the indoor environment can be better satisfied when adopting the strategy of no shading. At 15:00 and 16:00, the composite score reaches the maximum when the shading angle is 0° and 15°, which is mainly because a large amount of outdoor sunlight enters the room at these two moments, and, to reduce the phenomenon of excessive indoor illumination, moderate shading measures are taken to better meet the comprehensive indoor demand. According to the maximum comprehensive score of each time period, the regulation strategy of west-facing side window lighting under typical summer days is formulated as follows: 8:00~14:00, no shading; 14:00~15:00, 0° shading; and 15:00~16:00, 15° shading.

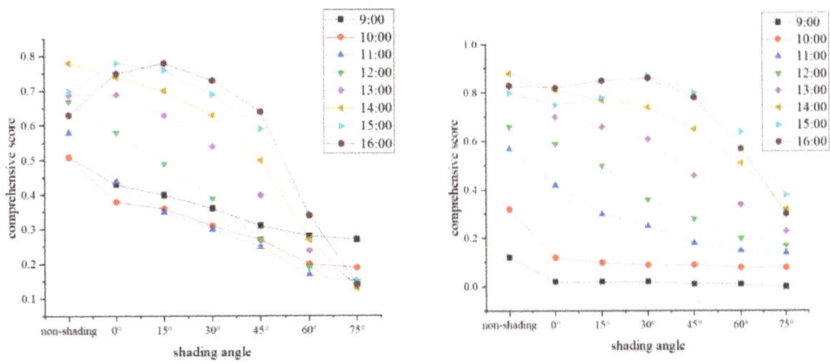

(a) (b)

Figure 12. West-facing side window light composite score: (**a**) summer conditions; and (**b**) winter conditions.

The composite scores under winter conditions are shown in Figure 12b, which shows that, similar to the summer conditions, because the light environment indices in winter under the west-facing side window lighting conditions have a greater weight than the rest of the indices, to prioritize meeting the quality of the indoor light environment, the highest composite scores are obtained when taking no shading measures at all times between 9:00 and 14:00 to better meet the demand for the indoor comprehensive environment. At 15:00 and 16:00, the indoor score reaches the maximum at a shading angle of 30°, which is mainly due to the phenomenon of excessive illumination in some indoor areas at that time in the afternoon, to reduce the phenomenon of excessive illumination to improve the quality of indoor lighting; compared with no shading, moderate shading measures can better satisfy the comprehensive needs of the indoor environment. According to the highest comprehensive score of each time period, the control strategy for the west-facing side window lighting under typical winter days is formulated as follows: 8:00~14:00, no sunshade; and 14:00~16:00, 30° sunshade.

3.5. Skylight Daylighting Temporal–Spatial Partition Control Strategy

In the form of skylight lighting, the weight of each index in different working conditions is shown in Figure 13. It can be seen that, under summer working conditions, the weight of the thermal environment indicators is significantly larger than that of the light environment indicators and comprehensive energy consumption indicators, indicating that the indoor thermal environment plays a more significant role in the comprehensive evaluation under the summer skylight lighting conditions, and more attention should be paid to the comfort of the indoor thermal environment. In winter conditions, the weight of the light environment index is greater than the rest of the indices, indicating that the quality of the indoor light environment should be prioritized in winter conditions to achieve a more reasonable comprehensive evaluation.

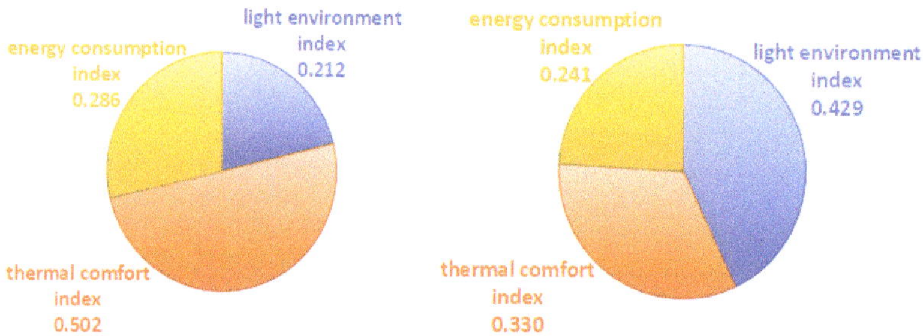

(a) (b)

Figure 13. Skylight daylight light indicator weights: (**a**) summer conditions; and (**b**) winter conditions.

The composite scores under summer conditions for skylight daylighting are shown in Figure 14a, which shows that different shading angles have higher composite scores at different moments compared to no shading, and that shading has an obvious improvement on the indoor thermal environment in summer for skylight daylighting. At the same time, it can be seen that, at 9:00, the overall score reaches the highest when the shading angle is 15°; at 10:00 and 11:00, the overall score reaches the highest when the shading angle is 30°; at 12:00 and 13:00, the overall score reaches the highest when the shading angle is 45°; and, in the afternoon, the overall score reaches the highest when the shading angle is 0° at each moment between 14:00 and 16:00. The control strategy for skylight daylighting under typical summer days were developed based on the highest composite scores for each time period as follows: 8:00~9:00, 15° shading; 9:00~11:00, 30° shading; 11:00~13:00, 45° shading; and 13:00~16:00, 0° shading.

The composite scores under winter conditions are shown in Figure 14b. Unlike summer, the composite scores for different shading angles under most moments of a typical day in winter are lower compared to those without shading, which is mainly because the light environment indices are more heavily weighted under winter conditions, and the absence of shading brings about better lighting conditions in the room most of the time, which leads to a better composite evaluation. The composite score under each moment from 9:00 to 13:00 is highest when there is no sunshade; at 14:00 and 15:00, the composite score is highest when the sunshade angle is 45°; and, at 16:00, the composite score is highest when the sunshade angle is 75°. The control strategies for skylight daylighting under typical summer days were developed based on the highest composite scores for each time period as follows: from 8:00 to 13:00, no shading; from 13:00 to 15:00, 45° shading; and from 15:00 to 16:00, 75° shading.

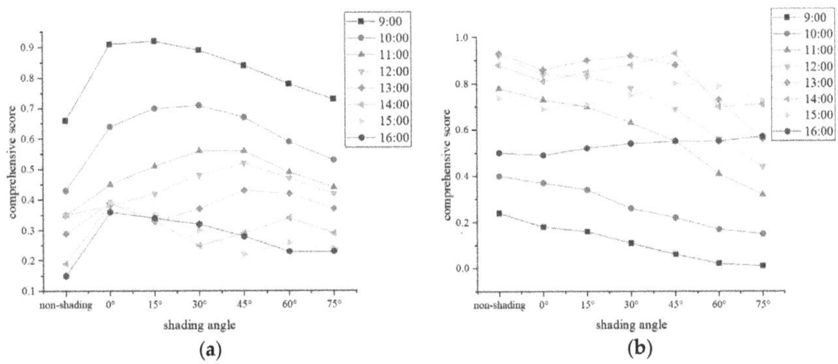

Figure 14. Skylight daylighting composite score: (**a**) summer conditions; and (**b**) winter conditions.

4. Conclusions and Outlook

4.1. Conclusions

This paper takes the college gymnasium as the research object, adopts the temporal–spatial partition regulation means of partition lighting regulation and temporal–spatial partition shading regulation, carries out the simulation research of the temporal–spatial partition regulation of luminous and thermal environment in gymnasium with different lighting forms, and establishes the comprehensive evaluation indices based on the results of the temporal–spatial partition simulation, conducts the comprehensive evaluation of the indoor luminous and thermal environment and energy consumption, and formulates the temporal–spatial partition regulation strategy. The main conclusions are as follows:

(1) The overall DA, UDI-a, and sDA indices of the different lighting forms of the stadium model have been improved to a certain extent after lighting zoning, and the stadium has an overall lighting evaluation that is more in line with the actual lighting needs after lighting zoning. After adopting zoned lighting control, the heating energy consumption of the different lighting forms of the gymnasium model has increased to some extent, but the cooling and lighting energy consumption has been greatly reduced, and the total energy consumption has been reduced. Zoned lighting control causes the energy efficiency of the gymnasium to be improved, especially in the form of skylight lighting; zoned lighting control after the total energy consumption was reduced by 4.85%; and the energy-saving effect of natural light is more significant, proving the necessity of zoned lighting control.

(2) Based on the temporal–spatial partition simulation results, the entropy weighting method is used to calculate the weights of the light environment, heat environment, and comprehensive energy consumption indices of the gymnasium in different lighting forms under different working conditions. In the form of the south-facing side window lighting, the weight of the light environment index in summer working conditions is 0.437, the weight of the heat environment index is 0.413, and the weight of the comprehensive energy consumption index is 0.150; the weight of light environment index in winter working condition is 0.291, the weight of heat environment index is 0.474, and the weight of comprehensive energy consumption index is 0.235. In the form of the westward side window lighting, the weights of the light environment indicators in summer working conditions are 0.702, the weight of the heat environment indicators are 0.152, and the weights of the comprehensive energy consumption indicators are 0.146; in winter working conditions, the weights of the light environment indicators are 0.587, the weight of the heat environment indicators are 0.224, and the weights of the comprehensive energy consumption indicators are 0.189. In the form of skylight lighting, the weight of the light environment index in summer working conditions is 0.212, the weight of the heat environment index is 0.502, and the weight of the comprehensive energy consumption index is 0.286; in winter working conditions, the weight of light environment index is 0.429,

the weight of heat environment index is 0.330, and the weight of comprehensive energy consumption index is 0.241.

(3) Through the comprehensive evaluation model to obtain the comprehensive indoor environment score under different conditions, based on the score in the partition lighting regulation based on the reasonable control of the shading angle in time, under the lighting form of south-facing side windows, the shading angle time-sharing control strategy in summer is as follows: 8:00~10:00, no shading; 10:00~14:00, 0°; and 14:00~16:00, no shading; and the shading angle time-sharing control strategy in winter is as follows: 8:00~16:00, no shading. Under the lighting form of the west-facing side windows, the time-sharing control strategy for the summer shading angle is as follows: 8:00~14:00, no shading; 14:00~15:00, 0°; and 15:00~16:00, 15°; and the time-sharing control strategy for winter shading angle is as follows: 8:00~14:00, no shading; and 14:00~16:00, 30°. In the form of top skylight lighting, the typical summer shading angle time-sharing control strategy is as follows: 8:00~9:00, 15°; 9:00~11:00, 30°; 11:00~13:00, 45°; and 13:00~16:00, 0°; and the winter shading angle time-sharing control strategy is as follows: 8:00~13:00, no shading; 13:00~15:00, 45°; and 15:00~16:00, 75°.

(4) The design of a building requires the integration of three key variables: lighting comfort, thermal comfort, and energy consumption. These variables have a direct impact on the energy efficiency of the building and the user's experience. Good lighting enhances the visual comfort of athletes and spectators, while a suitable thermal environment is related to human thermal comfort, and reducing energy consumption is key to achieving sustainable development. In practical application, comprehensive on-site research and demand analysis should be carried out to ensure that the regulation strategy can meet the actual use needs of specific buildings and the comfort requirements of users. Secondly, it is recommended that we adopt intelligent control system to realize the automatic control of shading facilities and the lighting system to improve the implementation efficiency and accuracy of the control strategy. In addition, the effectiveness of the regulation strategy is regularly evaluated and optimized to adapt to the changes in the external environment and the evolution of user needs.

4.2. Outlook

In this paper, the university gymnasium was the research object with which to carry out the research on the temporal–spatial partition light and heat environment regulation method; due to the actual conditions and time constraints, there are still some problems and shortcomings in this research:

(1) This paper is mainly based on relevant experience and specification requirements for the lighting level target illuminance division; subsequent to this, the actual high-space building zoning lighting control can be based on the actual research situation for the indoor target illuminance division of different areas, making it more in line with the actual needs.

(2) This paper only provides the temporal–spatial partition control means of partition lighting control and time-lapse external louver sunshade control, and, in the future, the rest of the sunshade forms and light and heat environment control means can be carried out in the temporal–spatial partition control research.

Author Contributions: Conceptualization, Y.W.; methodology, K.G. and S.L.; formal analysis, K.G. and S.L.; investigation, K.G., S.L. and M.J.; data curation, Y.W. and S.L.; writing—original draft preparation, K.G.; writing—review and editing, Y.W., K.G., S.L. and M.J.; supervision, Y.W.; project administration, Y.W.; funding acquisition, Y.W. All authors have read and agreed to the published version of the manuscript.

Funding: This work was supported by Central Guidance of Local Science and Technology Development Funds (236Z5202G).

Data Availability Statement: The data are contained within the article.

Conflicts of Interest: The authors declare no conflicts of interest.

References

1. Ekonomou, G.; Menegaki, A.N. The Role of the Energy Use in Buildings in Front of Climate Change: Reviewing a System's Challenging Future. *Energies* **2023**, *16*, 6308. [CrossRef]
2. Xie, S.; Zhong, K.; Zhu, H.; Ye, X.; Kang, Y. Effects of the position of windows on large-height spaces' warm-air heating systems. *J. Donghua Univ. (Nat. Ence)* **2015**, *41*, 376–381.
3. Eiz, H.; Mushtaha, E.; Janbih, L.; Rifai, R.E. The Visual and Thermal Impact of Skylight Design on the Interior Space of an Educational Building in a Hot Climate. *Eng. J.* **2021**, *25*, 187–198. [CrossRef]
4. Evuti, A.M.; Baba, Y.M.; Adeyemi, K. A Study of Energy Demand and Consumption in University of Abuja, Nigeria. *IOP Conf. Ser. Earth Environ. Sci.* **2020**, *476*, 012087. [CrossRef]
5. Elghamry, R.; Hassan, H. Impact of Window Parameters on the Building Envelope on the Thermal Comfort, Energy Consumption and Cost and Environment. *Int. J. Vent.* **2020**, *19*, 233–259. [CrossRef]
6. Solovyov, A.K. Daylight, Solar Radiation, Architectural Expression, and Energy Efficiency of Buildings. *Light Eng.* **2021**, *29*, 6–9. [CrossRef]
7. Katunsky, D.; Vertal, M.; Dolnikova, E.; Zozulakova, S.; Hutkai, K.; Dicka, Z. Mutual Interaction of Daylight and Overheating in the Attic Space in Summer Time. *Sustainability* **2022**, *14*, 15634. [CrossRef]
8. Kim, S.-K.; Ryu, J.-H.; Seo, H.-C.; Hong, W.-H. Understanding Occupants' Thermal Sensitivity According to Solar Radiation in an Office Building with Glass Curtain Wall Structure. *Buildings* **2022**, *12*, 58. [CrossRef]
9. Rabczak, S.; Bukowska, M.; Proszak-Miąsik, D.; Nowak, K. Influence of Shading on Cooling Energy Demand. *IOP Conf. Ser. Mater. Sci. Eng.* **2017**, *245*, 052046. [CrossRef]
10. te Kulve, M.; Schlangen, L.; Lichtenbelt, W.V.M. Interactions between the Perception of Light and Temperature. *Indoor Air* **2018**, *28*, 881–891. [CrossRef]
11. AL-Dossary, A.M.; Kim, D.D. A Study of Design Variables in Daylight and Energy Performance in Residential Buildings under Hot Climates. *Energies* **2020**, *13*, 5836. [CrossRef]
12. Nazari, S.; Keshavarz Mirza Mohammadi, P.; Sareh, P. A Multi-Objective Optimization Approach to Designing Window and Shading Systems Considering Building Energy Consumption and Occupant Comfort. *Eng. Rep.* **2023**, *5*, e12726. [CrossRef]
13. Hviid, C.A.; Nielsen, T.R.; Svendsen, S. Simple Tool to Evaluate the Impact of Daylight on Building Energy Consumption. *Sol. Energy* **2008**, *82*, 787–798. [CrossRef]
14. Krarti, M.; Erickson, P.M.; Hillman, T.C. A Simplified Method to Estimate Energy Savings of Artificial Lighting Use from Daylighting. *Build. Environ.* **2005**, *40*, 747–754. [CrossRef]
15. Li, D.H.W.; Cheung, A.C.K.; Chow, S.K.H.; Lee, E.W.M. Study of Daylight Data and Lighting Energy Savings for Atrium Corridors with Lighting Dimming Controls. *Energy Build.* **2014**, *72*, 457–464. [CrossRef]
16. Cao, S. Intelligent Lighting Control System in Large-Scale Sports Competition Venues. *Light Eng.* **2018**, *26*, 172–182. [CrossRef]
17. Zhang, L.; Meng, Q.L.; Li, J. Passive Energy-Saving Design in Hot and Humidity Area: A Case Study of Multi-Functional Stadium of Guangdong College of Pharmacy. *Appl. Mech. Mater.* **2012**, *121–126*, 715–719. [CrossRef]
18. Chow, S.K.H.; Li, D.H.W.; Lee, E.W.M.; Lam, J.C. Analysis and Prediction of Daylighting and Energy Performance in Atrium Spaces Using Daylight-Linked Lighting Controls. *Appl. Energy* **2013**, *112*, 1016–1024. [CrossRef]
19. Gomes, M.G.; Santos, A.J.; Rodrigues, A.M. Solar and Visible Optical Properties of Glazing Systems with Venetian Blinds: Numerical, Experimental and Blind Control Study. *Build. Environ.* **2014**, *71*, 47–59. [CrossRef]
20. Daum, D.; Morel, N. Assessing the Saving Potential of Blind Controller via Multi-Objective Optimization. *Build. Simul.* **2009**, *2*, 175–185. [CrossRef]
21. Yao, J. An Investigation into the Impact of Movable Solar Shades on Energy, Indoor Thermal and Visual Comfort Improvements. *Build. Environ.* **2014**, *71*, 24–32. [CrossRef]
22. Littlefair, P.; Ortiz, J.; Das Bhaumik, C. A Simulation of Solar Shading Control on UK Office Energy Use. *Build. Res. Inf.* **2010**, *38*, 638–646. [CrossRef]
23. Zhang, J.; Wu, Y.; Huang, Y.; Cheng, H.; Liu, D. Shading strategy and field investigation of indoor light and thermal environment of high-speed railway stations in hot summer areas. *Huazhong Archit.* **2021**, *39*, 41–49. [CrossRef]
24. Ansari, K.; Pont, U.; Mahdavi, A. Visual Comfort and Lighting Levels in Sport Facilities: A Case Study. In Proceedings of the CESB 2016: Central Europe Towards Sustainable Building 2016—Innovations for Sustainable Future, Prague, Czech Republic, 22–24 June 2016.
25. Eltaweel, A.; Su, Y. Parametric Design and Daylighting: A Literature Review. *Renew. Sustain. Energy Rev.* **2017**, *73*, 1086–1103. [CrossRef]
26. Yang, D.; Zhang, M.; Han, G. Reference building model for school gymnasium daylighting in Beijing-Tianjin Region. *China Illum. Eng. J.* **2017**, *28*, 115–122. [CrossRef]
27. *GB/T50034-2024*; Standard for Lighting Design of Buildings. China Architecture and Building Press: Beijing, China, 2024.
28. Wang, Y. Research on Daylighting Design of Hangzhou University Gymnasium Based on Dynamic Lighting Evaluation. Master's Thesis, Zhejiang University, Hangzhou, China, 2023. [CrossRef]
29. Wang, X. Research on Multi-Objective Optimization of Shading Louver System Based on Genetic Algorithm. Master's Thesis, Guangdong University of Technology, Guangzhou, China, 2022. [CrossRef]

30. Zhou, R.; Fan, F.; He, D.; Qiu, W. Integrated entropy weight method based on data stability and subjective preference in multi-attribute group decision-making. *Control Decis.* **2012**, *27*, 1169–1174. [CrossRef]
31. Ai, L.; Liu, S.; Ma, L.; Huang, K. A Multi-Attribute Decision Making Method Based on Combination of Subjective and Objective Weighting. In Proceedings of the 2019 5th International Conference on Control, Automation and Robotics (ICCAR), Beijing, China, 19–22 April 2019; pp. 576–580. [CrossRef]
32. Li, H.; Yang, L. Passive ultra-low energy consumption building evaluation based on entropy weight method. *J. Build. Energy Effic.* **2021**, *49*, 47–51.

buildings

MDPI

Article

Research on the Carbon Reduction Potential of the Life Cycle of Building Roofs Retrofit Designs

Dawei Mu, Wenjin Dai, Yixian Zhang *, Yixu Shen, Zhi Luo and Shurui Fan

College of Civil Engineering and Architecture, Hainan University, Haikou 570228, China;
daivemu@126.com (D.M.); 13118930402@163.com (W.D.); 18976332605@163.com (Y.S.); ralphzinc@163.com (Z.L.);
13919814783@163.com (S.F.)
* Correspondence: zhangyixian902@163.com

Abstract: This study examines existing buildings in Haikou in China under tropical island climate conditions. It presents three retrofit design models for greenhouses roofs (GHR), green roofs (GR) and photovoltaic roofs (PVR). The carbon cost of each retrofit roof model is calculated in the production and transportation phases of building materials, construction, and demolition. The changes in electricity consumption, water consumption, and plant carbon reduction are coupled to calculate the carbon reduction generated by each phase of the use of the retrofitted roofs. The carbon reduction per unit area for GHR, GR and PVR over the life cycle (20 years) is then comprehensively calculated. The life cycle carbon reduction per unit area is 262.57 kg·m^{-2} for GHR, 127.41 kg·m^{-2} for GR and 2567.12 kg·m^{-2} for PVR. Among the three retrofit methods, PVR has the greatest potential for reducing carbon emissions. The study can as a guide for implementing carbon reduction measures in tropical island areas. Domestic research on rooftop greenhouses also focuses on technology, yield, and energy consumption, mostly for northern regions with cold winters, and less research on rooftop greenhouses applied to regions with hot summers and warm winters. But domestic and foreign studies on the potential of rooftop greenhouses to reduce emissions have not yet been combined with plant cultivation of hydroelectric carbon emissions and plant carbon sequestration.

Keywords: carbon emissions of building; greenhouses roofs; green roofs; photovoltaic roofs; life cycle

1. Introduction

"Carbon Peak and Carbon Neutral" is one of the important strategic tasks in China, and according to the "2022 China Building Energy Consumption and Carbon Emission Research Report" issued by China Building Energy Conservation Association, the total carbon emission of the whole process of the national building in 2020 was 5.08 billion tons, accounting for 50.9% of the national carbon emission 50.9% [1]. As an important main body of carbon emissions, buildings should also be the main contributor to the realization of the dual-carbon target. Building roofs occupy 21% to 26% of all built-up areas [2] and have great potential for development.

Carbon emissions associated with the built environment dominate the total carbon emissions of society. Carbon emissions can be measured in three ways: firstly, by considering carbon dioxide alone; secondly, by the six gases identified by the Kyoto Protocol: carbon dioxide (CO_2), methane (CH_4), nitrous oxide (N_2O), hydrofluorocarbons (HFCs), perfluorocarbons (PFCs), and sulphur hexafluoride (SF6); and thirdly, by the framework of the Intergovernmental Panel on Climate Change (IPCC), which sets out a large number of GHG emissions. Of these, the IPCC framework is the most commonly used methodology and was established to facilitate reporting of carbon emissions consistent with the Kyoto Protocol. In this methodology, the impact of each GHG is expressed in terms of global warming potential (GWP) expressed in kilograms of carbon dioxide equivalent (kg CO2eq). The underlying framework for the assessment of carbon emissions from buildings, although it varies from country to country, is usually based on a well-established whole life cycle assessment (LCA) process [3].

LCA is often considered a 'cradle-to-cradle' approach, whereby a systematic assessment is carried out throughout the entire lifecycle of a product, e.g., raw material extraction, manufacturing, use, end-of-life treatment and recycling. The LCA (Life Cycle Assessment) method has been standardized by ISO [4,5] and can be used to evaluate the potential improvements in the ecological status of products, thereby enabling the design of more efficient and environmentally friendly complex products, such as entire buildings, building materials, and components. The literature shows that many studies have emphasized the role of LCA as a decision-making support tool for selecting low-impact building materials and components and designing green buildings. An emerging trend in the LCA literature is the commitment to low-energy consumption buildings, where the goal of reducing operational energy consumption implies an increase in embodied energy consumption due to the additional materials and components used in the building envelope and the technical equipment for energy-saving measures [6]. The main methods for assessing carbon emissions over the whole life cycle of a building are the process analysis method, the input–output method and the hybrid method. The process analysis method is a systematic approach to analysing the inputs and outputs of each process of a product. This method is highly encouraged by ISO standards due to its detailed and accurate process, but its analysis process requires a large amount of data from a wide range of unique sources, resulting in a significant amount of time and cost [7]. The process analysis method is suitable for calculating carbon emissions at the individual building level. Input–output methods to quantify direct and indirect emissions from large supply chains link final emissions to input–output economic data from economic sectors, often using publicly available data on industry inputs, emissions, energy use, and so on. This approach has become easier and faster due to the increasing availability of databases. However, it is less detailed and accurate than the process analysis approach, and its reliance on databases makes it impossible to calculate carbon emissions from atypical products. This method is suitable for macro-level studies in the construction industry. A hybrid approach combines the strengths of both approaches, using process analysis for detailed product information and input–output methods for general sector-level information. Although hybrid models have been shown to achieve specificity and system integrity, their accuracy and their respective use is controversial [8,9].

China has now introduced a national standard for calculating carbon emissions from buildings [10]. The standard applies to the calculation of carbon emissions at the design stage of civil buildings or the accounting of carbon emissions after construction. The regulations divide the carbon emissions of the whole life cycle carbon of the building into three stages: building operation, construction and demolition, and production and transport of building materials, and the carbon emissions in the standard are measured using the IPCC framework, which mainly adopts the carbon emission factors of various types of energy and materials for calculation.

The initial stage in developing a pertinent design methodology is to measure the carbon emissions of the edifice. Although it may vary from country to country. Research on carbon reduction and sequestration in the building sector is receiving both domestic and international attention to achieve carbon reduction goals. Song Zhixi et al. calculated the carbon emissions of Hangzhou West Railway Station throughout its life cycle. They concluded that using recyclable building materials during the production stage, as well as during the use phase of the building, and improving the efficiency of the HVAC and lighting systems, can reduce the carbon footprint per unit area [11]. Zhao Susu et al. calculated the life cycle carbon emissions of a reinforced concrete residential building in Jiangsu region. They proposed a carbon reduction strategy that involves improving the recycling rate of building materials during production, implementing green planting during use phase, and gradually demolition the building using small-scale machines during the demolition phase [12]. Ni Songyuan et al. proposed a strategy to reduce carbon emissions in the construction industry by producing synthetic biomass recycled aggregate using coconut shells to replace natural aggregate in concrete production. This would reduce carbon dioxide emissions during production and construction [13]. Roof retrofitting also means that environmental aspects can be improved: reducing the urban heat island effect [14].

Domestic and international research on building carbon reduction and sequestration has involved multiple fields and aspects, from policy guidance to model prediction, from carbon reduction calculation to carbon neutralization path, building carbon reduction and sequestration research has become increasingly rich and comprehensive. Carbon reduction measures can be implemented at all phases of a building's life cycle. Some carbon reduction technologies are already widely used, while others are still in the feasibility stage. There are few comparative studies on the life cycle of greenhouses roofs (GHR), green roofs (GR), and photovoltaic roofs (PVR) under tropical island climate conditions. On the basis of the roof of the reporting centre, as shown in Figure 1, GHR, GR, and PVR are established, respectively. GHR grows tomatoes in the greenhouse for carbon reduction and lays photovoltaic panels above it, a layer of plants in the GR, and PVR covered with photovoltaic panels. The aim of this paper is to establish a model for studying the carbon reduction potential of building GHR, RG, and PVR. This model will provide guidance for selecting carbon reduction strategies in engineering practice. This study establishes a model for the use of rooftop space for carbon reduction retrofitting of existing buildings applicable to tropical islands in the context of urban renewal. It also quantifies the carbon reduction potential of these three green roof retrofit approaches from a full life cycle perspective, which is then extended to the urban scale to explore the carbon reduction potential of building roofs after urban regeneration.

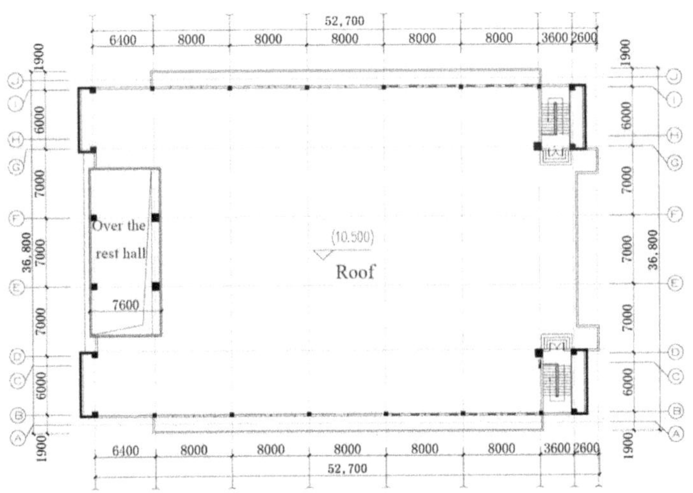

Figure 1. Roof plan.

The modelling will be carried out below, using the conference lecture hall as the base model, and adding three forms of building modelling on top of it, respectively, and then the analysis of the three buildings in which the creation of formulas is carried out. A careful calculation is presented in it. The fourth part is the core of this paper, i.e., the presentation and discussion of the research results. Through an in-depth analysis of the data, the essential features of the research problem are revealed and discussed in depth in relation to the theory. Meanwhile, the uncertainties and limitations of the results are objectively analysed to provide reference for further research. The main findings, innovations, and implications for future research are clarified. At the same time, the shortcomings of the study will be reflected upon to provide directions for improvement in subsequent research.

2. Materials and Methods

2.1. Building Roof Model

Based on a lecture hall, we established models for GHR, RG, and PVR, respectively. The building is a reinforced concrete frame structure, with a height of 10.95 m and two floors above ground. We used Software SketchUp 2017 to create the building model based on the building plan. The building has a flat roof with walls measuring 4 m in height on the south, north, and west sides, while the east wall is only 1.5 m high. The second-floor lounge on the west side features a glass curtain wall that is 6 m higher than the roof. The roof plan is illustrated in Figure 1. The basic model of the building and the building incorporates three design models for GHR, GR, and PVR, and the structure of the building is shown in Figure 2.

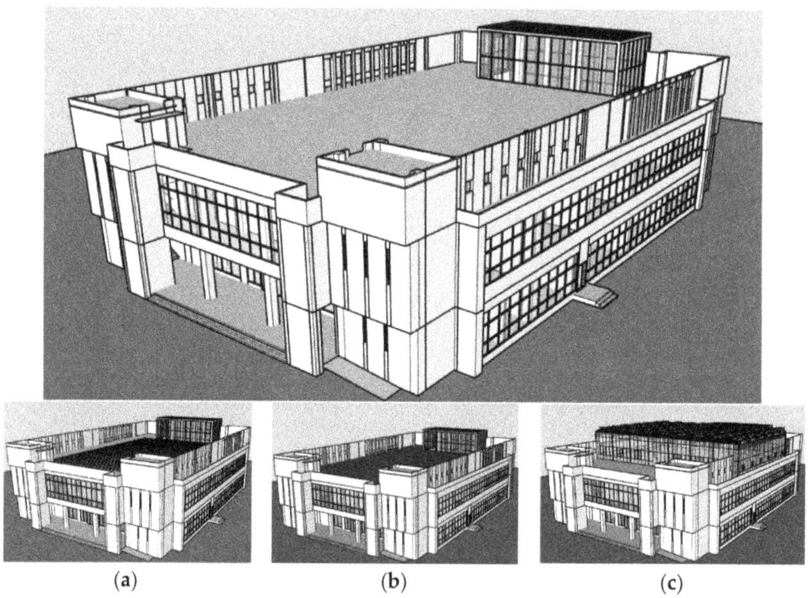

Figure 2. Lecture hall building model. (**a**) Photovoltaic roofs; (**b**) green roofs; (**c**) greenhouses roofs.

2.1.1. GHR Design

Haikou city is suitable for constructing GHR in a north–south orientation. The load-bearing members of the GHR are situated on the building columns and load-bearing beams below, ensuring enhanced structural safety. The modern greenhouse on the roof consists of two small roofs of 4 m each, i.e., a span of 8 m, an opening of 5 m, an eave height of 6 m, a ridge height of 6.80 m, and a truss height of 0.5 m. The east–west direction is 24 m, and the north–south direction is 35 m. The GHR has a surface area of 840 m^2, and the utilization rate of the roof space is about 60%. The main component types, materials and cross-section dimensions are shown in Figure 3 and Table A1. The steel selection is Q235. This GHR opted for side sliding windows combined with the top-hung segmental roof windows for natural ventilation.

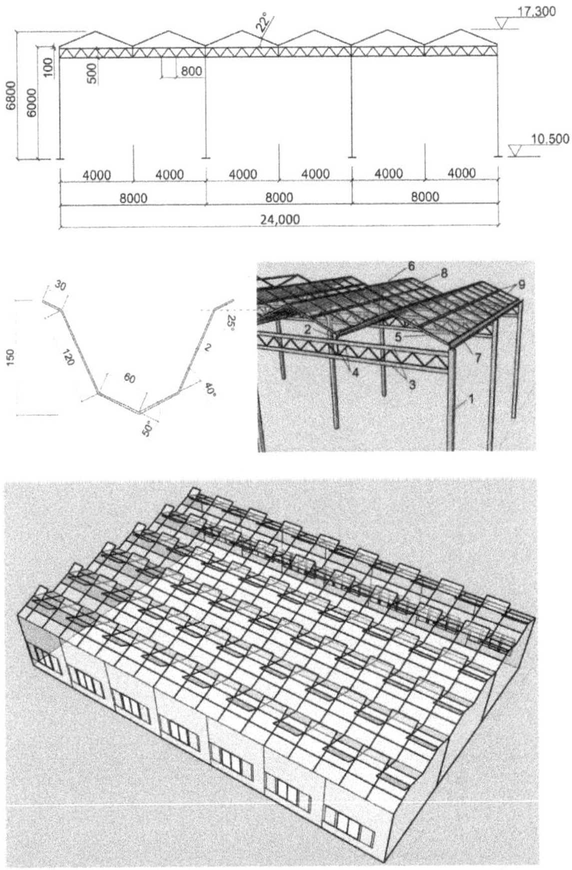

Figure 3. The elevation and dimensions of the GHR structure (mm). (The numbers 1–9 in the figure correspond to the component types in Table A1).

The crop cultivation planning in this study utilised single-layer cultivation beds with a width of 1.20 m for planting. A 0.6-m access space was left between the beds for crop management, and a 1-m-wide main access was created around the greenhouse to facilitate the transportation of production means. Additionally, a 4-m-wide area on the north side was allocated for equipment control and storage. This resulted in the placement of 17 cultivation beds, each with a width of 1.20 m and a length of 22 m, in the greenhouse. Figure 4 shows that the greenhouse planting area is 448.80 m^2. Tomato planting is carried out at a density of 30 cm × 50 cm, with 176 tomatoes planted in each cultivation bed, resulting in a total of 2992 tomatoes planted in the greenhouse. During the summer, high temperatures can cause the greenhouse to become too hot. Therefore, it is necessary to take cooling measures. This greenhouse uses the wet curtain-fan evaporation cooling system. As per the "greenhouse wet curtain-fan cooling system design specification" [15], it is recommended to install the wet curtain and fan on opposite walls, with a distance of 30–45 m between them. The height of the wet curtain should be between 0.60–2.40 m, and the air inlet at the wet curtain should be closable without removing it. In summary, the arrangement of wet curtains-fans in the greenhouse is shown in Figure 5. The distance between wet curtains and fans is 35 m. The wet curtains are 1 m above the ground, 2 m high, and 150 mm thick, and they are arranged continuously on the south side of the mountain wall, with the windows at the air inlet openings in the form of outwardly flap windows, and one fan is arranged every 4 m on the wall on the opposite side of the wall for a total of 6 sets.

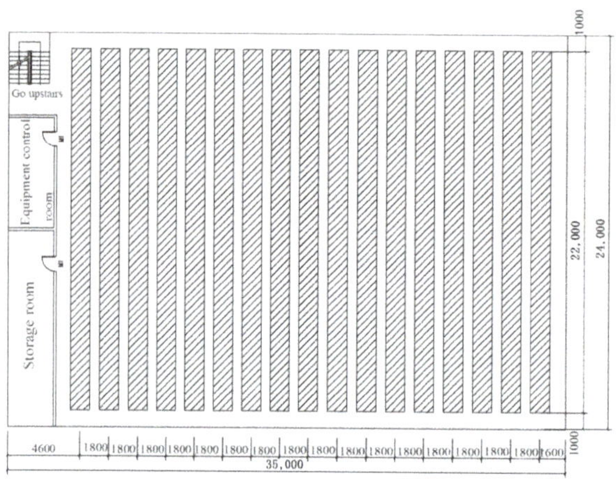

Figure 4. The planting plan for the GHR (mm).

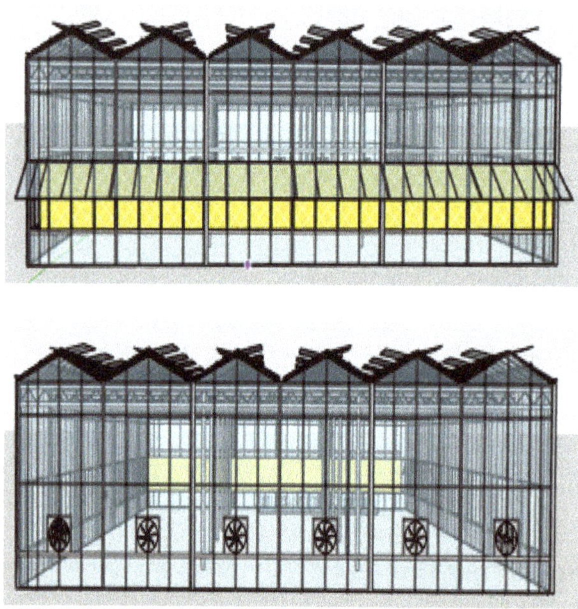

Figure 5. Wet curtain-fan arrangement.

The ventilation and cooling system in the greenhouse must be equipped with wet curtain fans, pumps, and gear motors and other equipment, as indicated in Table 1.

Table 1. Details of main greenhouse equipment.

	Equipment	Equipment Parameters	Quantities	Rated Power (kW)	Utilization	Annual Electricity Consumption (kW·h)
Wet curtain fan cooling system	Negative pressure fan	Diameter 1.38 m, airflow 56,000 $m^3 \cdot h^{-1}$	6	1.10	June to September, 11:00–15:00 daily	3220.80
	Circumfluence fan	Impeller diameter 400 mm, air volume 6450 $m^3 \cdot h^{-1}$	6	0.37	200 days per year at 3 h per day	1332
	circulating water pump	Dust 6 m with a flow rate of 13,000 $L \cdot H^{-1}$	3	0.10	June to September, 11:00–14:30 daily	128.10
Window opening system	Reduction motor	For wet curtain outward-facing windows, and roof casement systems	2 sets	0.75	Based on an average of four runs per day, each run will not exceed 10 min	365

The GHR glass requires design considerations for structural design, ventilation and cooling, planting, and the use of PV power generation glass instead of ordinary glass. The photovoltaic glass on the GHR needs to be designed for transmittance, model, and arrangement to enable energy production while growing crops. The design of the photovoltaic panels on the GHR differs from that of distributed photovoltaic power generation on the roofs of buildings. It is important to ensure that the photovoltaic panels on the GHR do not affect the photosynthesis of the crops inside. A suitable option for the panels is "power generation glass" made of cadmium telluride, which has a light transmittance of 40%. A total of 696 PV glass modules are arranged as shown in Figure 6. The product parameters for the selected PV glass are presented in Table 2.

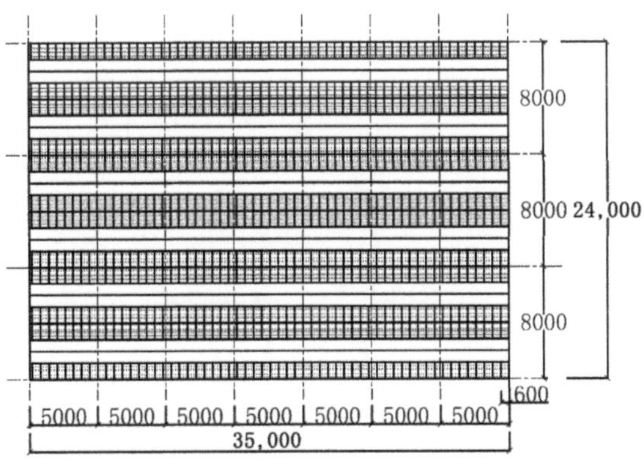

Figure 6. The arrangement of PV glazing on the GHR (mm).

Table 2. Product parameters for PV glass RK-T40.

Parameter Type	Specific Types	Specific Parameters
Mechanical parameters	Component size	1200 × 600 × 6.80 mm
	Framework	3.2 mm CdTe power generation glass + 0.4 mm EVA + 3.2 mm tempered glass
	Battery type	Cadmium telluride thin film (CdTe)
Photothermal performance parameters	Visible light transmittance	40%
	Heat transfer coefficient	5.76 $W{\cdot}m^{-2}k^{-1}$
	SHGC value	0.54
Electrical performance parameters	Power (Pmpp)	60 Wp
	Short-circuit current (Isc)	0.74 A
	Open circuit voltage (Voc)	123.80 V
System integration characterization parameters	Temperature coefficient (Pmpp)	−0.29%/°C
	Temperature coefficient (Isc)	+0.04%/°C
	Temperature coefficient (Voc)	−0.28%/°C

2.1.2. GR Design

When constructing the GR project, it is necessary to waterproof the roof again. For this purpose, we have selected the SBS waterproofing membrane as the material for the waterproof layer when constructing the additional layer of the GR. PVC sheets are used as a root barrier over the waterproofing layer to prevent plant roots from damaging the roof structure and to protect the waterproofing layer. Above the root barrier layer, high-impact polystyrene is used as a drainage and storage layer. This layer quickly drains rainwater to prevent accumulation on the roof. A filter layer should be placed between the drainage layer and the cultivation substrate. Its purpose is to prevent substrate particles from entering the drainage layer, ensuring the smoothness of the drainage layer and the quality of the cultivation substrate. Polyester fiber should be used for the filter layer. The top layer consists of a mixture of pumice, sand, clay, and organic matter. This study selected Sedum lineare Thunb as the plant for GR, which is currently widely used domestically and internationally. Figure 7 shows each additional material layer of the GR.

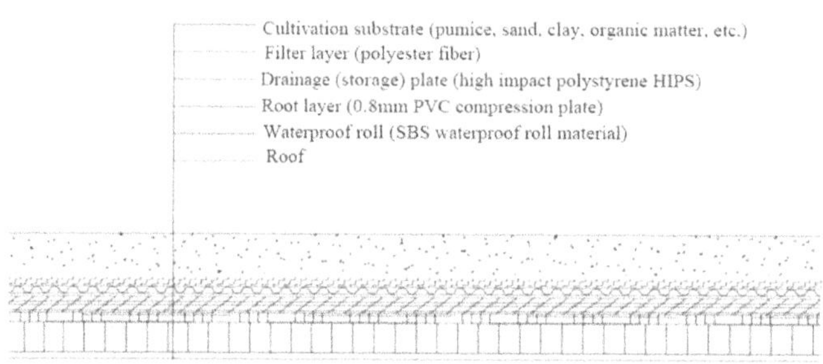

Figure 7. Supplementary materials for GR.

2.1.3. PVR Design

Haikou is located at the northern edge of the tropics, with high solar radiation throughout the year. The low latitude and the large altitude angle of the sun make it possible to have relatively long hours of direct sunlight. In Haikou, the annual sunshine hours in most areas can reach more than 2000 h, and even more than 2400 h in some areas. Especially in the dry season, there are more sunny days with high light intensity, providing good conditions for crop growth and solar energy utilisation. Sufficient light also makes Hainan's tropical plants grow luxuriantly, and the abundant solar energy resources also provide favourable natural conditions for the development of new energy industries such as photovoltaic power generation. This provides abundant solar energy resources for photovoltaic power generation, which allows photovoltaic panels to efficiently convert solar energy into electricity.

The PV power generation technology selected for the tropical island climate zone in this study is CdTe thin-film solar cells, CdTe thin-film cells have a low manufacturing cost, high conversion efficiency, and good low-light properties, and their temperature coefficient is about half of that of crystalline silicon solar cells [16], and lower temperature coefficients mean that their power generation efficiency is less affected by the temperature, so they are better suited for high-temperature regions. It has also been shown to be more environmentally friendly, with cadmium emissions much lower than conventional energy sources such as oil and coal and only half that of crystalline silicon solar cells, and emissions of other heavy metals in production and use lower than that of silicon solar cells [17,18], and its life cycle carbon emissions are also lower than that of other photovoltaic technologies [19,20]. CdTe thin film cells were selected in this study and the specific parameters are shown in Table 3.

Table 3. Product parameters for CdTe thin film battery.

Parameter Type	Specific Types	Specific Parameters
Mechanical parameters	Component size	1200 × 600 × 6.8 mm
	framework	3.20 mm CdTe power generation glass + 0.40 mm EVA + 3.20 mm tempered glass
	Battery type	Cadmium telluride thin film (CdTe)
Photothermal performance parameters	visible light transmittance	0
	heat transfer coefficient	5.76 $W{\cdot}m^{-2}k^{-1}$
	SHGC value	0.31
Electrical performance parameters	Power (Pmpp)	110 Wp
	Short-circuit current (Isc)	2.48 A
	Open circuit voltage (Voc)	63.00 V
System integration characterization parameters	Temperature coefficient (Pmpp)	−0.29%/°C
	Temperature coefficient (Isc)	+0.04%/°C
	Temperature coefficient (Voc)	−0.28%/°C

When selecting a fixed array of PV panels, the optimum tilt angle and array parameters can be calculated based on the latitude of the region. Refer to the "Photovoltaic Power Station Design Code" [21] for the optimal tilt angle reference value when selecting the tilt angle of the panels [20]. When arranging PV arrays, it is important to calculate the minimum spacing between them to avoid any decrease in power generation efficiency caused by surrounding walls or other arrays blocking sunlight. The calculation to determine the minimum distance is done using Equation (1).

$$\mathrm{d} = \mathrm{a}[\cos\beta + \frac{\sin\beta}{\tan(90^\circ - \delta - \phi)}] \quad (1)$$

The variables used in this context are: d for the minimum spacing between PV arrays, a for the width of the PV panels, β for the tilt angle of the PV panels, δ for the winter solstice equinox of 23.26°, and φ for the latitude of the region. The study arranged the PV panels in a fixed manner and calculated their array parameters using Equation (1). The latitude of the Haikou area is 20.03°, and the width of the PV panels is 600 mm. The tilt angle refers to the optimal tilt angle reference value of PV arrays in the "Specification for Design of Photovoltaic Power Stations" [21]. The recommended tilt angle for the Haikou area is the latitude minus 3, which equals $\beta = \varphi - 3 = 20.03° - 3° = 17.03°$. There must be a minimum distance of 739 mm between the arrays when arranging the PV array. The height of the wall on the south side of the roof is 4 m, and the wall is vertical ($\beta = 90°$, $\delta = 23.26°$, $\varphi = 20.03°$). The minimum distance between the PV arrays and the wall is 3.76 m. Equation (1) was also used in this calculation. After calculating the minimum spacing between the PV arrays, it was determined that the PV panels are arranged as shown in Figure 8. The distance between the PV arrays and the south side fence is set to 4 m, and the distance between the arrays is 800 mm, allowing for 35 rows. There is a 1-metre-wide aisle on both the left and right sides, with 30 PV panels laid in each row. The panels are arranged in 2 groups of 15 pieces each, with a distance of 800 mm between the groups. In total, there are 1050 PV modules.

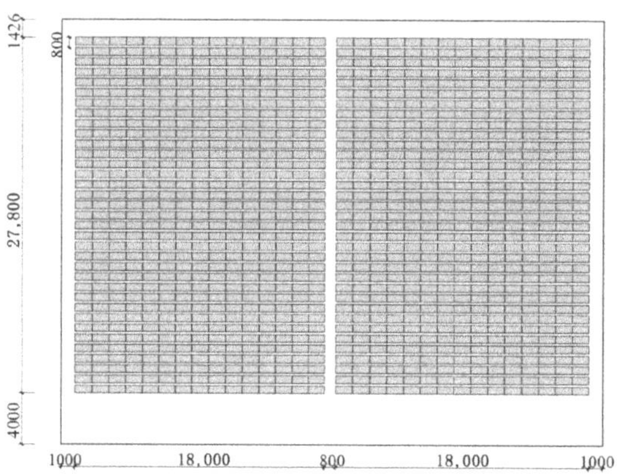

Figure 8. The PV arrangement (mm).

2.2. Mathematical Model

The calculation of the life cycle carbon emissions of the building can be materialization stage, use stage and demolition stage [22–25], and the life cycle carbon emissions of the building can be calculated by Equation (2).

$$E_{LC} = E_p + E_U + E_E \quad (2)$$

where E_{LC}, E_P, E_U and E_E are the carbon emissions of the life cycle of the building, the materialization phase, the use phase, and the demolition phase, respectively.

2.2.1. The Materialization Phase

During the material production and transportation stages, carbon emissions are primarily generated by the production and transportation of building materials required for constructing the GHR. These emissions are calculated in accordance with the "Standard for calculating building carbon emissions" [10] using Equation (3).

$$E_P = \Delta C_{sc} + \Delta C_{ys} = \sum_{i=1}^{n} M_i F_i + \sum_{i=1}^{n} M_i D_i T_i \quad (3)$$

where ΔC_{sc} represents the carbon emissions from the production phase of building materials. ΔC_{ys} represents carbon emissions at the material transport stage. The variable M_i represents the consumption of the i material, and F_i represents the carbon emission factor of the i material. The symbol D_i represents the average transportation distance of the i material, which is determined based on the actual distance. In cases where the actual distance is unclear, a default distance of 500 km may be used. The symbol T_i represents the carbon emission factor per unit weight for the transportation distance of the i material.

2.2.2. The Use Phase

Energy consumption is calculated using Equation (4).

$$E_U = \Delta C_M \quad (4)$$

The carbon emissions during the use phase of the building are represented by ΔC_M, calculated according to Equation (5). ΔE_a and ΔE_b represent the annual change in building electricity and water use during the operational phase of each scenario, EF_a and EF_b represent the carbon emission factors for electricity and water, which can be queried based on standards or publicly available carbon emission databases, and represent the annual carbon reductions brought about by the plants, and y represents the design life of each scenario (20 years).

$$\Delta C_M = (\Delta E_a EF_a + \Delta E_b EF_b - \Delta C_p)y \quad (5)$$

(1) Annual energy consumption

The energy consumption of the typical model during the building use stage was calculated by Software Designbuilder 6.1. The annual power generation of the PV system was calculated using Archelios Pro 2022R1 simulation software.

Equation (6) is used to calculate the annual change in electricity consumption brought to the building by each scenario in this paper.

$$\Delta E_a = E_{a_1} - E_{a_0} + E_{a_2} - ER_a \quad (6)$$

where ΔE_a is the annual change in electricity consumption brought about by each design for the main building; E_{a_0} is the annual energy consumption of the original building (Lecture Hall Building Model); E_{a_2} is the electricity consumed in the production of a typical model (GHR, GR, PVR); ER_a is the annual electricity production of the photovoltaic system; EF_a and EF_b are the carbon emission factors for electricity and tap water, which can be obtained from the standard or publicly available carbon emission databases.

(2) Annual water energy consumption

Irrigation water required for crop production during the use phase of a building, as well as water required to run some equipment, can add to the annual water consumption of a building. Irrigation water consumption can be based on plant water demand combined with irrigation methods, and cultivation methods for calculation can be based on local standards in the water quota provisions.

The water consumption required for the use of the equipment in the building can be calculated according to the relevant equipment specifications, and the water consumption of the wet curtain in the wet curtain fan system can be calculated according to the Design Code for Greenhouse Wet Curtain—Fan Cooling System [15].

$$V_s = \frac{C_s A_\omega}{1000} \quad (7)$$

$$U_p = \kappa_p Q_{ws} \quad (8)$$

where C_s is the minimum pool capacity per unit wet curtain area, A_ω is the wet curtain area, V_s represents the minimum volume of the pool. κ_p is the wet curtain per unit air intake, Q_{ws} is the evaporated water volume of the wet curtain fan system air intake, and U_p is the evaporated water demand of the wet curtain.

(3) Carbon sequestration by plants

The factors that influence the amount of carbon sequestered by plants in a building include the amount of carbon sequestered by the plants themselves and the amount of carbon emitted by vegetable crops that reduce food miles. This was calculated using Equation (9) with reference to the existing literature [26].

$$\Delta C_p = C_P + E_C \quad (9)$$

The amount of carbon sequestered by different plants C_P can be determined by consulting the relevant published literature on existing studies (with references), or experimentally. The carbon reduction E_C obtained by reducing the mileage can be calculated according to Equation (10).

$$E_C = \sum_{i=1}^{n} P_{V_i} J_i L_i \quad (10)$$

E_C, P_{V_i}, J_i, and L_i are variables used to calculate the carbon reduction achieved by reducing food miles. P_{V_i} represents the weight of vegetables transported using mode i, while J_i represents the amount of carbon dioxide emitted per unit weight of vegetables transported 1 km under mode i. According to statistics, the carbon dioxide emissions generated by transporting one tonne of food for one kilometre using maritime transport is 10 g, using land transport is 63 g, and using air transport is 1580 g [26]. L_i represents the mileage of transporting the food by mode i.

2.2.3. The Demolition Phase

Carbon emissions from the building demolition phase, which includes the construction and demolition phases, the waste removal phase, and the construction waste recycling phase, are calculated in Equation (11).

$$E_E = \Delta C_{jz} + \Delta C_{cc} \quad (11)$$

ΔC_{jz} represents the carbon emissions produced by the design aspect of the construction program, while ΔC_{cc} indicates the change in carbon emissions during the demolition phase.

(1) Building construction

Equation (12) calculates the carbon emissions resulting from the construction of the greenhouse component at this stage of building construction.

$$\Delta C_{jz} = \sum_{i=1}^{n} E_{jz,i} EF_i \quad (12)$$

ΔC_{jz} denotes the carbon emissions generated by the design part of the construction programme. $E_{jz,i}$ denotes the total energy usage of the i type, which is calculated based on the various energy usage of the machinery used in the construction. If there is no measured value, the energy usage can be estimated based on the energy consumption per unit shift of commonly used construction machinery. EF_i denotes the carbon emission factor of the i type of energy, which can be found in specifications and databases (calculation standard for building carbon emissions: [10,27]).

(2) Building demolition work

The change in carbon emissions during the demolition phase (ΔC_{cc}) is mainly divided into the demolition construction phase, the waste removal phase, and the recycling phase of the renewable construction waste generated during the demolition phase, and is calculated by Equation (13),

$$\Delta C_{cc} = C_{sg} + C_{wy} - C_{hs} \tag{13}$$

Among them, C_{sg} represents the demolition construction stage. The calculation method is the same as that of the construction stage. When the energy consumption during construction is not clear, it can be estimated based on 90% of the energy consumption in the construction stage [28]. C_{wy} represents the carbon emission generated in the waste transportation stage. The calculation method is the same as that of the building materials transportation stage. It is calculated based on the mode of transportation and the distance of transportation. When the distance of transportation is not clear, it can be taken as the default distance (500 km). The carbon emissions resulting from the recycling of renewable materials are represented by C_{hs}. The renewable building waste generated from demolition can be recycled. The carbon emissions from the recyclable part will be deducted from the carbon emissions generated during the demolition stage, taking into account the recycling rate of different materials.

3. Results and Discussion

3.1. Carbon Emissions of the Materialization Phase

3.1.1. GHR

The required building material consumption of the typical roof model and the quantity of building materials is calculated, as shown in Table A1.

The carbon emission factors of a hot dip galvanised welded straight seam square steel tube and carbon steel hot-dip galvanized coil are referred to in the "Calculation Standard for Carbon Emission in Buildings" [10]. The carbon emission factors of Tempered glass and 100% native aluminum profiles are referred to in the "Life-cycle energy consumption and carbon emission of typical decoration materials" [29]. The carbon emission factors of CdTe photovoltaic modules are shown in Table A1 [30].

The power generation of the GHR photovoltaic modules was simulated using the software archeliosPro 2020. The PV modules were arranged on the three-dimensional model according to Figure 6, and the total power generation of the PV modules was 973,700 kW·h (20 years).

Table A2 shows that the carbon emissions of the production stage of GHR building materials are ΔC_{sc} = 6.88 × 3.91 + 1.62 × 3.11 + 19.75 × 1.53 + 0.88 × 254 + 0.97 × 18 = 303.25 t.

There are 696 PV modules on the roof, each weighing 12 kg. The default transport distance of 500 km is used to calculate the weight of transport, mode of transport, and carbon emission factor per unit weight of transport distance, as shown in Table A2. The calculation yields ΔC_{ys} = (6.88 × 0.179 + 1.61 × 0.286 + 19.75 × 0.078 + 880.29 × 2.5 × 10^{-3} × 2.7 + 0.012 × 696 × 0.162) × 500 × 10^{-3} = 2.82 t.

$$E_P = \Delta C_{sc} + \Delta C_{ys} = 303.25 + 2.82 = 306.07 \text{ t}$$

3.1.2. GR

Table A3 displays the consumption of building materials and their corresponding carbon emission factors for constructing GR. The types of building materials and their consumption are calculated based on the weight per unit area of each material used in Figure 7 and the green area of 1400 m^2. The carbon emission factor is selected based on existing standards, published authoritative literature, and carbon emission factor databases. The carbon emission factors of SBS waterproofing membrane and polyester fibre are referred to in [31], while the carbon emission factor of PVC pressed board is referred to in the "Calculation Standard for Carbon Emission in Buildings" [10]. The carbon emission factor of high impact polystyrene is derived from Software Ecoinvent v2.2. The carbon emission in the production stage of GR building materials as,

$$\Delta C_{sc} = 6.17 + 11.45 + 4.08 + 2.88 = 24.58t$$

Table A3 shows the consumption of various building materials, with an average transport distance of 500 km used as the default. The carbon emissions from the transportation of materials stage are

$$\Delta C_{ys} = 1.27 \times 500 \times 10^{-3} = 0.63 \text{ t}.$$
$$E_P = \Delta C_{sc} + \Delta C_{ys} = 24.58 + 0.63 = 25.31t$$

3.1.3. PVR

Table A4 shows the calculated carbon emissions during the production stage of these materials.

The power generation of the PVR was simulated using Software archelisPRO, and the total power generation of the PV was 2045.9 MWh (20 years).

Fthenakis compiled a life cycle inventory of CdTe cell and PV module materials. The carbon emissions from the production phase of CdTe PV thin film cells were calculated to be 18 $gCO_2e/kW{\cdot}h$ [30]. The carbon emissions from the production phase of the building materials were ΔC_{sc} = 36.83 t.

The weight of each PV module is 12 kg, and 1050 PV modules are used in this model. The mode and distance of transport are shown in Table A4. The calculation yields the carbon emissions from the material transport phase as ΔC_{ys}=12 × 1050 × 10^{-3} × 66.02 × 10^{-3} = 0.83 t.

$$E_P = \Delta C_{sc} + \Delta C_{ys} = 36.83 + 0.83 = 37.66 \quad (14)$$

3.2. *Carbon Emissions of the Use Phase*

The calculation of the change in carbon emissions during the use phase. We firstly need to determine the annual change in electricity use (ΔE_a), the annual change in water use (ΔE_b), and the annual carbon reduction of plants in the greenhouse (C_p).

3.2.1. Annual Electricity Consumption

GHR

The energy consumption of the existing building model was simulated using Software Designbuilder to build the model. The software sets the geographic location and meteorological data, and builds the geometric model of the building based on the floor plan. The energy consumption simulation model is shown in Figure 9, and enters the material and structure of the envelope according to the construction drawings. The software then calculates the heat transfer coefficients of the envelope (see Table 4) and sets the thermal parameters of the building for internal use (see Table 5).

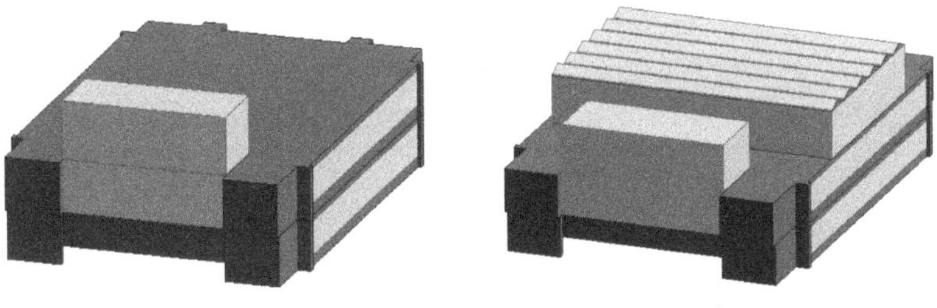

Figure 9. (**a**) Existing building energy simulation model; (**b**) GHR building energy simulation model.

Table 4. Material, construction, and thermal performance of enclosure structure.

	Structure Name	Structural Materials and Construction	Average Heat Transfer Coefficient U-Value ($W \cdot m^{-2}k^{-1}$)	Solar Heat Gain Coefficient SHGC Value
Opaque enclosure structure	Building facades	250 aerated concrete blocks, strength class A5.0, dry density class B06, with special masonry mortar, special plaster mortar inside and outside 10 mm.	0.69	—
	Building roofing	100-thick non-slip floor tiles + waterproof mortar hook joints + 20-thick polymer mortar paving base + 10-thick low strength grade mortar isolation layer + waterproofing layer (2.0 + 2.0-thick double-layer modified bitumen polyethylene-foam waterproofing roll-roofing) + 20-thick 1:3 cement mortar leveling + heat insulation layer (60-thick molded polystyrene foam board) + 30-thick lightweight aggregate concrete 2% slope layer + 150-thick reinforced concrete Roof panel	0.37	—
Transparent enclosure structure	Glass curtain wall	12-mm tempered laminated glass	3.17	0.77
	Exterior windows, exterior doors	3-mm aluminum alloy Low-E glass	4.51	0.78
	Greenhouse covering materials	5-mm tempered glass	5.84	0.84
	PV glass for GHR	PV glass RK-T40 (Table 2)	5.76	0.54

Table 5. Thermal parameter settings.

Internal Use Functions of Buildings	Office	GHR
Density of personnel	10 square meters per person	1 person
Electrical load	12 W·m^{-2}	0 W·m^{-2}
Lighting load	10 W·m^{-2}	0 W·m^{-2}
Ventilation	Mechanical ventilation	Natural ventilation
Minimum fresh air volume standard	30 cubic meters/hour. People	——
Natural ventilation airflow	——	20 times/hour
Air conditioning cooling control temperature	24–28 °C	——
Air conditioning COP	3.00	——

The annual electricity consumption of the main building in the modern GHR building model E_{a_1} is 698,692 kW·h. This is 31,046 kW·h more than the annual electricity consumption of the ordinary building model, E_{a_0}, which is an increase of 2.6%. The increased energy consumption is solely due to air-conditioning and refrigeration.

The GHR annual electricity consumption E_{a_2} is primarily calculated based on the electricity usage of its main equipment. Table 1 displays the electricity consumption of said equipment. Additionally, E_{a_2} is estimated to be 5046 kW·h.

The PV system on top of the GHR was simulated using Software archelios PRO 2022R1 to determine its annual electricity production ER_a. The results showed an average annual production ER_a of 48,685 kW·h.

The GHR changes the annual electricity consumption of the existing building $\Delta E_a = 698{,}692 - 667{,}646 + 5046 - 48{,}685 = -12{,}593$ kW·h

GR

The GR building model was used to calculate the annual change in electricity use (ΔE_a) and the annual carbon sequestration by plants (C_p). Using Software Designbuilder, we modelled the energy consumption (E_{a_1}) of the main building after the installation of the GR. Tables 4 and 5 show the thermal properties of the main building envelope, as well as the setting of thermal parameters. Table 6 displays the basic parameters of the GR substrate layer and plant layer.

Table 6. GR substrate layer and plant layer.

Parameters	Numerical Value
Thermal conductivity	0.30 W·m^{-1}K^{-1}
Soil density	500 kg·m^{-3}
Specific heat of soil	1000 J·kg^{-1}K^{-1}
Blade emissivity	0.95
Minimum stomatal resistance of the blade	100 s·m^{-1}
Maximum water content of the soil	0.50 m^3·m^{-3}
Minimum water content of the soil	0.01 m^3·m^{-3}
Initial volumetric moisture content of soil	0.20 m^3·m^{-3}
Soil thickness	0.10 m
Plant height	0.30 m
Leaf Area Index LAI	5.00

According to the above parameters to establish the energy consumption model, the simulation results show that the annual electricity consumption of the main building of the

GR building model E_{a_1} = 65,707 kW·h compared with the annual electricity consumption of the ordinary building model E_{a_0} = 667,646 kW·h can be seen, the GR can be reduced through the reduction of the main building cooling energy consumption to make the building reduce the electricity consumption by 10,571 kW·h, then the annual change in electricity consumption $\Delta E_a = -10{,}571$ kW·h.

PVR

During the use phase, carbon emissions are primarily calculated based on changes in electricity consumption, which includes changes in PV power generation and building energy consumption. The average annual photovoltaic power generation, calculated using Software Archelios PRO 2022R1, is 102,295 kW·h.

The impact of PVR on the energy consumption of the building was simulated using Software Designbuilder. The energy consumption model of the existing building was used as a basis for the PVR energy consumption model. This involves placing the PV modules on the roof of the building and setting the material structure and thermal performance parameters of the PV modules according to Table 3. The thermal performance and parameters of the main building are shown in Tables 4 and 5. The simulation results indicate that E_{a_1} = 661,927 kW·h. The PVR-mounted modules can reduce the annual electricity consumption of the building by 5719 kW·h by decreasing the cooling energy consumption of the main building, as shown by the comparison with the existing building model.

The annual change in electricity consumption $\Delta E_a = -102{,}295 - 5719 = -108{,}014$ kW·h.

The performance of the PVR modules will continue to degrade and remain stable for approximately 20 years.

3.2.2. Annual Water Consumption

The model for the annual variability of water use in the GHR includes the use of irrigation water and wet curtain water. Researchers have conducted extensive research on the water demand of greenhouse tomatoes in different growing environments [32–34] to develop a scientifically sound irrigation plan based on precision irrigation. This study calculates the daily water consumption for drip irrigation of tomatoes in a modern greenhouse with a tropical island climate and drip irrigation technology. The selected water consumption rate is 0.4 L per plant, based on a previous study [35]. Based on the planting plan for the GHR, it is possible to plant 2992 tomato plants, resulting in an annual irrigation water consumption of 436.83 t ($2992 \times 0.4 \times 365$).

The wet curtain fan system in the greenhouse uses 854.86 t of water annually. Additionally, after the installation of the GHR, there is an annual increase in water usage for the building of ΔE_b = 1291.69 t.

3.2.3. Carbon Reduction by Plants

(1) The study analysed the PV rate of tomatoes during four growth periods: seedling, flowering, fruiting, and fruiting at the end. An estimation module was established and a simulation model of the system dynamics was used to calculate that tomato plants can sequester 0.3 kg of carbon per plant during a 130 day growth cycle, assuming no pests or diseases occur [36]. In this study, a total of 2992 tomato plants were planted in the greenhouse, and the annual carbon sequestration of plants in the greenhouse was calculated to be $C_p = 0.3 \times 2992 \times 365 \div 130 \times 10^{-3} = 2.52$ t.

The average value of food miles for vegetables in Haikou City is approximately 1000 km, with 960 km being the distance of land transport and 40 km being the distance of sea transport. The annual carbon emissions from growing tomatoes in the greenhouse can be calculated by multiplying the weight of the transport, which is the annual production of tomatoes, by the distance travelled to the greenhouse. Therefore, the annual carbon emissions can be reduced by growing tomatoes in the GHR and reducing the number of food miles travelled. Based on the calculation, the reduction in annual carbon emissions is $6 \times 2992 \times 10^{-3} \times (960 \times 63 + 40 \times 10) \times 10^{-6} = 1.09$ t. The greenhouse grows a total of 2992 tomato plants, with an expected average annual yield of 6 kg per plant.

The carbon reduction achieved by plants in the GHR is 3.61 t. The main structure of the GHR has a design service life of 20 years, as established in this study's typical model of the GHR. The main structure of the greenhouse in the establishment of the typical model of the GHR in this study, the change in the building's carbon emissions during use as,

$$E_U = \Delta C_M = \left(-12,593 \times 10^{-3} + 1291.69 \times 10^{-3} - 3.61\right) \times 20 = -298.82t$$

(2) Sedum plants were selected to cover the GR, and the results of existing studies show that Sedum plants for GR can sequester 375 g of carbon per square metre per year for the whole GR system [37]. In this study, this literature reference value is selected, and the calculation can be obtained that the annual carbon sequestration of plants in the building model of GR $C_p = 375 \times 1400 \times 10^{-6} = 0.53$ t. when establishing the typical model of GR, the design service life of the GR is 20 years, which can be calculated to obtain that the GHR change in the building's carbon emissions during use as,

$$E_U = \Delta C_M = \left(-10,571 \times 0.917 \times 10^{-3} - 0.53\right) \times 20 = -204.5t$$

When modelling a typical GHR, the main structure of the greenhouse is calculated to give the change in building carbon emissions from the GHR during use as,

$$E_U = \Delta C_M = \left(-108,014 \times 0.917 \times 10^{-3}\right) \times 20 = -1981.0t$$

3.3. *Carbon Emissions of the Demolition Phase*

3.3.1. GHR

The carbon emissions generated from the construction of the GHR are shown in Table 7, in which the total electrical energy usage is estimated to be $164.31 \times 1 + 60.27 \times 21 = 1430$ kW·h according to the energy consumption per unit shift of the commonly used construction machinery (calculation standard for building carbon emissions: [10]), and the carbon emission factor of the electrical energy of the Haikan Province is 0.917 $kgCO_2e \cdot kW \cdot h^{-1}$ [27], then the carbon emission generated during the construction process is $\Delta C_{jz} = 1.31$ t. The energy consumption of the GHR demolition construction phase is calculated as 90% of the energy consumption during the building construction process [28], $\Delta C_{sg} = 1.3 \times 90\% = 1.18$ t.

The transport distance for outbound transport of waste is still taken as the default distance of 500 km, and the carbon emission factors for the weight of waste, mode of transport, and transport distance per unit weight are shown in Table 7, $\Delta C_{wy} = 5.33 \times 500 \times 10^{-3} = 2.67$ t.

The steel, glass, and aluminium alloy profiles generated by the demolition are recyclable materials, combined with their recycling rate [38], the carbon emissions of the recyclable portion are subtracted from the greenhouse carbon emissions, and the types of recyclable materials and their recycling rate, etc., are shown in Table 8, i.e., $C_{hs} = 232.97$ t. Therefore, in the GHR during the demolition phase, the carbon emission generated is −229.12 t.

Table 7. Carbon emissions from construction of GHR.

Construction Machinery	Performance Specifications	Number of Machine Shifts	Energy Consumption Type	Energy Consumption per Unit Shift (kW·h)	Carbon Emission Factor ($kgCO_2e{\cdot}kW{\cdot}h^{-1}$)	Carbon Emission (t)
Tower crane	Improve quality by 400 tons	1	Electric energy	164.31	0.92	0.15
AC welding machine	Capacity 21 kV·A	21	Electric energy	60.27	0.92	1.16
			Total			1.30

Table 8. Carbon emissions from GHR waste transport during the demolition phase and the recycling phase.

Category of Waste Material	Weight (t)	Waste Transportation Method	Carbon Emission Factor ($kgCO_2e{\cdot}t^{-1}km^{-1}$)	Distance (km)	Recycling Utilization Rate	Carbon Emissions Generated During the Production Phase (t)	Carbon Emission (t)
Steels	8.49	Heavy diesel truck (with a load capacity of 10 tons)	0.16	500	90%	31.91	−28.03
Glass	19.75	Heavy diesel truck (with a load capacity of 30 tons)	0.08		10%	−30.23	−2.25
Aluminium alloy	5.94	Medium diesel truck (with a load capacity of 8 tons)	0.18		90%	−223.59	−200.70
Photovoltaic module	8.35	Heavy diesel truck (with a load capacity of 10 tons)	0.16	——	0	0	1.35
			Total				−229.12

The GHR demolition phase is,

$$E_E = \Delta C_{jz} + \Delta C_{cc} = 1.31 - 229.12 = -227.81\text{t}$$

3.3.2. GR

The selection of commonly used construction machinery is shown in Table 9, the carbon emission generated in the construction process is $\Delta C_{jz} = 152.49 \times 10^{-3} = 0.152$ t, and the carbon emission generated in the demolition process is calculated, where the energy consumption in the demolition construction phase is 90% of the energy consumption in the construction process, i.e., $\Delta C_{sg} = 0.152 \times 90\% = 0.14$ t.

Table 9. Carbon emissions from construction of GR.

Construction Machinery	Performance Specifications	Number of Machine Shifts	Type of Energy Consumption	Energy Consumption per Unit Shift (kW·h)	Carbon Emission Factor ($kgCO_2e{\cdot}kW{\cdot}h^{-1}$)	Carbon Emission (kg)
Tower crane	Improve quality by 60 tons	1	Electricity	166.29	0.98	152.49

The waste transport distance is still taken as the standard distance of 500 km, and the transport mode for medium-sized diesel lorries (load 8 t) is calculated in the same way as for the construction materials transport stage, from Table A3 material weight outbound (4.06 + 1.57 + 1.16 + 0.28) × 500 × 0.179 × 10^{-3} = 0.63 t. If there are no recycled construction materials, then the recycling stage C_{hs} for 0, $\Delta C_{cc} = 0.63 + 0.14 = 0.77t$.

Calculations give the carbon emissions from the GR and demolition phases as,

$$E_E = \Delta C_{jz} + \Delta C_{cc} = 0.15 + 0.77 = 0.92t$$

3.3.3. PVR

The carbon emissions from building construction and demolition are shown in Table 10, and the carbon emissions generated in the construction process are ΔC_{jz} = 0.93 t.

Table 10. Carbon emissions during the construction phase of PVR.

Construction Machinery	Performance Specifications	Number of Machine Shifts	Type of Energy Consumption	Energy Consumption per Unit Shift (kW·h)	Carbon Emission Factor ($kgCO_2e{\cdot}kW{\cdot}h^{-1}$)	Carbon Emissions (t)
Tower crane	Improve quality by 400 tons	1	Electricity	164.31	0.98	0.15
AC welding machine	Capacity 21 kV·A	14	Electricity	60.27	0.92	0.77

Carbon emissions from demolition take 90% of the carbon emissions generated during the construction phase, i.e., 0.93 × 90% = 0.84 t. The carbon emissions generated during the waste transportation phase are the same as those generated during the transportation phase of building materials, i.e., 0.83 t, $\Delta C_{cc} = 0.84 + 0.83 = 1.67t$.

Excluding the recycling of PV modules, the carbon emissions from demolition of PVR are as follows,

$$E_E = \Delta C_{jz} + \Delta C_{cc} = 0.93 + 1.67 = 2.60t$$

3.4. Carbon Emissions of the Life Cycle of Building Roofs Retrofit

Table A5 summarises the data for GHR, GR, and PVR. The life cycle carbon reductions of three types of building roofs retrofit are, in descending order, PVR > GHR > GR. This paper evaluates the comparison of the carbon emissions of three roofing models using the LCA approach. In this text, the reasons for the largest carbon reduction of PVR are analyzed as follows: It is likely that the rooftop photovoltaic system can directly convert solar energy into electrical energy with a relatively high energy conversion efficiency. In contrast, GHR and GR mainly reduce carbon emissions indirectly by means of adjusting indoor temperature, absorbing carbon dioxide, etc., and their carbon reduction effects are relatively limited.

In the paper by Ref. [39] the LCA approach was used to study the different contributions of two roof (GHR, PVR) forms, comparing two different roof resource production systems, constructing a model and implementing assessment calculations to analyse the environmental costs, resource recycling benefits, and sensitivities, and exploring nine major influences in the study, Climate Change, Ozone Depletion, Terrestrial Acidification, Freshwater Eutrophication, Photochemical Oxidation Formation, Terrestrial Ecotoxicity, Water Depletion, Metal Depletion, and Fossil Depletion. In their study, PVR was more environmentally friendly in all impact categories except terrestrial ecotoxicity, water consumption, and metal depletion, which were the only categories that benefited neither GHT nor PVR [40]. Ref. [41] highlighted that the most significant shortcoming of the PVR lies in its inability to tackle problems related to the disposal and maintenance of photovoltaic components once their service life comes to an end. In this particular study, only the complete 20-year service life of the PV panels was taken into consideration throughout the entire life cycle. However, it failed to account for the carbon emissions that could be triggered by unforeseen circumstances. For instance, in case of damage occurring in the middle of the period, maintenance work would be required, and this process was overlooked in terms of its potential carbon footprint.

In this paper, a comparison was made among three roof models with the aim of addressing carbon emissions. It was found that the PVR model exhibited the lowest carbon emissions. However, an analysis of the heat island effects that each of these models might induce during the power generation process was lacking. In contrast, ref. [8] integrated PV panels with GR in their research. The PV panels contributed to the buildings by generating electricity, yet they also had the potential to exacerbate the urban heat island effect. Their study demonstrated that a well-irrigated GR could mitigate this adverse impact. They further proposed setting region-specific conditions for separately introducing GR and PV systems to conduct comparative studies. Meanwhile, this paper focused on comparing the carbon emissions from power generation among the three distinct roof models. In this paper, carbon emissions were also calculated using the LCA method to compare the economic benefits of the three types of roofs, but some of the reasons for this were not taken into account, for example, the implementation of GHR reduces the transport and consumption of food transport, and reduces the expenditure on fertilisers, nevertheless, the study also confirms that all the economic benefits of PVR are indeed the best. PVR applications have numerous advantages: they reduce the dependence on fossil fuel power generation, decrease the energy losses related to power transmission and distribution, and do not require land for installation. However, various environmental, social, economic, and technical limitations associated with the retrofit uses of rooftops have been emphasized. In the case of GHR, some concerns at the social level have been expressed, as consumers may be reluctant to use soilless cultivation techniques. In addition, health risks may arise from air pollution or irrigation with improperly treated wastewater [42]. In this paper, these factors are not considered in GHR, but it is true that growing in GHR creates situations such as wastewater contamination, compared to PVR which does not create risks such as affecting people's health [33]. The researchers also conducted a study [43] on the comparison between PVR systems and rooftop agriculture, specifically GR and GHR. They put forward a comprehensive comparison between PVR technology and rooftop agricultural technology. The research findings indicated that, under simulated conditions and with respect to the selected comparison criteria, food production was demonstrated to be more advantageous than energy production, not only for system owners but also for local communities. When evaluated holistically, the GHR exhibited significant edges over

GR and solar PV systems, as both GR and solar PV systems imposed certain impacts on operators and local communities. In this paper, further comparisons were predominantly centered around carbon emissions, where PVR held an absolute upper hand. Nevertheless, aspects such as food production, the utilization of food within cities, and the impact on surrounding communities were not delved into any further.

4. Conclusions

This study establishes models of GHR, PVR, and GR for green retrofitting of existing buildings using roof space in tropical island climate. Based on the models, the carbon cost of retrofitted roofs in the stages of production and transportation of building materials, construction, and demolition were calculated according to the "Standard for calculating building carbon emissions" [10].

For the use of GHR, GR, and PVR to renovate the roof space of existing buildings, the carbon costs invested in the production and transport of building materials, construction, and the demolition phase are 78.26 t, 26.13 t, and 40.26 t, respectively, with greenhouses requiring the highest carbon costs.

Combining the energy production and consumption in the use phase (20 years), it is calculated that the GHR produces changes in electricity consumption of 251.86 MWh, changes in water consumption of 25,834 t, and the carbon reduction of plants of 72.2 t in the existing buildings during the 20-year period, and the total carbon reduction during the use phase is 298.82 t. The GR produces changes in electricity consumption of 211.42 MWh, carbon reduction of plants of 10.6 t, and a total reduction of 204.5 t; PVR produces changes in electricity consumption of 2160.28 MWh, and carbon reduction of 1981.0 t.

The life cycle calculation of the carbon reduction per unit area of each roof renovation method can be obtained, the carbon reduction per unit area of the GR with a design area of 1400 m^2 is 127.41 $kg{\cdot}m^{-2}$, the carbon reduction per unit area of the PVR with a photovoltaic module surface area of 756 m^2 is 2567.12 $kg{\cdot}m^{-2}$, and the carbon reduction per unit area of the GHR with a design area of 840 m^2 is 262.57 $kg{\cdot}m^{-2}$.

The above conclusions reflect that the carbon reduction benefits of PVR as a carbon reduction scheme are the most significant. The benefits of PVR in terms of reducing carbon emissions by producing clean energy instead of traditional thermal power generation are very significant, and PV power generation is a sustainable form of energy that can provide mankind with a long-term supply of clean energy. In addition, roof-mounted PV panels are equivalent to installing a shading layer on the roof of the building. PVR can also reduce carbon emissions by reducing the energy consumption of the main building. Considering factors such as shading, the utilisation rate of roof space for PV power generation in this study is not high, but as PV technology continues to develop and PV modules become more adaptable, more space will be available for PVR.

It is evident that there are still some deficiencies in this study that require further investigation. The paper has only carried out preliminary theoretical analyses and screening of suitable roofs based on factors that can be quantified uniformly, such as building height, roof area, slope, and light shading. In addition, numerous other factors impede the feasibility of roof retrofitting. These include building function, roof ownership, and other planning, policy, and regulatory conditions. Furthermore, the loading and roof conditions of some old buildings that are no longer suitable for retrofitting due to safety concerns, and the effects of typhoons and other inclement weather on retrofitted roofs must be considered. These factors should be taken into account in the context of the practical application of the study. This study employs the calculation results of an existing building model and extrapolates them to the urban scale to assess the carbon reduction potential. However, this approach is not general enough and more typical building models should be selected for the calculation. Alternatively, a more elaborate reasoning model should be established, and a few buildings of general sizes should be selected for experiments. Finally, experiments and software simulations should be used for comparisons.

Author Contributions: Conceptualization, W.D. and Y.Z.; methodology, D.M. and Y.S.; software, Y.S.; validation, Y.Z. and D.M.; formal analysis, W.D.; investigation, W.D. and Y.S.; resources, D.M.; data curation, W.D.; writing—original draft preparation, Y.S.; writing—review and editing, Y.Z. and W.D.; visualization, W.D.; supervision, W.D.; project administration, Z.L. and S.F.; funding acquisition, D.M. and Y.Z. All authors have read and agreed to the published version of the manuscript.

Funding: This project is supported by the National Natural Science Foundation of China (No. 52068017), the Natural Science Foundation of Hainan Province (No. 521RC501), the Key Research and Development Project of Hainan Province (No. ZDYF2022SHFZ353), and the Natural Science Foundation of Hainan Province (No. 623RC454).

Data Availability Statement: The data that support the findings of this study are available from the corresponding author upon reasonable request.

Conflicts of Interest: The authors declare no conflict of interest.

Appendix A. Life Cycle Inventory of Scenarios in This Study

Table A1. List of GHR building materials consumption in the materialization stage.

Category of Building Materials	Component Type	Material Type	Section Size (mm)	Cross Sectional Area	Unit Mass ($kg{\cdot}m^{-1}$ or $kg{\cdot}m^{-2}$)	Length or Area (m or m^2)	Quantity	Dosage (kg or m^2)	Total (kg or m^2)
Hot dip galvanised welded straight seam square steel tube	Column	Square steel	160 × 160 × 3	18.45	14.49	6	32	2782	6877
	Strut	Rectangular	60 × 30 × 3	4.21	3.30	0.10	56	18	
	Trusses	Rectangular	90 × 50 × 3	8.04	6.31	8	48	2423	
	Truss beam	Square steel	25 × 25 × 1.5	1.31	1.03	0.60	480	296	
	Arch rod	Rectangular	60 × 30 × 3	4.21	3.30	2.1	96	665	
	Purlins	Rectangular	60 × 30 × 3	4.21	3.30	35	6	693	
Carbon steel hot-dip galvanized coil	Gutter	Galvanized	Figure 3	—	15.70	0.42 × 35	7	1615.53	1615.53
Tempered glass	Covering	Tempered	5	—	12.50	1580.20	—	19,752.50	19,752.50
100% native aluminum profiles	Metal frame	Aluminum	70 × 60 × 2.50	7.42	—	0.26 × 6 0.26 × 32 0.26 × 35 0.26 × 0.80 0.26 × 0.40 $\sqrt{2^2+0.8^2} \times 0.26$	130 8 40 12 24 432	880.29	880.29

Table A2. List of GHR carbon emissions during the materialization stage and transportation during the materialization phase.

Category of Building Materials	Consumption of Building Materials/Photovoltaic Module Power Generation	Carbon Emission Factor	Transportation Method	Dosage	Density ($t{\cdot}m^{-3}$)	Distance (km)	Transportation Carbon Emissions Factor ($kgCO_2e{\cdot}t^{-1}km^{-1}$)	Carbon Emissions (t)
Hot dip galvanised welded straight seam square steel tube	6877 kg	3.91 $kgCO_2e{\cdot}kg^{-1}$	Medium diesel truck (with a load capacity of 8 tons)	6.88 t	—	500	0.18	27.51

Table A2. *Cont.*

Category of Building Materials	Consumption of Building Materials/ Photovoltaic Module Power Generation	Carbon Emission Factor	Transportation Method	Dosage	Density ($t \cdot m^{-3}$)	Distance (km)	Transportation Carbon Emissions Factor ($kgCO_2e \cdot t^{-1}km^{-1}$)	Carbon Emissions (t)
Carbon steel hot-dip galvanized coil	1615.53 kg	3.11 $kgCO_2e \cdot kg^{-1}$	Light diesel truck (with a load capacity of 2 tons)	1.61 t	——		0.29	5.25
Tempered glass	19,752.50 kg	1.53 $kgCO_2e \cdot kg^{-1}$	Heavy diesel truck (with a load capacity of 30 tons	19.75 t	——		0.08	31.01
100% native aluminum profiles	880.29 m^2	254 $kgCO_2e \cdot m^{-2}$	Medium diesel truck (with a load capacity of 8 tons)	880.29 × 2.5 × 10^{-3} m^3	2.70		0.18	224.11
CdTe photovoltaic modules	973,700 kW·h	18 $gCO_2e \cdot kW \cdot h^{-1}$	Heavy diesel truck (with a load capacity of 10 tons	12 kg × 696	——		0.16	18.19
			Total					306.07

Table A3. List of GR building materials and their carbon emission factors in the materialization stage and transportation during the materialization phase.

Category of Building Materials	Consumption of Building Materials (kg)	Carbon Emission Factor ($kgCO_2e \cdot kg^{-1}$)	Type of Transportation	Weight (t)	Distance (km)	Carbon Emission Factor ($kgCO_2e \cdot t^{-1}km^{-1}$)	Carbon Emissions (t)
SBS waterproofing membrane	4060	1.52	Medium diesel truck (with a load capacity of 8 tons)	4.06	500	0.18	6.53
PC pressing board	1568	7.30		1.57			11.59
High impact polystyrene HIPS	1165	3.50		1.16			4.18
Polyester fiber	280	10.28		0.28			2.91
			Total				25.21

Table A4. Carbon emissions from production during the materialization phase of PVR building materials and PV module transportation during the materialization stage.

Category of Building Materials	Power Generation (MWh)	Carbon Emission Factor ($kgCO_2e/kW \cdot h$)	Type of Transport	Quantity	Weight (kg)	Distance	Carbon Emission Factor ($kgCO_2e \cdot t^{-1}km^{-1}$)	Carbon Emissions (t)
Photovoltaic module	2046	0.02	Heavy diesel truck (with a load capacity of 18 tons)	1050	12	500 km	0.13	37.66
			Sea freight			80 km	0.02	

Table A5. Carbon emissions of the life cycle of building roofs retrofit.

Life Cycle	Content	Unit	GHR	GR	PVR
Materialization phase	ΔC_{sc}	t	303.25	24.58	36.83
	ΔC_{ys}	t	2.82	0.63	0.83
	E_P	t	306.07	25.21	37.66
Use phase	$\Delta E_a EF_a$	t	230.96	193.87	1980.98
	$\Delta E_b EF_b$	t	4.34	0	0
	ΔC_p	t	72.20	10.60	0
	E_U	t	−298.82	−204.50	−1981.00
Demolition phase	ΔC_{jz}	t	1.31	0.15	0.93
	ΔC_{sg}	t	1.18	0.14	0.84
	ΔC_{wy}	t	2.67	0.63	0.83
	ΔC_{cc}	t	−229.12	0.77	1.67
	ΔC_{hs}	t	232.98	0	0
	E_E	t	−227.81	0.92	2.60
Carbon emissions		t	−220.56	−178.37	−1940.74

References

1. China Building Energy Conservation Association; Chongqing University Urban and Rural Construction and Development Research Institute. Research Report on China's Building Energy Consumption and Carbon Emissions). *Architecture* **2023**, 57–69.
2. Getter, K.L.; Rowe, D.B. Role of extensive green roofs in sustainable development. *Hortscience* **2006**, *41*, 1276–1285. [CrossRef]
3. Golubchikov, O.; Deda, P. Governance, technology, and equity: An integrated policy framework for energy efficient housing. *Energy Policy* **2012**, *41*, 733–741. [CrossRef]
4. *ISO 14040:2006*; Environmental Management–Life Cycle Assessment–Principles and Framework. ISO: Geneva, Switzerland, 2006.
5. *ISO 14044:2006*; Environmental Management—Life Cycle Assessment—Requirements and Guidelines. ISO: Geneva, Switzerland, 2006.
6. Sharma, A.; Saxena, A.; Sethi, M.; Shree, V. Life cycle assessment of buildings: A review. *Renew. Sustain. Energy Rev.* **2011**, *15*, 871–875. [CrossRef]
7. Cabeza, L.F.; Rincón, L.; Vilariño, V.; Pérez, G.; Castell, A. Life cycle assessment (LCA) and life cycle energy analysis (LCEA) of buildings and the building sector: A review. *Renew. Sustain. Energy Rev.* **2014**, *29*, 394–416. [CrossRef]
8. Pomponi, F.; Lenzen, M. Hybrid life cycle assessment (LCA) will likely yield more accurate results than process-based LCA. *J. Clean. Prod.* **2018**, *176*, 210–215. [CrossRef]
9. Yang, Y.; Heijungs, R.; Brandão, M. Hybrid life cycle assessment (lca) does not necessarily yield more accurate results than process-based lca. *J. Clean. Prod.* **2017**, *150*, 237–242. [CrossRef]
10. *GB/T 51366-2019*; Ministry of Housing and Urban Rural Development Calculation Standard for Building Carbon Emissions. China Construction Industry Press: Beijing, China, 2019. Available online: http://www.jzlj.org.cn/Item/Show.asp?m=1&d=8065 (accessed on 20 August 2024).
11. Song, Z.; Li, Y.; Ma, G.; Luo, X. Research on the whole life cycle carbon reduction of hangzhou west railway station hub buildings. *Chin. Foreign Archit.* **2022**, *12*, 28–34. [CrossRef]
12. Zhao, S.; Zhu, J.; Wang, Z.; Xu, K.; Zhu, K.; Chen, X. Research on building carbon emission calculation and emission reduction strategies based on lca: Taking a residential project as an example. *Constr. Econ.* **2023**, *44*, 371–378. [CrossRef]
13. Ni, S.; Liu, H.; Li, Q.; Quan, H.; Gheibi, M.; Fathollahi-Fard, A.M.; Tian, G. Assessment of the engineering properties, carbon dioxide emission and economic of biomass recycled aggregate concrete: A novel approach for building green concretes. *J. Clean. Prod.* **2022**, *365*, 132780. [CrossRef]
14. Wang, W.; Shu, J. Urban renewal can mitigate urban heat islands. *Geophys. Res. Lett.* **2020**, *47*, e2019GL085948. [CrossRef]
15. *NY/T 2133-2012*; Design Rule for Fan-Pad Cooling System in Greenhouse. Ministry of Agriculture of the People's Republic of China: Beijing, China, 2012.
16. Qin, Z.; Yu, P.; Wei, N. Current situation and trend analysis of cdte thin film solar cells at home and abroad. *Glass* **2021**, *48*, 9–13. [CrossRef]
17. Fthenakis, V.M. Life cycle impact analysis of cadmium in cdte pv production. *Renew. Sustain. Energy Rev.* **2004**, *8*, 303–334. [CrossRef]
18. Liu, X.; Yang, X. Full-cycle cadmium emission analysis of cdte photovoltaic in China's national environment. *Sci. Bull.* **2013**, *58*, 12.
19. Liu, F.; van den Bergh, J.C.J.M. Differences in CO_2 emissions of solar pv production among technologies and regions: Application to china, eu and usa. *Energy Policy* **2020**, *138*, 1–11. [CrossRef]
20. Miller, I.; Gençer, E.; Vogelbaum, H.S.; Brown, P.R.; Torkamani, S.; O'Sullivan, F.M. Parametric modeling of life cycle greenhouse gas emissions from photovoltaic power. *Appl. Energy* **2019**, *238*, 760–774. [CrossRef]

21. *GB 50797-2012*; China Electricity Council, Ministry of Housing and Urban-Rural Development of the People's Republic of China. Design Code for Photovoltaic Power Station. China Planning Press: Beijing, China, 2012.
22. Robati, M.; Daly, D.; Kokogiannakis, G. A method of uncertainty analysis for whole-life embodied carbon emissions (CO_2-e) of building materials of a net-zero energy building in australia. *J. Clean. Prod.* **2019**, *225*, 541–553. [CrossRef]
23. Schmidt, M.; Crawford, R.H.; Warren-Myers, G. Integrating life-cycle ghg emissions into a building's economic evaluation. *Build. Cities* **2020**, *1*, 361–378. [CrossRef]
24. Yu, M.; Robati, M.; Oldfield, P.; Wiedmann, T.; Crawford, R.; Nezhad, A.A.; Carmichael, D. The impact of value engineering on embodied greenhouse gas emissions in the built environment: A hybrid life cycle assessment. *Build. Environ.* **2020**, *168*, 106452. [CrossRef]
25. Zhang, X.; Wang, F. Life-cycle assessment and control measures for carbon emissions of typical buildings in china. *Build. Environ.* **2015**, *86*, 89–97. [CrossRef]
26. Zhao, M. Research on the Methodology for Assessing the Agricultural Potential of Rooftops of Existing Urban Buildings. Master's Thesis, Tianjin University, Tianjin, China, 2018. [CrossRef]
27. *CECS 374:2014*; China Academy of Architectural Design and Research Building Carbon Emission Measurement Standard. China Planning Press: Beijing, China, 2014. Available online: https://www.zlglpt.com/book/book_view.aspx?id=2715 (accessed on 20 August 2024).
28. Wang, Y. Whole Life Cycle Carbon Emission Study of Industrialized Prefabricated Buildings. Ph.D. Thesis, Southeast University, Nanjing, China, 2016.
29. Yu, H.; Zeng, J.; Zhao, M. Life-cycle energy consumption and carbon emission of typical decoration materials. *Build. Sci.* **2014**, *30*, 21–25. [CrossRef]
30. Fthenakis, V.; Kim, H.C. Energy use and greenhouse gas emissions in the life cycle of cdte photovoltaics. *MRS Online Proc. Libr.* **2005**, *895*, 306. [CrossRef]
31. Tao, T.; Wei, W.; Wu, Y.; Yu, T.; Zhu, Q. *Study on Carbon Costs and Carbon Benefits of Green Roofs Based on Life Cycle Assessment*; Poster Session Presentation: Beijing, China, 2013.
32. Fan, J.B. Experimental study on water demand by drip irrigation in greenhouse tomato. *Jilin Water Resour.* **2016**, *31*, 14–16. [CrossRef]
33. Feng, C.; Zhang, Y.; Fang, X. Study of tomato water demand regulation under the condition of drip irrigation. *J. Anhui Agric. Sci.* **2015**, *52*, 49–50. [CrossRef]
34. Sun, L.; Sun, J.S.; Liu, H.; Liu, Z.G.; Zhang, J.Y. Water requirement rules of tomato in sunlight greenhouse. *J. Irrig. Drain.* **2008**, *27*, 51–54.
35. Harmanto; Salokhe, V.M.; Babel, M.S.; Tantau, H.J. Water requirement of drip irrigated tomatoes grown in greenhouse in tropical environment. *Agric. Water Manag.* **2005**, *71*, 225–242. [CrossRef]
36. Zhang, J. Construction of Organic Vegetable Disease Prediction Model and Carbon Sequestration Estimation Model Based on System Dynamics and Internet of Things Technology. Master's Thesis, Zhejiang Agriculture and Forestry University, Hangzhou, China, 2015.
37. Getter, K.L.; Rowe, D.B.; Robertson, G.P.; Cregg, B.M.; Andresen, J.A. Carbon sequestration potential of extensive green roofs. *Environ. Sci. Technol.* **2009**, *43*, 7564–7570. [CrossRef]
38. Gong, X. Research on Building Dismantling and Material Reuse Technology. Master's Thesis, Tianjin University, Tianjin, China, 2010.
39. Houchmand, L.J.; Martí, M.M.; Gassó-Domingo, S. Photovoltaics and green roofs: Holistic analysis in built environments. *Renew. Sustain. Energy Rev.* **2025**, *207*, 114987. [CrossRef]
40. Corcelli, F.; Fiorentino, G.; Petit-Boix, A.; Rieradevall, J.; Gabarrell, X. Transforming rooftops into productive urban spaces in the mediterranean. An lca comparison of agri-urban production and photovoltaic energy generation. *Resour. Conserv. Recycl.* **2019**, *144*, 321–336. [CrossRef]
41. Agarwal, U.; Rathore, N.S.; Jain, N.; Sharma, P.; Bansal, R.C.; Chouhan, M.; Kumawat, M. Adaptable pathway to net zero carbon: A case study for techno-economic & environmental assessment of rooftop solar pv system. *Energy Rep.* **2023**, *9*, 3482–3492. [CrossRef]
42. Specht, K.; Siebert, R.; Hartmann, I.; Freisinger, U.B.; Sawicka, M.; Werner, A.; Thomaier, S.; Henckel, D.; Walk, H.; Dierich, A. Urban agriculture of the future: An overview of sustainability aspects of food production in and on buildings. *Agric. Hum. Values* **2014**, *31*, 33–51. [CrossRef]
43. Benis, K.; Turan, I.; Reinhart, C.; Ferrão, P. Putting rooftops to use—A cost-benefit analysis of food production vs. Energy generation under mediterranean climates. *Cities* **2018**, *78*, 166–179. [CrossRef]

buildings

MDPI

Article

Evaluating Housing Quality, Energy Performance, Indoor Environment, and User Satisfaction in Renovated Locked-In Owner-Occupied Homes in Belgium

Leontien Bielen [1], Katleen Van den Broeck [2], Alexis Versele [1] and Hilde Breesch [1,*]

[1] Building Physics and Sustainable Design, Department of Civil Engineering, KU Leuven, Gebroeders De Smetstraat 1, 9000 Ghent, Belgium
[2] Research Group Social and Economic Policy and Social Inclusion, KU Leuven, Parkstraat 47, P.O. Box 5300, 3000 Leuven, Belgium
* Correspondence: hilde.breesch@kuleuven.be

Abstract: Approximately 3% of home-owners in Flanders (Belgium) are "locked-in", residing in substandard dwellings with limited financial means for improvement. Under the European Urban Innovative Actions project ICCARus, 84 dwellings of locked-in home-owners in Ghent (Belgium) underwent renovations with an average budget of EUR 29,000. Their impact on housing quality, energy performance, indoor environmental quality (IEQ), and user satisfaction was assessed. The housing quality was evaluated using the Flemish Housing Code, and energy performance was calculated. A monitoring campaign in selected dwellings evaluated the IEQ. Interviews documented home-owners' satisfaction. The results revealed a significant enhancement in housing quality, with post-renovation compliance reaching 83%. Energy performance notably improved, from a pre-renovation F-label prevalence (31% of the cases) to a post-renovation majority achieving a C-label (55%). The average indoor temperature in all rooms increased after renovation. A moderate correlation between enhancement in the living room and bathroom temperature and budget for energy performance measures was noted. The CO_2 concentration in the majority of the homes remained unchanged. A total of 61% of the participants reported increased satisfaction with their dwelling post-renovation, with more pronounced improvement in winter. This enhanced satisfaction was correlated with improved energy performance.

Keywords: renovation; locked-in home-owners; housing quality; energy performance; indoor environmental quality

1. Introduction

Approximately 11% of residential buildings in Flanders, Belgium, equivalent to 310,000 dwellings, are categorized as being in poor or very poor condition [1]. Specifically, nearly half (47%) of privately rented dwellings exhibit inadequate housing quality [2]. Furthermore, the European Union's Renovation Wave advocates for energy-efficient upgrades in 35 million buildings, constituting 65% of residences constructed before 1970, aiming to reduce net greenhouse gas emissions by 55% by 2030 and achieve carbon neutrality by 2050 [3]. In alignment with these goals, Flanders targets elevating the Energy Performance Certificate (EPC) of all dwellings (new and renovated) to the highest level (i.e., A), requiring primary energy use below 100 kWh/m^2 [4]. To meet this objective, the annual rate of deep renovations (to EPC level A) must increase to 3% yearly [4]. The projected cost of aligning the Flemish housing stock with the Flemish Housing Code [5], stipulating basic safety, health, housing quality standards, and energy requirements, was estimated at EUR 137–145 billion in 2019 [6].

Recent research in Flanders [1,7] underscores a higher demand for energy-efficient renovations in dwellings occupied by vulnerable households. The 2013 Flemish Housing Survey [8] reveals that 118,000 home-owners (i.e., 4% of the Flemish housing market) face challenges combining housing quality and affordability issues. Among these home-owners, 70% fall within the three lowest income quintiles, constituting the "locked-in home-owners" group, i.e., residents in poorly maintained dwellings with limited financial means for improvement.

Research on the housing shortage in Flanders indicates that the proportion of locked-in owners in the Belgian cities of Antwerp and Ghent averages 4.3% [9,10]. In Ghent alone, approximately 6000 owner–occupiers fall into this category. A significant concern is that 24.2% of the residents allocate at least 30% of their family income to housing expenses [11].

Compared to other owner-occupied houses or the broader housing stock in Flanders, residences of locked-in owners exhibit notably higher rates of quality issues. Structural and non-structural problems in roofs and walls, moisture-related concerns, and window defects are approximately twice as prevalent in these homes. Issues with electrical installations, heating systems, indoor environment, and basic comfort are typically three to four times more frequent in the houses of locked-in owners compared to the overall housing stock. Additionally, the risk of carbon monoxide intoxication is four times higher in these homes [10].

Considering the socio-economic conditions of locked-in owners and the substantial renovation requirements, there are doubts, as outlined by [10] regarding the ability of locked-in owners to undertake quality renovations and align their houses with the minimum requirements set by the Flemish Housing Code [5] (which in the meantime changed to the Flemish Codex on Housing with a different evaluating code) Without adequate support, it remains challenging for this group to execute renovation works and address structural defects due to limited financial resources.

In an effort to renovate the residences of locked-in owners and enhance their living conditions, the City of Ghent initiated the ICCARus ("Gent knapt op") project, with support from the European Urban Innovative Actions (UIA) initiative [12]. The project provided financial assistance, a maximum of EUR 30,000 per locked-in owner, for the implementation of renovation works. Notably, this funding is reimbursed to public authorities upon any official property ownership changes, ensuring the sustainability of the initiative (e.g., sale, disposal, donation, inheritance). This innovative approach allows for the continual reuse of funds for new renovation projects in the future, extending the impact beyond a specific group. Operational from 1 November 2018 to 31 October 2022, the project facilitated various renovation measures, including the renewal of heating or electrical systems, insulation works, refurbishment of external joinery, and moisture treatment in 84 dwellings across the City of Ghent, Belgium. This paper aims to assess the impact of these renovation measures in dwellings owned by locked-in owners in terms of energy performance, indoor environment, housing quality, and user satisfaction.

Existing monitoring studies showed that renovation measures can have an attributable impact on reducing the energy demand for heating in a dwelling [13–15]. Hamilton et al. [13] examined the change in gas use from 2005 to 2007 in a cohort of 168,998 gas-heated English dwellings. They showed that retrofits do have an impact on reducing energy demand by 4.2% per year. The retrofit measures associated with the largest change in energy demand were the installation of a condensing gas boiler (−5.2%) and cavity wall insulation (−3.8%). Lambie [14] monitored the impact of renovations in 18 Belgian dwellings from 4 different projects. She identified the most efficient renovation measures from this monitoring campaign. A roof renovation combined with some small additional measures regarding the building envelope was found to be efficient. A floor renovation did not result in significant improvements. In addition, a complete renovation of the building envelope without changing the building geometry was found to be efficient. Du et al. studied energy retrofits of 37 buildings in Finland and 15 buildings in Lithuania. The heating energy consumption decreased after renovation by an average of 24% and 49% in the Finnish and Lithuanian buildings, respectively.

The impact of energy renovations in residential buildings on the indoor environment was studied, among others, by [14–20]. Lambie [14] noticed in the 18 monitored Belgian dwellings that the global thermal comfort temperature in winter, on average, increased by 0.8 °C for small renovations and 2.5 °C for medium and deep renovations. She also noticed that a balanced mechanical ventilation system with heat recovery is an efficient measure to obtain good post-retrofit indoor air quality. In addition, Broderick et al. [16] monitored thermal parameters and indoor air pollutants in 15 Irish social dwellings in the living room and the main bedroom before and after an energy upgrade during the winter periods of 2015 and 2016. The living room temperatures increased from 17.8 to 18.6 °C, but this change was not statistically significant. The indoor relative humidity did not significantly change after the retrofit and was still within the recommended range of between 30 and 60%. They noticed increases in pollutant concentrations (formaldehyde, PM2.5, and TVOC) that were correlated with lower building air exchange rates post-retrofit. Moreover, Földváry et al. [17] evaluated the impact of energy renovations on IAQ, air exchange rates, and occupant satisfaction in Slovak naturally ventilated multifamily buildings. Temperature, relative humidity, and CO_2 concentrations were measured in 94 apartments during one week in the winter. They noticed that the indoor air temperature was significantly lower in the original buildings than in the renovated ones. However, the relative humidity was similar in the renovated and the non-renovated buildings. The average CO_2 concentration in every apartment for the whole measurement period was higher in the renovated than in the non-renovated buildings (1290 ppm vs. 1100 ppm) but not statistically significant. IAQ, by means of NO_2, TVOC, formaldehyde, and CO_2, was further investigated in 20 apartments before and after its renovation. CO_2 and formaldehyde concentrations were significantly higher, and the air change rates were significantly lower in the renovated buildings. Du et al. [15] studied the impact of energy retrofit on IAQ, including formaldehyde and microbial content. Concentrations of formaldehyde tended to be lower after the retrofits in Finnish buildings, whereas an opposite trend was seen in Lithuanian buildings. Hassan et al. [20] studied the impact of deep renovation in 12 Irish homes on indoor air pollutants, including PM2.5, CO_2, and formaldehyde. CO_2 concentrations decreased, but PM2.5 ($p < 0.0001$) and formaldehyde ($p < 0.05$) concentrations increased post-retrofit due to inadequate ventilation, the presence of a wood stove, and occupant activities. Fisk et al. [21] studied the empirical data from evaluations of the influence of residential energy efficiency retrofits of dwellings with low-income occupants on indoor environmental quality conditions. Data were extracted from 36 studies described in 44 papers plus two reports, including the above-mentioned studies of Broderick et al. [16] and Földváry et al. [17]. Nearly all reviewed studies were performed in Europe or the United States. They concluded that the average indoor temperature during winter typically increased after the retrofit, usually by less than 1.5 °C. Moreover, the formaldehyde concentrations tended to increase after retrofits that did not add mechanical ventilation in the entire dwelling. Hassan et al. [20] concluded from a sample of 12 Irish homes that the homes were significantly ($p < 0.00001$) warmer post-retrofit. Pungercar et al. [19] also noticed an increased indoor temperature of up to 2.5 °C during the heating period in a retrofit of a single-family house in Germany. Willand et al. [18] synthesized the results of 28 energy efficiency improvement programs in residential buildings. All reviewed studies found the dwellings to be warmer after the retrofit interventions.

Regarding the changes in housing quality and health before and after the energy renovation measures, Fisk et al. [21] concluded that, based on occupant reports, dampness, and mold almost always decreased after retrofit. This decrease in reported problems with mold was confirmed by the synthesis of Willand et al. [18]. In addition, Maidment et al. [22] systematically quantified the impact of household energy efficiency measures on health and wellbeing. Thirty-six studies (in total of more than 33,000 participants) were meta-analyzed. On average, the energy efficiency interventions led to a small but significant improvement in the health of the residents. Significant health benefits were identified for children, people with poor health, and vulnerable groups in general. The increased temperature and lowered relative humidity were also found with benefits for cardiovascular (subjective and objective) and respiratory health [18]. Significant reductions in fungal and bacterial concentrations were found after renovation in Finland and Lithuania [15].

Differences in user satisfaction before and after renovation in residential buildings were determined by [17,18,21,23,24]. Földváry et al. [17] surveyed the perceived air quality, sick building syndrome symptoms, and airing habits of the occupants. They noted that energy renovation was associated with lower occupant satisfaction with IAQ. In addition, self-reported headache, itchy eyes, dry skin and fatigue were more prevalent after renovation than before. Willand et al. [18] also noted better self-reported respiratory conditions of the occupants. Pedersen et al. [23] investigated the tenant perception of IEQ changes after the renovation of a multifamily housing area in Sweden with 323 dwellings. A post-renovation survey showed that the renovation increased the proportion of residents who are satisfied with thermal comfort and indoor air quality in winter. Also, Coggins et al. [24] showed that in deep energy retrofitted Irish residences, the majority of the home-owners were satisfied with indoor temperature post-retrofit compared to pre-retrofit. They were also mostly satisfied with perceived IAQ and natural light post-retrofit. Satisfaction with these factors significantly improved ($p < 0.005$) post-retrofit. Fisk et al. [21] also reviewed self-reported thermal comfort and health and subjectively reported that thermal comfort (statistically significant), thermal discomfort, non-asthma respiratory symptoms, general health, and mental health (statistically significant) nearly always improved after retrofit.

The above-mentioned studies reveal trends in the impact of energy renovation on energy use, indoor conditions, and user satisfaction in residential buildings. However, locked-in home-owners, i.e., residents in poorly maintained dwellings with limited financial means for improvement, have not been targeted yet in a large-scale field study. This paper's contribution lies in unraveling the influence of limited energy renovation measures for this specific type of home-owners via 40 field studies in Ghent (Belgium).

The paper's structure unfolds in the following sequence. It commences with an overview of the data collection process, followed by a detailed description of the field studies. This encompasses information on housing type, floor area, dwelling inhabitants, pre-renovation owner satisfaction, and the specific renovation measures undertaken. The results chapter quantitatively evaluates the impact of the implemented renovation measures on housing quality, energy performance, indoor environment, and owner satisfaction. Finally, the paper concludes with a discussion of findings and an overarching conclusion.

2. Data Collection Method

2.1. Introduction

The data collection encompasses both quantitative and qualitative aspects. A housing quality assessment aligned with the Flemish Housing Code was conducted through housing controls, and a monitoring campaign was initiated to measure the renovation's impact on the indoor environment (temperature and indoor air quality).

In addition, qualitative data were obtained through pre- and post-renovation interviews, capturing insights into owners' satisfaction with their dwelling and comfort levels during summer and winter.

A total of 84 locked-in owners participated in the project, with 36 contributing to both pre- and post-renovation interviews. In total 54 dwellings are considered involved in the financial analysis of the renovation measures. The housing quality from 36 dwellings and the energy performances from 28 dwellings are documented both pre- and post-renovation. Table 1 provides a comprehensive overview of participant distribution in this research.

Table 1. Number of participants involved in different parts of the research.

Part of Research Project	Amount of Dwellings/ Participants
Analysis of renovation measures	84
Financial analysis of renovation measures (invoices of the renovation measures available)	54
Housing quality analysis	
• Pre-renovation	82
• Post-renovation	36
Analysis energy performance	
• Pre-renovation	39
• Post-renovation	28
Pre- and post-renovation interviews	36
Indoor environment monitoring	10

2.2. Quantitative Data

The total renovation budget invested per dwelling is based on available invoices and data from the technical supervisors. This analysis includes only dwellings for which invoices were available, i.e., 54 dwellings. Based on the available invoices of the implemented renovation measures, an estimate was made of the average investment per measure. Renovation measures were grouped into 4 categories (i.e., increasing housing comfort, improving energy performance, decreasing safety risks, and improving accessibility), and most implemented measures were analyzed.

Energy performance certificates (EPCs) were issued by an external auditor before and after renovation [25]. This certificate describes the energy performance of a property in terms of annual primary energy use per m^2 floor area. Moreover, it gives some additional data about the houses, such as the average usable floor area. These data were also considered in this study.

The indoor environment was measured in ten houses before renovation during 5.5 weeks in the winter of 2019/2020 (11 January 2020–18 February 2020) and one year later after renovation over the same period. The Belgian climate is classified as zone 4A: Mixed—Humid under the ASHRAE climate classification [26]. The monthly average temperatures in January and February 2020 were equal to 5.9 °C and 7 °C, respectively, warmer than normal. Most daily temperatures deviate less or larger than 5 °C compared to the normalized values from 1991 to 2020 [27]. On the contrary, the monthly average temperatures in January and February 2021 were normal, i.e., 3 °C and 5.3 °C. A cold week with a deviation larger than 5 °C from the normalized values was noticed in February 2021.

In the living areas of the houses, operative temperature, relative humidity (RH), and CO_2 concentration were measured with HOBO MX1102 sensors with a time interval of 10 min. In the bathroom and bedroom, temperature and relative humidity were measured with HOBO U12-013 sensors, also with a time interval of 10 min. All monitoring devices were located at a height of approximately 1.5 m and far from heating sources. The accuracy of these monitoring devices is summarized in Table 2.

Table 2. Accuracy of the used IEQ sensors (Onset, HOBO MX1102a—Onset, sd).

Parameter	HOBO MX1102	HOBO U12-013
Temperature Accuracy	±0.21 °C	±0.35 °C from 0 °C to 50 °C
RH Accuracy	±2% (20% to 80%) to a maximum of ±4.5% and ±6% below 20% and above 80%	±2.5% (10% to 90%) to a maximum of ±3.5% ±3.5%
CO_2 Accuracy	±50 ppm ±5% of reading at 25 °C	N/A

2.3. Qualitative Data

Before and after renovation, the housing quality was examined through a technical assessment by the housing control service of the City of Ghent using the Flemish Housing Code [5]. The Flemish Housing Code defines the health, safety, and quality requirements that every house in Flanders should meet, such as stability of roof, walls, and floors, safe heating system without risk of carbon monoxide intoxication, safe stairs, no moisture problems, minimal sanitary facilities, supply of drinkable water, minimum roof insulation and double glazing. All this information was collected through the technical assessment checklist. Penalty points are given for each missing requirement. A property allocated more than 15 penalty points is labeled as a property of inadequate quality. The selection criteria to join the project were as follows. Only houses of inadequate quality could benefit from the initiative. At the same time, the feasibility of obtaining below 15 penalty points with a maximum of EUR 30,000 for renovation measures was required.

Both prior to and three to six months following the renovation, a structured interview was conducted to collect residents' perspectives on various subjective aspects, encompassing their satisfaction with their living spaces and the comfort experienced during the summer and winter seasons [28]. The residents were asked to rate their satisfaction on a five-point scale ranging from "Not at all satisfied" to "Very satisfied". 65 pre-renovation and 46 post-renovation interviews with locked-in owners were conducted. The evaluation focused on individuals who actively participated in interviews both before and after the renovation, involving a dataset of 36 home-owners.

3. Description of Field Studies and Implemented Renovation Measures

3.1. General Information

In this research project, 94% of the case study houses are terraced houses, with only 6% comprising apartments, specifically focusing on the renovation of their private areas. The houses have an average usable floor area of 124 m^2. Among them, the smallest house covers 69 m^2, while the largest spans a usable floor area of 296 m^2 before renovation and 302 m^2 after renovation. Typically, participating households consist of two residents. The maximum household size observed was eight, comprising two adults and six children, while the minimum occupancy was represented by a single resident.

3.2. Dwelling Characteristics Before Renovation

The examination of the building envelope concentrated on assessing the roof and glazing properties. A notable finding revealed that over half of the home-owners inhabited residences where the glazing in the living room failed to comply with the Flemish Housing Code's requirement for double-glazed windows. Specifically, in 13 houses (equivalent to 19%), over 80% of the glazing in the living room consisted of single glazing.

Additionally, more than half of the surveyed houses (69%) exhibited roofs lacking full insulation. In 30 houses, a striking 80% or more of the roof lacked insulation. These figures surpass the regional average in Flanders, where roof insulation is (partially) present in 83% of houses, as reported by [29]

A significant concern emerged from the findings, revealing that more than half of the houses (58%) featured aging heating units, surpassing 12 years in operation. These units predominantly comprised gas or oil boilers, with only occasional instances of condensing boilers. This aging infrastructure heightened the potential risk of carbon monoxide (CO) intoxication. Furthermore, in 21 residences (or 30%), outdated appliances for domestic hot water production were identified, contributing to the overall risk.

Data from the Great Housing Survey [8] indicated that 15% of owner-occupied houses in Flanders equipped with combustion-based heating systems were identified as at risk of CO intoxication [30]. Notably, among locked-in owners, this risk has escalated to 39%, as reported by [10].

Concerns regarding safety hazards in the electrical installations of the surveyed houses are notable. In 31 houses, there existed an imminent risk of acute electrocution or fire, with some integral electricity installations deemed obsolete. Another 25 houses, constituting approximately one third of the total, had unsafe electrical installations, although these issues were deemed fixable with limited interventions. Only in 13 of the participating houses was the electrical installation considered to be without any safety risks.

Alarming is the fact that among the surveyed houses, a substantial 81% presented potential safety risks such as electric shocks, fire hazards, or electrocution. This percentage significantly surpasses the regional average in Flanders, where the prevalence of such safety risks is reported at 21% [2].

Furthermore, 85% of the participants reported facing moisture-related issues in their homes. Among them, 36 houses (52%) had challenges with rising or condensing moisture, localized leaks, or leaking gutters. Additionally, 23 houses (33%) exhibited severe moisture damage from structural defects in the roof, façade, or floor. Notably, both of these figures surpass the regional averages in Flanders, where limited and severe moisture problems were found in only 16% and 4% of houses, respectively [2].

It is worth noting that these percentages align more closely with the experiences of locked-in home-owners in Flanders, where 26% reported damage due to condensation problems, and 31% reported damage resulting from infiltrating or rising moisture [10].

3.3. Owner's Satisfaction Before Renovation

The baseline interviews show that 52% of the locked-in owners are satisfied with their house (very satisfied or rather satisfied), which is much lower than the average for households in Ghent in 2017 (City Monitor) [11], where 82% of residents were (very or rather) satisfied with their house. It should be kept in mind that the descriptions here are at the household level, while the analyses in the City Monitor are at the resident level. However, this cannot explain the large difference, and therefore, it is concluded that satisfaction with the house is much lower among the participants than what is average for Ghent. These answers are in line with the condition of the dwelling as described above and the need for renovation, which is a prerequisite for participation in this project.

The reasons given by locked-in owners for their lack of satisfaction are mostly related to the poor (structural) condition of the house. In particular, the poor comfort in the houses is cited (cold in winter) as well as moisture problems. Furthermore, lack of space, quality of sanitation, and comfort are also mentioned, as well as external factors related to the neighborhood.

3.4. Implemented Renovation Measures

3.4.1. Classification of Renovation Measures

Priority renovation measures were determined individually for each house, guided by a comprehensive technical assessment. These renovation measures were subsequently classified into four categories, as depicted in Figure 1:

- Category 1: Housing comfort measures
- Category 2: Measures with a potential increase in energy performance
- Category 3: Safety measures
- Category 4: Accessibility measures

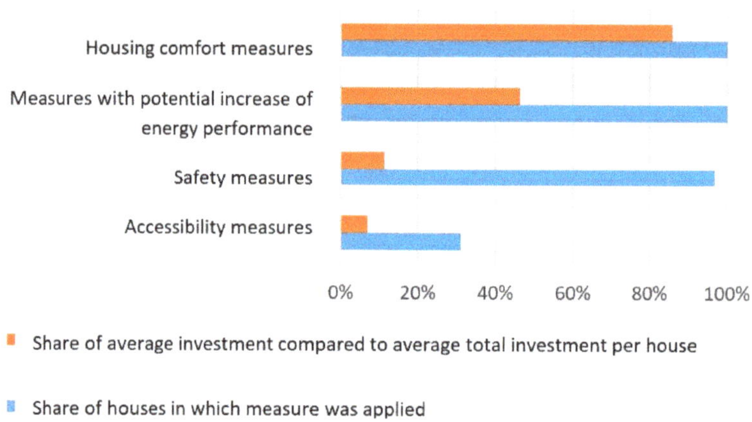

Figure 1. Share of houses in which measure was applied and share of average investment compared to average total investment per house.

Nearly all undertaken actions contribute to enhancing housing comfort (category 1). Investments that fall outside or have a limited impact in this category include engineering costs, container rentals, or demolition works. Measures with a potential impact on energy performance (category 2) encompass the installation of insulation in the building envelope, the replacement or enhancement of heating and ventilation systems, and the renovation of windows and doors, among others.

Safety measures (category 3) encompass the inspection and/or complete renewal of the electrical installation, structural enhancements to improve house stability, installation of smoke detectors, and stability calculations conducted by an engineer. In the context of accessibility (category 4), specific measures include the installation of shower handles and stair lifts, the adaptation of bathrooms to improve accessibility, the replacement of bathtubs with showers, and the substitution of steps with ramps.

Renovation measures can belong to several categories. For example, façade insulation or renewing the heating system will improve both the energy performance and housing comfort of the residents. Another example is the complete renewal of the electricity that will make the house safer and offer better housing comfort.

Further detailed analysis of 54 dwellings for which the invoices were available, the most frequently implemented specific measures are as follows (see Figure 2):

- Installation of smoke detectors (category 3)
- Update or integral renewal and inspection of electricity installation (category 3)

- Renewal of external joinery (including glazing) (both in façades and roof) (categories 1 and 2)
- Insulation, complete renovation, or finishing of the roof (categories 1 and 2)
- Installation of a new heating system or repair/maintenance of the existing heating system (categories 1 and 2)
- Injection or treatment against moisture (categories 1 and 2)

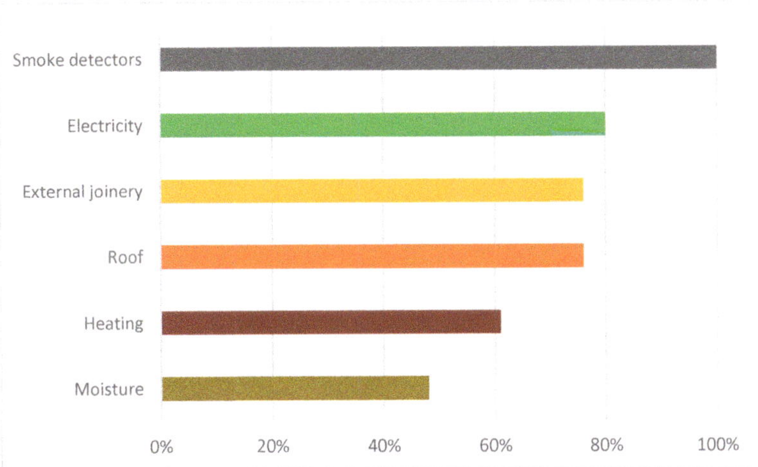

Figure 2. Most implemented renovation measures (the relative amount of dwellings, in %), based on analysis of 54 dwellings.

In adherence to the compulsory regulations outlined in the Flemish Housing Code [5], smoke detectors were universally installed in all houses. The electricity systems underwent repair or renewal in 80% of the houses. Furthermore, renewal of external joinery and roof works were undertaken in over 75% of the houses. This is in line with a study conducted by the Flemish Energy and Climate Agency [29], focusing on energy awareness and behavior of Flemish households, substantiating the prevailing trend of prioritizing roof insulation. In addition, a substantial 61% of the houses underwent a renewal of the heating system. Additionally, almost half of the houses required injection or treatment against moisture.

3.4.2. Financial Analysis of the Renovation Measures

The total renovation budget invested per dwelling averages EUR 29,000, with a minimum of EUR 4045 and a maximum of EUR 37,135. Figure 1 illustrates the share of houses in which a measure was applied, along with the share of the average investment relative to the average total investment per house.

Measures of energy performance and housing comfort were implemented in all dwellings. On average, measures in the context of housing comfort (category 1) took up 86% of the total budget, while 46% of the total budget was spent on energy performance (category 2). Because several measures were included in several categories, the sum of the average investments per category does not correspond to the average renovation budget. The safety measures for residents (category 3) averaged EUR 3267 per house. Despite being implemented in 97% of houses, the budget allocation for these critical measures remains relatively low, constituting only 11% of the total average renovation budget. For accessibility measures (category 4), an average of EUR 2012 was invested per dwelling, representing 7% of the average total renovation budget.

The budget allocated for specific renovation measures, excluding smoke detectors, is illustrated in Figure 3. The largest expenditure per dwelling was dedicated to roof renovations, followed by the replacement of external joinery and the heating system.

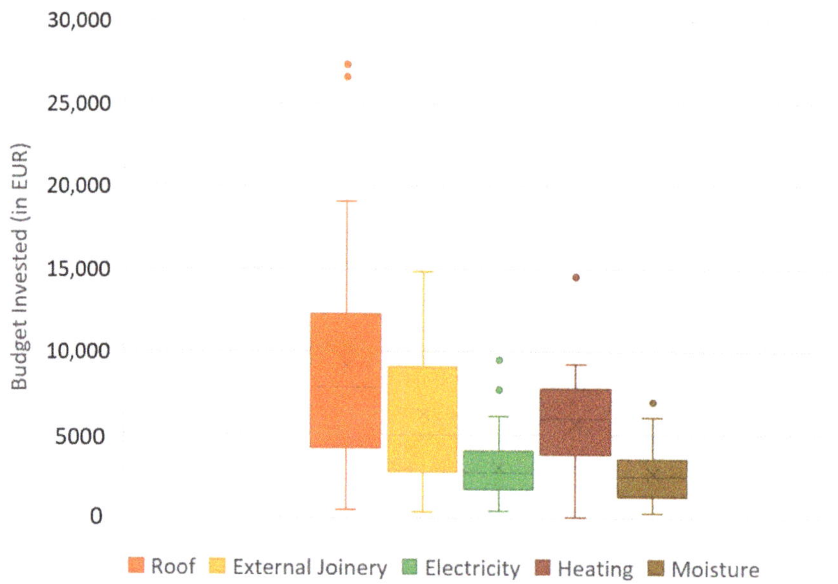

Figure 3. Budget invested per renovation measure (in EUR), based on analysis of 54 dwellings.

For roof renovations, an average of EUR 9161 was spent per dwelling, with a median expenditure of EUR 7841. The cost of these roof works showed significant variability, ranging from a minimum of EUR 456 to a maximum of EUR 27,323 for a single dwelling, which is almost the entire available budget.

In the houses in which windows and/or doors were replaced, the average investment was EUR 6273, with a median of EUR 498. Renovating or repairing the heating system costs an average of EUR 5596 per dwelling, with a median of EUR 5979. Although electrical work was required in most houses (approximately 80%), it accounted for a relatively small portion of the budget, with an average expenditure of EUR 3024 and a median of EUR 2708. Moisture treatment or injection was necessary for about half of the dwellings, with a rather small average investment of EUR 2617 and a median of EUR 2434.

3.4.3. Renovation Measures in the Subset of the Dwellings Selected for IEQ Monitoring

The impact of the renovation measures on indoor environmental quality (IEQ) was monitored in a subset of 10 dwellings. Table 3 details the specific measures implemented in these selected houses, including insulation, external joinery, heating systems, and ventilation systems. The table distinguishes between the three rooms where measurements were conducted: the living room, bathroom, and bedroom. This breakdown allows for an assessment of whether the implemented measures had an effect on the IEQ measurements. Cells are color-coded to indicate the presence of specific measures, with any additional relevant details provided in the accompanying text within the table.

Table 3. Overview of the implemented renovation measures in the 10 selected houses for IEQ monitoring. Color-coded cells indicate the presence of specific measures, with any additional relevant details provided in the accompanying text.

House	Insulation			External Joinery			Heating			Ventilation		
	Living room	Bathroom	Bedroom	Living Room	Bathroom	Bedroom	Living Room	Bathroom	Bedroom	Living Room	Bathroom	Bedroom
A			roof									
B			roof				radiators	radiators	radiators			
C			roof				central heating + radiators					
D							central heating + radiators					
E												
F	roof											
G		roof										
H		roof	roof				central heating + radiators					
I							central heating + radiators					
J	roof											

Insulation was installed in 9 out of 10 houses (excluding dwelling D), primarily focusing on roof insulation. In some cases, this involved insulating the sloped roof of the main volume, potentially affecting the indoor climate of the bedrooms. In other homes, it involved insulating the flat roofs of annex buildings, which could impact the indoor climate of the bathroom or living room. Façade insulation in dwelling I may influence the indoor climate of the entire house. Exterior joinery was replaced and/or repaired in eight houses (excluding dwellings G and I), mainly involving the replacement of several windows, which could affect all the rooms where measurements were taken. In seven houses (excluding dwellings A, G, and J), central heating was installed, existing heating systems were repaired, and/or radiators were installed. In four houses (E, F, I, J), a mechanical extraction ventilation system was installed in the bathroom.

4. Results

This section analyses the housing quality, energy performance, and indoor environmental quality post-retrofit. It also explores the link between implemented renovation measures and the satisfaction of the locked-in home-owners.

4.1. Housing Quality Assessment

Housing quality was assessed before and after renovation through a technical assessment conducted by the housing control service of the City of Ghent in accordance with the Flemish Housing Code [5]. Figure 4 shows the distribution of the penalty scores before and after renovation. In addition, Table 4 compares the penalty score for the 10 selected dwellings for the IEQ monitoring campaign pre- and post-renovation. Prior to the renovation, the dwellings received penalty scores ranging from 18 to 120 points, with an average of 62 points. This is notably higher than the average penalty score for inadequate dwellings in Flanders, which, according to the Flemish Housing Survey 2013, was 28 points for dwellings exceeding the inadequacy threshold of 15 points [30].

A significant difference was observed between the pre- and post-renovation penalty scores from Figure 4, tested via a t-test for unequal variance, with $p = 3 \times 10^{-13} < 0.05$. Following the renovations, a significant improvement was observed, with 30 out of 36 houses (83%) meeting the requirements of the Flemish Housing Code, defined as a score below 15 penalty points. In 13 houses (36%), all penalty points were successfully eliminated. However, a minority of houses did not meet the post-renovation requirements. This was due to some owners opting to undertake minor works themselves to reduce costs, which were not included in the project scope.

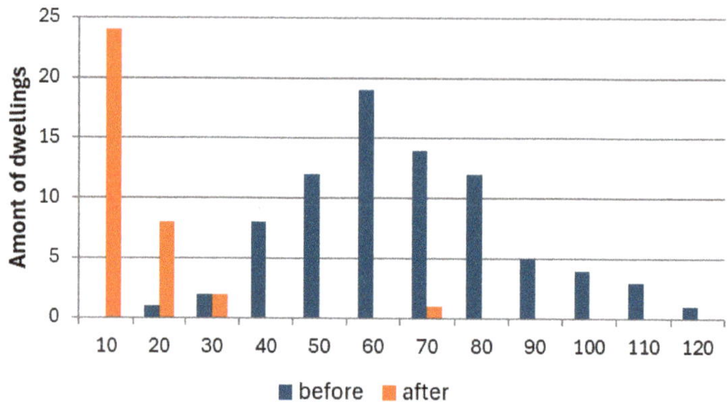

Figure 4. Distribution of the penalty scores assessing housing quality pre- and post-renovation.

Table 4. Overview of the thermal characteristics before renovation, housing quality (penalty score), and energy performance (EPC-label) in the 10 selected houses for IEQ monitoring before and after renovation.

	Thermal Characteristics Before Renovation		Housing Quality Penalty Score (-)		EPC-Label	
House	**Roof Insulation**	**Single Glazing in Living Room**	**Before**	**After**	**Before**	**After**
A	>80% not insulated	partly	78	9	- [1]	A
B	Partly	>80%	38	10	F	C
C	>80% not insulated	partly	85	0	- [1]	C
D	>80% not insulated	>80%	74	0	E	C
E	Partly	>80%	26	0	F	C
F	Partly	>80%	93	3	D	B
G	>80% not insulated	partly	61	12	F	- [1]
H	>80% not insulated	no	98	15	F	- [1]
I	Partly	partly	93	3	D	- [1]
J	Partly	no	51	12	F	C

[1] Data not available.

Improvement in housing quality due to renovation is in line with existing studies showing a decrease in dampness and mold problems after retrofitting [18,21].

4.2. Energy Performance Evaluation

4.2.1. Analysis of the Energy Performance Certificates (EPC)

The energy performance of the houses was assessed using the Flemish Energy Performance Certificate (EPC) [25] both before and after renovation, as shown in Figure 5. The energy performance of the 10 selected houses is also presented in Table 4. Table 5 presents the energy performance certificates along with their corresponding theoretical primary energy use. In addition, Figure 6 compares the annual primary energy use per usable floor area for 28 dwellings pre- and post-renovation.

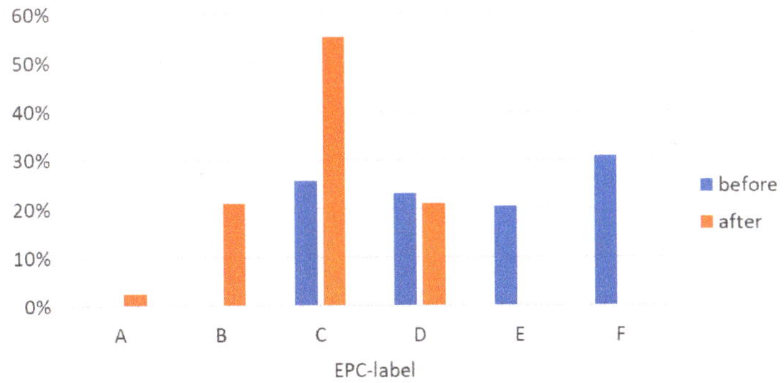

Figure 5. Energy performance certificates (EPC) before and after renovation (explanation about EPC in 5).

Table 5. Theoretical primary energy use per year and per m^2 usable floor area in kWh/(m^2·a) and the corresponding Energy Performance Certificate [31].

EPC	Primary Energy Use (kWh/(m^2·a))
A+	less than 0
A	between 0 and 100
B	between 100 and 200
C	between 200 and 300
D	between 300 and 400
E	between 400 and 500
F	more than 500

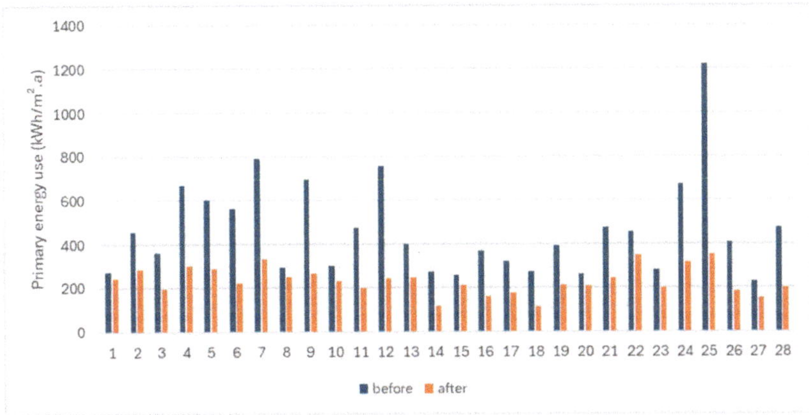

Figure 6. Overview of theoretical annual primary energy use per floor area per dwelling pre- and post-renovation.

Before renovation, 31% of the houses were classified in the worst category, i.e., F. Notably, none of the dwellings achieved an A+, A, or B rating, indicating that all had a calculated primary energy use exceeding 200 kWh/m^2 per year. Following renovation, the calculated primary energy use of all dwellings decreased significantly, calculated via a t-test with a p-value = $8 \times 10^{-6} < 0.05$. The majority of the houses (55%) achieved a C rating, 21% obtained a B, and 3% received an A.

Moreover, the correlation between the renovation budget invested for measures with a potential increase in energy performance and the change in annual primary energy use per floor area per dwelling was investigated via Pearson correlation. The correlation was found to be weak to moderate, with a Pearson correlation coefficient of 0.3.

Despite these improvements, 97% of the dwellings did not meet the Flemish 2050 objective of achieving a primary energy use of less than 100 kWh/m^2 per year (i.e., an A rating). This shortfall is primarily due to the extremely poor housing quality before renovation, which necessitated prioritizing measures to improve basic housing quality and reduce safety risks within the constraints of the limited budget.

4.2.2. Estimation by the Home-Owners

In post-renovation interviews, home-owners were asked to estimate their energy use following the renovation. Interestingly, 55% of participants estimated that their energy use remained the same or even increased. They cited several reasons for this perception:

- More rooms are in use after renovation or have heating systems, leading to an increase in the heated area.
- Additional (electrical) appliances were installed, likely increasing electricity consumption.
- An increase in visitors post-renovation, which they believe may contribute to higher energy consumption.

However, many owners based their energy use estimates on their energy bills. Given the rising energy prices, it is possible that those who believed their energy use had increased might actually have seen higher costs due to price peaks, even if their actual energy use remained constant or decreased.

Conversely, 45% of the participants estimated a decrease in their energy use after renovation. They attributed this to improvements such as insulation, new windows, upgraded heating systems, and sealing of gaps and cracks.

These findings align with other studies, which suggest that the actual impact of energy-saving investments often falls short of theoretical predictions [32]. The actual reduction in energy use is often (much) lower. The reduction in energy use is frequently lower than anticipated for several reasons. One factor is the "rebound effect", where increased efficiency leads to increased consumption in other areas [33]. Additionally, research by [34] found that energy use predictions often overestimate consumption in poorly insulated homes while underestimating it in energy-efficient ones. Furthermore, miscalculations of expected returns, failure to account for contextual influences, and reliance on average scenarios can all contribute to discrepancies. The feedback provided to users about their energy use and the differences between short-term and long-term effects also play a role [32].

The interviews also explored participants' behavior regarding energy use. Half of the respondents (51%) described themselves as "economical with energy, but could still save in some areas". Financial motivations were a key driver for energy efficiency. Some participants reported less energy-efficient behaviors, attributing this to their children, while others prioritized thermal comfort. Notably, the renovation did not significantly change most owners' attitudes toward energy use, with 77% stating their behavior remained the same and only 18% reporting a more conscious approach.

4.3. Indoor Environmental Quality Assessment

The impact of the renovation measures on the indoor environmental quality was monitored in a subset of 10 dwellings. Indoor temperatures and CO_2 concentration in the living room, bathroom, and main bedroom were analyzed before and after renovation during 4 weeks in winter.

4.3.1. Indoor Temperature

Figure 7 compares the indoor temperature variations in the living room before and after renovation. The average living room temperature in all the houses increased during the measurement campaign from 18.9 °C before renovation to 20.6 °C after renovation. Taking into account that outdoor temperatures after renovation were significantly lower, it can be concluded that the renovation measures had a positive impact on the average indoor temperature. However, the difference in average temperature per dwelling pre- and post-renovation is not statistically significant, tested via a paired t-test for average values, with a p-value of $0.24 > 0.05$. This result has to be interpreted with precautions due to the small sample size (i.e., 8 dwellings).

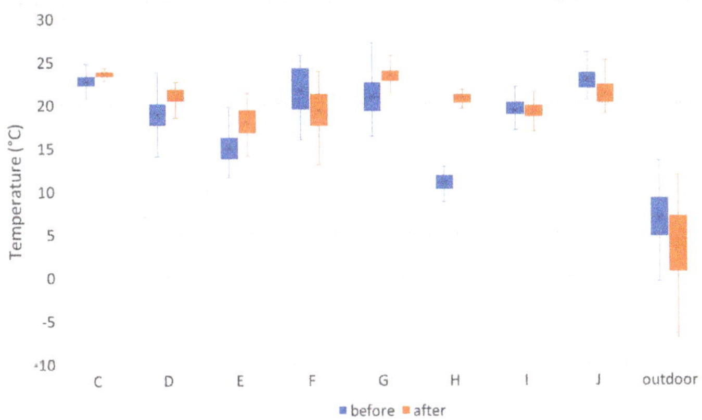

Figure 7. Indoor temperature in the living room before and after renovation.

In 75% of the dwellings, the indoor temperature in the living room increased (or stayed the same) after renovation. Only in dwellings F and J is a decrease in temperature noticed. The differences in living room, bathroom, and bedroom temperature per dwelling are significant (via t-test) with a p-value < 0.05. In dwelling H, very low temperatures before renovation were measured. The average temperature here increased from 10.8 °C to 20.6 °C after renovation. This living room was rarely used and thus not heated because it was in too poor condition. The bedroom was used as a living room. In the living room on the ground floor, the exterior joinery was replaced, central heating was installed, and interior finishing was provided. As a result, the living room was reoccupied and heated after renovation.

Moreover, large fluctuations in temperature were measured in houses C, D, F, and G on a daily basis before renovation. These may indicate poor insulation of the houses, as depicted in Table 4. After renovation, these fluctuations in temperature were significantly reduced, except in dwelling F.

Thermal comfort in the living room was assessed by checking the compliance with the minimum operative temperature for the winter season of comfort category II from EN 16798-1 [35], i.e., 20 °C (see Figure 8). As the occupancy was not monitored, this graph has to be handled with caution. Most of the dwellings comply after renovation with this minimum temperature. However, dwelling E and I still show low temperatures in the living room after renovation, despite placing insulation in both E and I and replacing the external joinery (only in E).

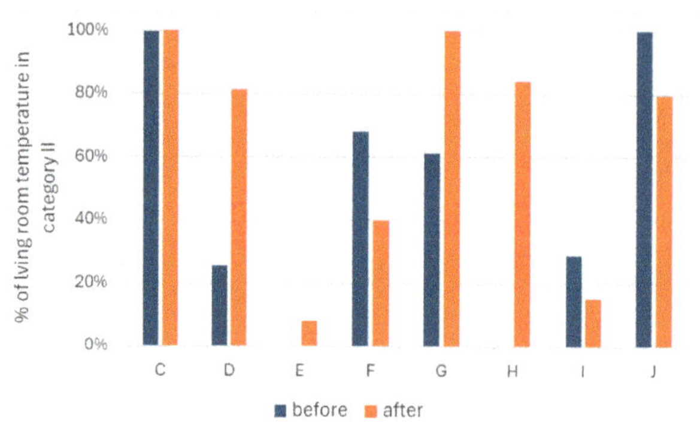

Figure 8. Share of living room temperatures within category II from EN 16798-1 before and after renovation.

Furthermore, Figure 9 illustrates the measured temperatures in the bathroom before and after renovation. The average bathroom temperature increased from 16.6 °C before renovation to 18.8 °C afterward. While most houses experienced a rise in temperature post-renovation, in three houses (F, I, J), the temperature slightly decreased, mirroring the temperature changes observed in the living room. However, the difference in average temperature per dwelling pre- and post-renovation is not statistically significant, tested via a paired t-test for average values, with a p-value of $0.089 > 0.05$. This result has to be interpreted with precautions due to the small sample size (i.e., 10 dwellings).

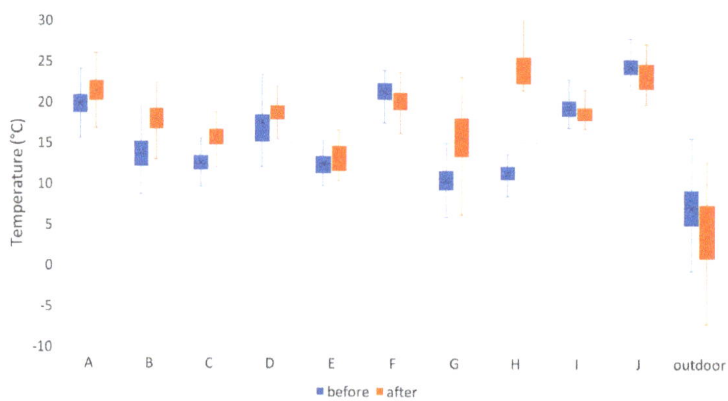

Figure 9. Indoor temperature in the bathroom before and after renovation.

House H exhibited the most significant temperature difference in the bathroom between pre- and post-renovation measurements. Prior to the renovation, the bathroom lacked insulation and heating, resulting in a low average temperature. During the renovation, the bathroom roof was insulated, a new skylight was installed, and a radiator was connected to the central heating system. These substantial temperature differences may indicate either changes in user behavior or reduced heat loss in the room with consistent user behavior. Given that user behavior remained largely unchanged (with only one owner maintaining similar usage of the room before and after renovation), the significant increase in average temperature can be attributed primarily to the renovation efforts.

In all dwellings, except for dwellings A and D, an increase in bedroom temperature was recorded after renovation (see Figure 10). The average bedroom temperature increased from 15.9 °C before renovation to 17.8 °C afterward. The difference in average temperature per dwelling pre- and post-renovation is statistically significant, tested via a paired t-test for average values, with a p-value of $0.023 < 0.05$.

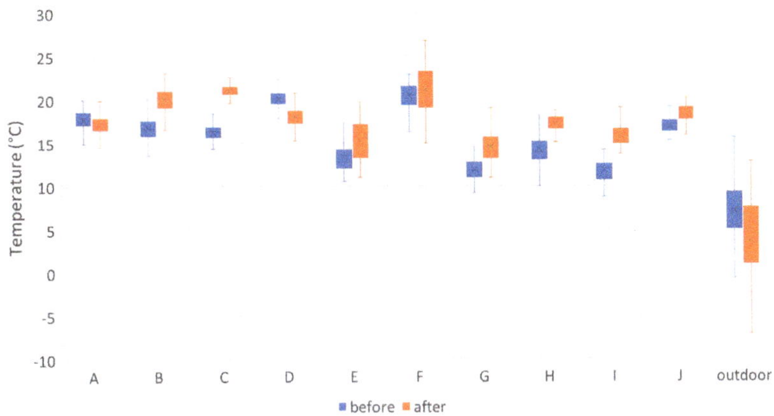

Figure 10. Analysis of bedroom temperature before and after renovation.

The most significant temperature increases were observed in dwellings C and I, likely due to the installation of roof insulation and/or the replacement of external joinery. In dwelling D, despite the replacement of external joinery, a decrease in average bedroom temperature was observed. This can be attributed to a change in how the bedroom was used after the renovation: it was used exclusively for sleeping at night, while during the day, the residents primarily occupied the living room.

Moreover, the correlation between the renovation budget invested for measures with a potential increase in energy performance and the change in median temperature in the living room, bathroom, and main bedroom was investigated via Pearson correlation. The correlation was found to be moderate in the living room as well as the bathroom, with a Pearson correlation coefficient of 0.46 and 0.42, respectively. Correlation in the bedroom was noted to be weak, with a correlation coefficient of 0.28. It can be concluded that the renovations caused an improvement in indoor temperatures.

The increase in indoor temperature in most of the dwellings due to renovation is confirmed by existing field studies [14,16–18,21].

4.3.2. CO_2-Concentration

During the measurement campaign, CO_2 concentration was monitored in the living rooms of eight houses before and after renovation to assess the indoor air quality (see Figure 11). The average CO_2 concentration increased from 680 ppm before renovation to 807 ppm after renovation. However, the difference in average CO_2 concentration per dwelling pre- and post-renovation is not statistically significant, tested via a paired t-test for average values, with a p-value of $0.31 > 0.05$. This result has to be interpreted with precautions due to the small sample size (i.e., 10 dwellings).

In five of the dwellings (F, G, H, I, J), CO_2 concentration remained relatively unchanged. However, a big increase in CO_2 concentration was observed in dwellings C and D, while in one house (E), this parameter decreased. The differences in living room CO_2 concentration per dwelling are significant (via t-test) with a p-value < 0.05.

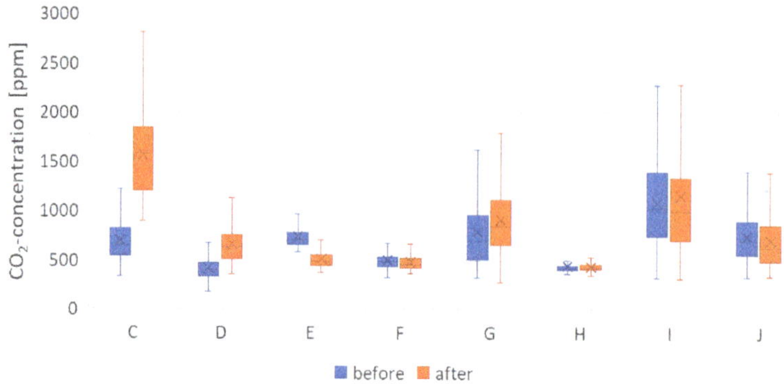

Figure 11. CO_2 concentration in living room before and after renovation.

The rise in CO_2 concentration in dwellings C and D can be attributed to the renewal of external joinery, which increased the airtightness of the buildings. In both cases, the occupants spent most of their time at home in the living room, with their behavior remaining largely consistent across both measurement periods. Consequently, the increased airtightness due to new windows explains the higher average CO_2 concentration after renovation. In dwelling C, the frequent visits from (grand)children, who were sleeping in the living room, further contributed to the (very) high CO_2 concentration even at night.

None of the dwellings had a mechanical ventilation system in the living room, either before or after renovation. However, when the external joinery was replaced, windows with ventilation grilles were installed.

Similar findings have been reported in previous research. In renovation projects involving multifamily houses in Latvia, Slovakia, and Ireland, indoor air quality also deteriorated post-renovation due to increased airtightness without the installation of new ventilation systems [16,17,36].

4.4. Satisfaction of the Home-Owners

The renovation works aimed at enhancing housing quality, energy performance, and safety may lead to higher levels of occupant satisfaction. The relationship between the types of renovation measures implemented and the resulting changes in housing quality, energy performance, and satisfaction levels is explored in this section.

This analysis focuses on cases where interviews were conducted both before and after the renovation and where detailed information on the specific renovation measures implemented is available. This includes data from 36 dwellings, as outlined in Table 1.

Table 6 presents the changes in satisfaction with the dwelling and comfort levels during summer and winter. The data indicate that a majority of participants (61%) reported increased satisfaction with their dwelling post-renovation. However, some participants expressed unchanged or even decreased satisfaction.

Satisfaction with summer comfort improved for about half of the participants, while improvements in satisfaction with winter comfort were more pronounced. Before the renovation, difficulty in keeping the house warm during winter was a common source of dissatisfaction. The renovation measures, which focused on enhancing energy performance and indoor comfort during the heating season, likely contributed to the observed increase in winter comfort satisfaction. This is in line with the study of Fisk et al. [21] who concluded a statistically significant increase in self-reported thermal comfort after retrofit.

Table 6. Change in satisfaction with dwelling and comfort during summer/winter.

% After Renovation	Dwelling	Comfort During Summer	Comfort During Winter
Less satisfied with . . .	11%	14%	9%
As (dis)satisfied with . . .	28%	34%	29%
More satisfied with . . .	61%	51%	63%
Observations [a]	36	35	35

[a] Observations with interview before and after the renovation and where information is available on type of renovation works implemented.

Table 7 presents the results of a probit regression analysis where the dependent variables are zero or one, related to whether the level of satisfaction increased (1) or not (0 = remained the same or decreased) after the renovation. Measures on energy performance or housing comfort are not included as they were implemented in all dwellings and, therefore, do not offer the potential to look at differences between those who did or did not implement. However, this type of work can be found among the answers to the open questions on explaining the level of satisfaction after the renovations, which include kitchen renovation, new windows, insulation, etc.

Table 7. Link between improved satisfaction and renovation (effects) (via probit regression).

Dependent Variable = 0 (No Improvement in Satisfaction) or 1 (Improved Satisfaction) (Only Coefficients Shown)	Improved Satisfaction with Dwelling	Improved Satisfaction with Comfort During Summer	Improved Satisfaction with Comfort During Winter
1. Type of work implemented			
Roof	−0.60	0.73	0.52
Exterior joinery	0.17	0.17	0.93
Electricity	−0.43	0.92	0.23
Heating	0.31	0.65	1.07 *
Moisture	0.23	−0.26	−0.78
Observations	36	35	35
2. Depth of renovation works			
Number of type of works	−0.14	0.41	0.24
Observations	36	35	35
3. Housing quality			
Decrease in penalty score	−0.01	0.02	0.01
Observations	26	25	25
4. Energy performance			
Decrease in kWh	0.001	0.005 *	0.02 *
Observations	19	18	18

* Significant at 5% (coefficient is significantly different from zero).

No significant direct relationship is noticed between changes in satisfaction with the dwelling and the type of works implemented (Table 7 (1)), the depth of renovation measured by the number of types of works implemented (Table 7 (2)), the improvement in housing quality measured by a decrease in the quality penalty score (Table 7 (3)) or the improvement in energy performance measured by the decrease in expected energy use (kWh) (Table 7 (4)).

Only the following effects are noted to be significant: the magnitude of the improvement in energy performance does have a positive effect on satisfaction with comfort both in summer- and wintertime, with a larger effect in wintertime. The implementation of work on the heating system is also shown to have a strong effect on the change in satisfaction with comfort in wintertime.

5. Conclusions and Limitations

A total of 84 locked-in home-owners in Ghent (Belgium) participated in the European Urban Innovative Actions project "Gent knapt op", which aimed to enhance the housing quality and living conditions of these residents who lived in bad quality housing but lacked the resources to renovate their homes. Each participant received a maximum renovation budget of EUR 30,000, along with social and technical support throughout the renovation process. This study evaluated the impact of these renovations on housing quality, energy performance, indoor environment, and user satisfaction in these field studies.

5.1. Limitations

The limitations of this study are the following:

- This study focuses on a specific renovation project in Ghent, Belgium, involving approximately 40 out of 6600 locked-in owners. Therefore, the conclusions should be interpreted with caution.
- Due to practical constraints, different numbers of dwellings (and sometimes different dwellings) were investigated across various aspects of the study, making it challenging to generalize the conclusions.
- The analysis of energy performance is based solely on calculations, not on measured energy use before and after renovations.
- The impact of user behavior on the IEQ monitoring results could not be excluded.
- Due to delays in implementing the renovation measures, the planned time between the completion of renovations and the distribution of questionnaires was reduced from the original minimum of 6 months to less than 3 months. This may have influenced the occupants' satisfaction.

5.2. Conclusions

The main findings of this study are summarized below:

- The total renovation budget invested per dwelling averages EUR 29,000. The largest expenditure per dwelling was dedicated to roof renovations, replacement of external joinery, and renewal of the heating system. The most frequently implemented renovation measures included smoke detectors (all), renewal of electricity system (80% of the homes), external joinery (75%), roof renovation (75%), renewal of heating system (61%), and moisture treatment (50%).
- The housing quality was significantly enhanced because of the implemented renovation measures: 83% of the homes met the requirements of the Flemish Housing Code (i.e., a penalty score below 15) post-renovation compared to none before renovation.
- The energy performance of the houses significantly improved, with a calculated primary energy use decreasing from over 200 kWh/(m^2·a) (corresponding to a C, D, E, or F-label) to under 300 kWh/(m^2·a) (corresponding to an A, B, C, or D-label). However, only 45% of the residents estimated a decrease in their energy use after renovation. The correlation between the renovation budget for energy performance measures and the change in theoretical annual primary energy use was found to be weak to moderate.
- The renovations positively affected the indoor environment. The average indoor temperature in the living room, bedroom, and bathroom increased after renovation with 2.7 °C, 2.2 °C and 1.9 °C, respectively. However, only the difference in average temperature per dwelling in the bedroom was found to be statistically significant. The correlation between the renovation budget for energy performance measures and the change in median temperature in the living room and bathroom was found to be moderate, while this was noted to be weak in the bedroom.

- The average CO_2 concentration increased from 680 ppm before renovation to 807 ppm after renovation. However, the difference in average CO_2 concentration per dwelling pre- and post-renovation is not statistically significant. The CO_2 concentration in the majority of the homes remained unchanged. However, two dwellings showed a significant increase due to replacement of the windows and increased airtightness.
- The majority of the participants (61%) reported increased satisfaction with their dwelling post-renovation, with more pronounced improvement in winter than summer comfort. A statistically significant effect was noticed between the magnitude of improvement in energy performance and satisfaction with comfort in both winter and summer. A strong effect was also observed between the works on the heating system and the change in satisfaction with the comfort in wintertime.

Author Contributions: Conceptualization, L.B., K.V.d.B. and A.V.; Methodology, L.B., K.V.d.B., H.B. and A.V.; Formal Analysis, L.B., H.B. and K.V.d.B.; Writing—original draft preparation, L.B., K.V.d.B. and H.B.; Writing—review and editing, L.B., K.V.d.B., A.V. and H.B.; Supervision, H.B. and A.V.; Funding Acquisition, K.V.d.B. and A.V. All authors have read and agreed to the published version of the manuscript.

Funding: This research was funded by European Regional Development Fund—Urban Innovation Actions: ICCARus (Gent knapt op)—Improving housing Conditions for CAptive Residents in Ghent.

Data Availability Statement: Supporting data are saved by the research consortium but due to the ethical aspects, they cannot be shared publicly. Using the consent statement of the City of Ghent, participants gave permission to share their data with the research consortium. An agreement between the partners of this project guarantees their privacy.

Acknowledgments: This research was possible thanks to a collaboration between the City of Ghent, the Public Center for Social Welfare of Ghent, KU Leuven, Ghent University, AP Hogeschool Antwerp, de Energiecentrale, SIVI, Samenlevingsopbouw and Domus Mundi. It was part of the ICCARus project and was co-financed by the European Fund for Regional Development through the Urban Innovative Actions (UIA) initiative.

Conflicts of Interest: The authors declare no conflict of interest.

References

1. Heylen, K.; Vanderstraeten, L. *Wonen in Vlaanderen Anno 2018*; Steunpunt Wonen: Leuven, Belgium, 2019.
2. Winters, S.; Ceulemans, W.; Heylen, K.; Pannecoucke, I.; Vanderstraeten, L.; Van den Broeck, K.; De Decker, P.; Ryckewaert, M.; Verbeeck, G. *Wonen in Vlaanderen Anno 2013. De Bevindingen uit Het Grote Woononderzoek 2013 Gebundeld*; Steunpunt Wonen: Leuven, Belgium, 2015.
3. European Commission. *2030 Climate Targets*; European Commission: Brussels, Belgium, 2023.
4. Vlaamse Regering. *Langetermijnstrategie Voor de Renovatie van Vlaamse Gebouwen*; Vlaamse Regering: Vlaamse, Belgium, 2020.
5. Vlaamse Regering. Besluit van de Vlaamse Regering tot uitvoering van de Vlaamse Codex Wonen van 2021. In *Besluit Vlaamse Codex Wonen*; Vlaamse Regering: Vlaamse, Belgium, 2021.
6. Ryckewaert, M.; Van den Houte, K.; Vanderstraeten, L. *Inschatting van de Renovatiekosten om het Vlaamse Woningpatrimonium aan te Passen aan de Woningkwaliteits—En Energetische Vereisten*; Steunpunt Wonen: Leuven, Belgium, 2019.
7. Heylen, K. *Doelgroepen Sociale Huur en Specifieke Segmenten op de Woningmarkt*; Steunpunt Wonen: Leuven, Belgium, 2019.
8. Heylen, K. *Grote Woononderzoek 2013. Deel 2. Deelmarkten, Woonkosten en Betaalbaarheid*; Steunpunt Wonen: Leuven, Belgium, 2015.
9. De Decker, P.; Meeus, B.; Pannecoucke, I.; Schillebeeckx, E.; Verstraete, J.; Volckaert, E. *Woonnood in Vlaanderen: Feiten/Mythen/Voorstellen*; Garant Uitgevers nv: Antwerp, Belgium, 2015.
10. Vanderstraeten, L.; Ryckewaert, M. *Noodkopers, Noodeigenaars en Captive Renters in Vlaanderen. Nadere Analyses Op Basis Van Het GWO2013*; Steunpunt Wonen: Leuven, Belgium, 2019.
11. Stad Gent. Gent in Cijfers. 2023. Available online: https://gent.buurtmonitor.be/ (accessed on 2 January 2025).
12. European Regional Development Fund. ICCARus (Gent Knapt Op). 2024. Available online: https://www.uia-initiative.eu/en/uia-cities/ghent-call3#:~:text=Housing-,ICCARus%20(Gent%20knapt%20op)%20-%20Improving%20housing%20Conditions%20for%20CAptive,,%20financially,%20administrative%20and%20socially (accessed on 2 January 2025).

13. Hamilton, I.G.; Summerfield, A.J.; Shipworth, D.; Steadman, J.P.; Oreszczyn, T.; Lowe, R.J. Energy efficiency uptake and energy savings in English houses: A cohort study. *Energy Build.* **2016**, *118*, 259–276. [CrossRef]
14. Lambie, E. Evaluation of the Impact of Energy Renovation Measures: Experimental Study of Residential Buildings. Ph.D. Thesis, KU Leuven, Leuven, Belgium, 2021.
15. Du, L.; Leivo, V.; Prasauskas, T.; Täubel, M.; Martuzevicius, D.; Haverinen-Shaughnessy, U. Effects of energy retrofits on Indoor Air Quality in multifamily buildings. *Indoor Air* **2019**, *29*, 686–697. [CrossRef] [PubMed]
16. Broderick, A.; Byrne, M.; Armstrong, S.; Sheahan, J. A pre and post evaluation of indoor air quality, ventilation, and thermal comfort in retrofitted co-operative social housing. *Build. Environ.* **2017**, *122*, 126–133. [CrossRef]
17. Földváry, V.; Bekö, G.; Langer, S.; Arrhenius, K.; Petrás, D. Effect of energy renovation on indoor air quality in multifamily residential buildings in Slovakia. *Build. Environ.* **2017**, *122*, 363–372. [CrossRef]
18. Willand, N.; Maller, C.J.; Ridley, I. Towards explaining the health impacts of residential energy efficiency interventions—A realist review. Part 1: Pathways. *Soc. Sci. Med.* **2015**, *133*, 191–201. [CrossRef] [PubMed]
19. Pungercar, V.; Zhan, Q.; Xiao, Y.; Musso, F.; Dinkel, A.; Pflug, T. A new retrofitting strategy for the improvement of indoor environment. *Energy Build.* **2021**, *241*, 110951. [CrossRef]
20. Hassan, H.; Mishra, A.K.; Wemken, N.; O'Dea, P.; Cowie, H.; McIntyre, B.; Coggins, A.M. Deep energy renovations' impact on indoor air quality and thermal comfort of residential dwellings in Ireland-ARDEN project. *Build. Environ.* **2024**, *259*, 111637. [CrossRef]
21. Fisk, W.; Singer, B.C.; Chan, W.R. Association of residential energy efficiency retrofits with indoor environmental quality, comfort, and health: A review of empirical data. *Build. Environ.* **2020**, *180*, 107067. [CrossRef]
22. Maidment, C.; Webb, T.L.; Jones, C.R.; Hathway, A. The impact of household energy efficiency measures on health: A meta-analysis. *Energy Policy* **2013**, *65*, 583–593. [CrossRef]
23. Pedersen, E.; Gao, C.; Wierzbicka, A. Tenant perceptions of post-renovation indoor environmental quality in rental housing: Improved for some, but not for those reporting health-related symptoms. *Build. Environ.* **2021**, *189*, 107520. [CrossRef]
24. Coggins, A.M.; Wemken, N.; Mishra, A.K.; Sharkey, M.; Horgan, L.; Cowie, H.; Bourdin, E.; McIntyre, B. Indoor air quality, thermal comfort and ventilation in deep energy retrofitted Irish dwellings. *Build. Environ.* **2022**, *219*, 109236. [CrossRef]
25. Flemish Energy and Climate Agency (VEKA). *Langetermijnstrategie Voor de Renovatie van Vlaamse Gebouwen*; VEKA: Brussels, Belgium, 2020.
26. *ANSI/ASHRAE Standard 169-2013*; Climatic Data for Building Design Standards, 8400. ANSI/ASHRAE: Peachtree Corners, GA, USA, 2013.
27. RMI. Available online: https://www.meteo.be/nl/klimaat/klimaat-van-belgie/klimatologisch-overzicht/2021/jaar (accessed on 30 January 2025).
28. Van den Broeck, K.; Bielen, L.; Ferny, L.; Maelstaf, H.; Van Roy, K.; Versele, A. *Gent Knapt Op-Sociale Impact Voor de Deelnemers*; Ghent University: Gent, Belgium, 2022.
29. Flemish Energy and Climate Agency (VEKA). *REG 2019: Energiebewustzijn en—Gedrag van Vlaamse Huishoudens*; Flemish Energy and Climate Agency: Brussels, Belgium, 2019.
30. Vanderstraeten, L.; Ryckewaert, M. *Grote Woononderzoek 2013. Kwaliteitscontrole van de in-en Uitwendige Woningscreening*; Steunpunt Wonen: Leuven, Belgium, 2015.
31. Vlaamse Overheid. Energieprestatiecertificaat. Available online: https://www.vlaanderen.be/energieprestatiecertificaten-epcs (accessed on 2 January 2025).
32. Van den Broeck, K. *Effecten van Energetische Investeringen in Woningen. Literatuuronderzoek*; Steunpunt Wonen: Leuven, Belgium, 2020.
33. Van Hove, M.; Delghust, M.; Janssens, A. *Analyse Naar de Haalbaarheid van Statistische Modellen die Energiegebruik in Woningen Kunnen Voorspellen op Basis van Gebouwparameters*; Ghent University: Gent, Belgium, 2021.
34. Majcen, D.; Itard, L.; Visscher, H. Theoretical vs. actual energy consumption of labelled dwellings in the Netherlands: Discrepancies and policy implications. *Energy Policy* **2013**, *54*, 125–136. [CrossRef]
35. *EN 16798-1*; Energy Performance of Buildings-Ventilation for Buildings—Part 1: Indoor Environmental Input Parameters for Design and Assessment of Energy Performance of Buildings Addressing Indoor Air Quality, Thermal Environment, Lighting and Acoustics—Module M1. CEN: Brussels, Belgium, 2019.
36. Dimdina, I.; Krumins, E.; Lesinskis, A. Indoor Air Quality in Multi-Apartment Buildings before and after Renovation. *Constr. Sci.* **2014**, *16*, 4. [CrossRef]

MDPI AG
Grosspeteranlage 5
4052 Basel
Switzerland
Tel.: +41 61 683 77 34

Buildings Editorial Office
E-mail: buildings@mdpi.com
www.mdpi.com/journal/buildings

www.ingramcontent.com/pod-product-compliance
Lightning Source LLC
Chambersburg PA
CBHW040807250926
51850CB00007BA/1101
* 9 7 8 3 7 2 5 8 7 8 2 9 1 *